Plumbing
Level Three

Trainee Guide
Fourth Edition

PEARSON

Boston   Columbus   Indianapolis   New York   San Francisco   Upper Saddle River
Amsterdam   Cape Town   Dubai   London   Madrid   Milan   Munich   Paris   Montreal   Toronto
Delhi   Mexico City   São Paulo   Sydney   Hong Kong   Seoul   Singapore   Taipei   Tokyo

NCCER
President: Don Whyte
Director of Product Development: Daniele Stacey
Plumbing Project Manager: Chris Wilson
Senior Manager: Tim Davis
Quality Assurance Coordinator: Debie Ness

Desktop Publishing Coordinator: James McKay
Permissions Specialist: Megan Casey
Production Specialist: Megan Casey
Editor: Chris Wilson, James Bassett

Writing and development services provided by S4Carlisle Publishing Services, Dubuque, IA
Project Manager: Michael B. Kopf
Writers: Paul Lagasse and Jack Klasey
Art Development: S4Carlisle Publishing Services

Permissions Specialists: Kim Schmidt, Karyn Morrison
Media Specialist: Genevieve Brand
Copy Editor: Michael H. Toporek

Pearson Education, Inc.
Director, Global Employability Solutions: Jonell Sanchez
Head of Associations: Andrew Taylor
Editorial Assistant: Douglas Greive
Program Manager: Alexandrina B. Wolf
Project Manager: Janet Portisch
Operations Supervisor: Deidra M. Skahill
Art Director: Diane Ernsberger
Directors of Marketing: David Gesell, Margaret Waples
Field Marketers: Brian Hoehl, Stacey Martinez

Composition: NCCER
Printer/Binder: LSC Communications
Cover Printer: LSC Communications
Text Fonts: Palatino and Univers

Credits and acknowledgments for content borrowed from other sources and reproduced, with permission, in this textbook appear at the end of each module.

PEARSON

ISBN-13:   978-0-13-340424-1
ISBN-10:   0-13-340424-2

Preface

To the Trainee

Most people are familiar with plumbers who come to their home to unclog a drain or install an appliance. In addition to these activities, however, plumbers install, maintain, and repair many different types of pipe systems. For example, some systems move water to a municipal water treatment plant and then to residential, commercial, and public buildings. Other systems dispose of waste, provide gas to stoves and furnaces, or supply air conditioning. Pipe systems in generation plants carry the steam that powers turbines. Pipes are also used in manufacturing plants, such as wineries, to move material through production processes.

Plumbers and their associated trades constitute one of the largest construction occupations, holding about 420,000 jobs. The occupation continues to grow, and nearly 229,000 job openings are expected to be available by 2020. Plumbers are also among the highest paid construction occupations.

New with *Plumbing Level Three*

This fourth edition of *Plumbing Level Three* presents a new instructional design and features a streamlined teaching order to better prepare you for your career as a plumber. In addition, *Plumbing Level Three* is now produced in full color, and contains updated artwork, detailed figures, and the latest tools and technology available to the plumbing trade. Throughout this title, "Going Green" features emphasize sustainability and environmentally constructive procedures.

This edition of *Plumbing Level Three* has two new modules. *Sizing and Protecting the Water Supply System* has information on how to size water supply piping and information on backflow prevention. *Service Plumbing* focuses on servicing fixtures, valves, appliances, and piping systems.

Your instructor also has been provided with a new set of tools to better ensure your success as an apprentice plumber. As you continue with your third year of training in the plumbing craft, we wish you the best and hope that you'll continue your training beyond this curriculum. As many of the craftspeople employed in this trade will tell you, there are numerous opportunities awaiting those with the skills and the desire to move forward in the plumbing profession.

We invite you to visit the NCCER website at **www.nccer.org** for information on the latest product releases and training, as well as online versions of the *Cornerstone* magazine and Pearson's NCCER product catalog.

Your feedback is welcome. You may email your comments to **curriculum@nccer.org** or send general comments and inquiries to **info@nccer.org**.

NCCER Standardized Curricula

NCCER is a not-for-profit 501(c)(3) education foundation established in 1996 by the world's largest and most progressive construction companies and national construction associations. It was founded to address the severe workforce shortage facing the industry and to develop a standardized training process and curricula. Today, NCCER is supported by hundreds of leading construction and maintenance companies, manufacturers, and national associations. The NCCER Standardized Curricula was developed by NCCER in partnership with Pearson, the world's largest educational publisher.

Some features of the NCCER Standardized Curricula are as follows:

- An industry-proven record of success
- Curricula developed by the industry for the industry
- National standardization providing portability of learned job skills and educational credits
- Compliance with the Office of Apprenticeship requirements for related classroom training (*CFR 29:29*)
- Well-illustrated, up-to-date, and practical information

NCCER also maintains a Registry that provides transcripts, certificates, and wallet cards to individuals who have successfully completed a level of training within a craft in NCCER's Curricula. *Training programs must be delivered by an NCCER Accredited Training Sponsor in order to receive these credentials.*

Special Features

In an effort to provide a comprehensive, user-friendly training resource, we have incorporated many different features for your use. Whether you are a visual or hands-on learner, this book will provide you with the proper tools to get started in the Plumbing trade.

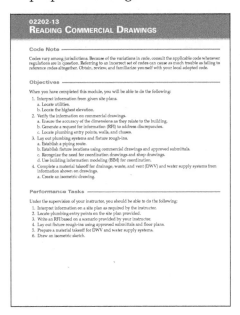

Color Illustrations and Photographs

Full-color illustrations and photographs are used throughout each module to provide vivid detail. These figures highlight important concepts from the text and provide clarity for complex instructions. Each figure reference is denoted in the text in *italics* for easy reference.

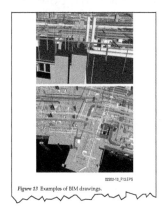

Figure 13 Examples of BIM drawings.

Introduction

This page is found at the beginning of each module and lists the Objectives, Performance Tasks, Trade Terms, and Required Trainee Materials for that module. The Objectives list the skills and knowledge you will need in order to complete the module successfully. The Performance Tasks give you an opportunity to apply your knowledge to the real-world duties that heavy equipment operators perform. The list of Trade Terms identifies important terms you will need to know by the end of the module. Required Trainee Materials list the materials and supplies needed for the module.

Notes, Cautions, and Warnings

Safety features are set off from the main text in highlighted boxes and are organized into three categories based on the potential danger of the issue being addressed. Notes simply provide additional information on the topic area. Cautions alert you of a danger that does not present potential injury but may cause damage to equipment. Warnings stress a potentially dangerous situation that may cause injury to you or a co-worker.

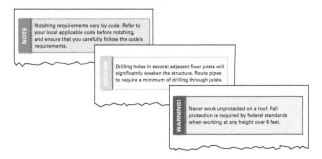

Special Features

Features provide a head start for those entering the Plumbing field by presenting technical tips and professional practices from operators in various disciplines. These features often include real-life scenarios similar to those you might encounter on the job site.

A Museum for Architecture

The Athenaeum of Philadelphia is a museum of American architecture. The museum, which has collected the work of about 1,000 American architects, has 150,000 drawings, 50,000 photographs, and many architectural documents.

Going Green

Going Green looks at ways to preserve the environment, save energy, and make good choices regarding the health of the planet. Through the introduction of new construction practices and products, you will see how the "greening of America" has already taken root.

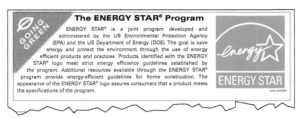

The ENERGY STAR® Program

ENERGY STAR® is a joint program developed and administered by the US Environmental Protection Agency (EPA) and the US Department of Energy (DOE). The goal is save energy and protect the environment through the use of energy efficient products and practices. Products identified with the ENERGY STAR® logo meet strict energy efficiency guidelines established by the program. Additional resources available through the ENERGY STAR® program provide energy-efficient guidelines for home construction. The appearance of the ENERGY STAR® logo assures consumers that a product meets the specifications of the program.

Did You Know?

The Did You Know? features offer hints, tips, and other helpful bits of information from the trade.

Did You Know?

Copper is one of the most plentiful metals. It has been in use for thousands of years. In fact, a piece of copper pipe used by the Egyptians more than 5,000 years ago is still in good condition. When the first Europeans arrived in the new world, they found the Native Americans using copper for jewelry and decoration. Much of this copper was from the region around Lake Superior, which separates northern Michigan and Canada.

One of the first copper mines in North America was established in 1664 in Massachusetts. During the mid 1800s, a new site

Step-by-Step Instructions

Step-by-step instructions are used throughout to guide you through technical procedures and tasks from start to finish. These steps show you not only how to perform a task but also how to do it safely and efficiently.

Step 1	Drive a stake on one side of the excavation trench.
Step 2	Drive a second stake on the opposite side of the trench, in line with the first stake. If a length of pipe has already been positioned in the trench, drive the stakes beside the hub end of the pipe.
Step 3	Place the batter board in position, and level it using a builder's level.
Step 4	For ease in calculating grade, locate the batter board at a convenient height above the pipe. General-purpose levels are suitable for this task, though the results will not be as accurate a...

Trade Terms

Each module presents a list of Trade Terms that are discussed within the text and defined in the Glossary at the end of the module. These terms are denoted in the text with **bold, blue type** upon their first occurrence. To make searches for key information easier, a comprehensive Glossary of Trade Terms from all modules is located at the back of this book.

sic parts and in a wide variety of styles. You can interchange these basic parts and produce an almost endless variety of drains. The basic parts are the drain body, grate (dome), deck clamp, clamping ring or flashing clamp, drain receiver (bearing pan), and extensions.

The basic component of all drains, the drain body, funnels the water into the piping system. Usually made of cast iron or plastic, it is designed to connect to the roof, floor, or foundation. Special installations may require stainless steel or porcelain-enameled cast iron. Examples of floor drains are shown in *Figure 1*.

Floor and area drains often contain a separate sediment bucket to prevent solids from entering the piping (see *Figure 2*). The drain body connects

Review Questions

Review Questions reinforce the knowledge you have gained and are a useful tool for measuring what you have learned.

Review Questions

1. The abbreviation DWV stands for _____.
   a. digital water valve
   b. drain, water, and vent
   c. drain, waste, and vent
   d. double water volume

2. Which of the following statements is true?
   a. It is more difficult to route DWV piping around water supply piping than vice-versa.
   b. The grade used for water supply piping is not important.
   c. The grade used for DWV piping is not important.
   d. It is more difficult to route water supply piping around DWV piping than vice-versa.

3. To calculate the load factor for a DWV system, you can use information contained in the _____.
   a. *Plumber's Handbook*
   b. applicable local code
   c. floor plans
   d. materials takeoff

4. Drainage and waste piping systems must be installed _____.
   a. below the frost line
   b. at a slope greater than 30°
   c. at a specified, constant grade
   d. by a master plumber

5. A general-purpose level can be used to provide _____.
   a. the same accuracy as a plumb bob or laser level
   b. a precise calculation of grade
   c. an accurate estimate of distance between level points
   d. a very rough approximation of grade

6. How can a general-purpose level be modified to make it more accurate for grade measurements?
   a. Extend its length with a wood lath.
   b. Attach a suitably sized wood block to one end.
   c. Add a piece of rebar to one

7. What can a transit level do that a builder's level cannot?
   a. Tilt upward to measure angles from the vertical
   b. Provide digital readouts
   c. Rotate horizontally to measure angles
   d. Be properly leveled on a sloping site

8. A vernier scale, used to make more precise readings, is found on a _____.
   a. laser level
   b. stadia rod
   c. steel tape
   d. builder's level

9. When setting up a builder's level, the leveling screws should be _____.
   a. backed out fully, then adjusted individually
   b. turned in opposite directions
   c. rotated counterclockwise
   d. tightened as far as possible, then backed off as necessary

10. Convert 8¾" to decimals of a foot.
    a. 0.872
    b. 0.688
    c. 0.693
    d. 0.698

11. Which of the following statements about the cold beam laser level is *not* true?
    a. The light beam is perfectly straight.
    b. It projects a beam of ultraviolet light.
    c. The light beam does not expand appreciably.
    d. You should not look directly into the light beam, even though it is considered harmless.

12. To protect a pipe that will pass through a concrete footing, install a(n) _____.
    a. insulation boot
    b. sleeve
    c. buffer
    d. expansion joint

NCCER Standardized Curricula

NCCER's training programs comprise more than 80 construction, maintenance, pipeline, and utility areas and include skills assessments, safety training, and management education.

Boilermaking
Cabinetmaking
Carpentry
Concrete Finishing
Construction Craft Laborer
Construction Technology
Core Curriculum:
    Introductory Craft Skills
Drywall
Electrical
Electronic Systems Technician
Heating, Ventilating, and
    Air Conditioning
Heavy Equipment Operations
Highway/Heavy Construction
Hydroblasting
Industrial Coating and Lining
    Application Specialist
Industrial Maintenance
    Electrical and Instrumentation
    Technician
Industrial Maintenance
    Mechanic
Instrumentation
Insulating
Ironworking
Masonry
Millwright
Mobile Crane Operations
Painting
Painting, Industrial
Pipefitting
Pipelayer
Plumbing
Reinforcing Ironwork
Rigging
Scaffolding
Sheet Metal
Signal Person
Site Layout
Sprinkler Fitting
Tower Crane Operator
Welding

Maritime

Maritime Industry Fundamentals
Maritime Pipefitting
Maritime Structural Fitter

Green/Sustainable Construction

Building Auditor
Fundamentals of Weatherization
Introduction to Weatherization
Sustainable Construction
    Supervisor
Weatherization Crew Chief
Weatherization Technician
Your Role in the Green
    Environment

Energy

Alternative Energy
Introduction to the Power
    Industry
Introduction to Solar
    Photovoltaics
Introduction to Wind Energy
Power Industry Fundamentals
Power Generation Maintenance
    Electrician
Power Generation I&C
    Maintenance Technician
Power Generation Maintenance
    Mechanic
Power Line Worker
Power Line Worker: Distribution
Power Line Worker: Substation
Power Line Worker:
    Transmission
Solar Photovoltaic Systems
    Installer
Wind Turbine Maintenance
    Technician

Pipeline

Control Center Operations,
    Liquid
Corrosion Control
Electrical and Instrumentation
Field Operations, Liquid
Field Operations, Gas
Maintenance
Mechanical

Safety

Field Safety
Safety Orientation
Safety Technology

Management

Fundamentals of Crew
    Leadership
Project Management
Project Supervision

Supplemental Titles

Applied Construction Math
Careers in Construction
Tools for Success

Spanish Translations

Basic Rigging
    (Principios Básicos de
    Maniobras)
Carpentry Fundamentals
    (Introducción a la
    Carpintería, Nivel Uno)
Carpentry Forms
    (Formas para Carpintería,
    Nivel Trés)
Concrete Finishing, Level One
    (Acabado de Concreto,
    Nivel Uno)
Core Curriculum:
    Introductory Craft Skills
    (Currículo Básico:
    Habilidades Introductorias del
    Oficio)
Drywall, Level One
    (Paneles de Yeso, Nivel Uno)
Electrical, Level One
    (Electricidad, Nivel Uno)
Field Safety
    (Seguridad de Campo)
Insulating, Level One
    (Aislamiento, Nivel Uno)
Ironworking, Level One
    (Herrería, Nivel Uno)
Masonry, Level One
    (Albañilería, Nivel Uno)
Pipefitting, Level One
    (Instalación de Tubería
    Industrial, Nivel Uno)
Reinforcing Ironwork, Level One
    (Herreria de Refuerzo,
    Nivel Uno)
Safety Orientation
    (Orientación de Seguridad)
Scaffolding
    (Andamios)
Sprinkler Fitting, Level One
    (Instalación de Rociadores,
    Nivel Uno)

Acknowledgments

This curriculum was revised as a result of the farsightedness and leadership of the following sponsors:

ABC Southern California Chapter
Ivey Mechanical Company, LLC
Lake Mechanical Contractors, Inc.
Lee Company

Putnam Career and Technical Center
Sundt
Wat-Kem Mechanical, Inc.
Worth & Company, Inc.

This curriculum would not exist were it not for the dedication and unselfish energy of those volunteers who served on the Authoring Team. A sincere thanks is extended to the following:

Doug Allen
Jonathan Byrd
Stanley Cordova
Steve Guy

David Hoover
Jan Prakke
Harry Rimbey

Brad Sims
Brent Thompson
Ray Thornton

NCCER Partners

American Fire Sprinkler Association
Associated Builders and Contractors, Inc.
Associated General Contractors of America
Association for Career and Technical Education
Association for Skilled and Technical Sciences
Carolinas AGC, Inc.
Carolinas Electrical Contractors Association
Center for the Improvement of Construction
  Management and Processes
Construction Industry Institute
Construction Users Roundtable
Construction Workforce Development Center
Design Build Institute of America
GSSC – Gulf States Shipbuilders Consortium
Manufacturing Institute
Mason Contractors Association of America
Merit Contractors Association of Canada
NACE International
National Association of Minority Contractors
National Association of Women in Construction
National Insulation Association
National Ready Mixed Concrete Association
National Technical Honor Society
National Utility Contractors Association

NAWIC Education Foundation
North American Technician Excellence
Painting & Decorating Contractors of America
Portland Cement Association
SkillsUSA®
Steel Erectors Association of America
U.S. Army Corps of Engineers
University of Florida, M. E. Rinker School of
  Building Construction
Women Construction Owners & Executives, USA

Contents

Module One
Applied Math

Reviews math concepts, including weights and measures, area and volume, temperature, pressure, and force. Also describes the six simple machines: inclined planes, levers, pulleys, wedges, screws, and wheels and axles.(Module ID 02301-14; 17.5 Hours)

Module Two
Sizing and Protecting the Water Supply System

Teaches techniques for sizing water supply systems, including calculating system requirements and demand, developed lengths, and pressure drops. Reviews the factors that can reduce efficiency of water supply piping. Introduces different backflow prevention devices and explains how they work, where they are used, and how they are installed in water supply systems. (Module ID 02312-14; 30 Hours)

Module Three
Potable Water Supply Treatment

Explains how to disinfect, filter, and soften water supply systems. Discusses how to troubleshoot water supply problems, flush out visible contaminants from a plumbing system, and disinfect a potable water plumbing system. (Module ID 02303-14; 15 Hours)

Module Four
Types of Venting

Reviews the different types of vents that can be installed in a DWV system and explains how they work. Also teaches design and installation techniques. (Module ID 02305-14; 20 Hours)

Module Five
Sizing DWV and Storm Systems

Explains how to calculate drainage fixture units for waste systems. Reviews how to size drain, waste, and vent (DWV) systems; storm drainage systems; and roof storage and drainage systems. (Module ID 02306-14; 20 Hours)

Module Six
Sewage Pumps and Sump Pumps

Discusses the installation, diagnosis, and repair of pumps, controls, and sumps in sewage and storm water removal systems. (Module ID 02307-14; 12.5 Hours)

Module Seven
Corrosive-Resistant Waste Piping

Discusses corrosive wastes and reviews related safety issues and hazard communications. Discusses how to determine when corrosive-resistant waste piping needs to be installed, as well as how to correctly select and properly connect different types of piping. (Module ID 02308-14; 7.5 Hours)

Module Eight
Compressed Air

Explains the principles of compressed air systems and describes their components and accessories. Reviews installation and periodic servicing of air compressor systems. (Module ID 02309-14; 10 Hours)

Module Nine

Service Plumbing

Covers the troubleshooting and repair of fix-tures, valves, and faucets in accordance with code and safety guidelines. Explains how to diagnose and repair water supply and drainage piping, water heaters, and other appliances and fixtures. Describes the effects of corrosion, freezing, and hard water on plumbing systems. (Module ID 02311-14; 27.5 Hours)

Glossary

Index

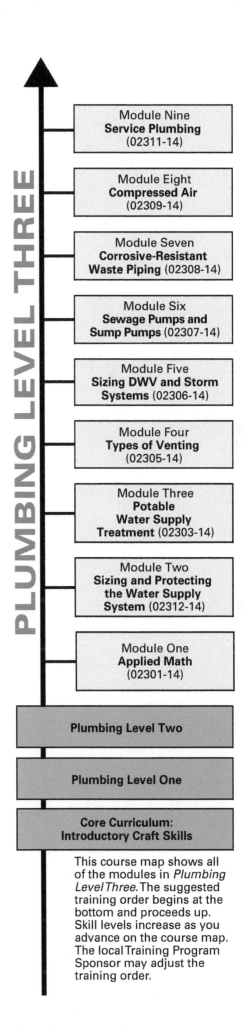

PLUMBING LEVEL THREE

Module Nine
Service Plumbing
(02311-14)

Module Eight
Compressed Air
(02309-14)

Module Seven
Corrosive-Resistant Waste Piping (02308-14)

Module Six
Sewage Pumps and Sump Pumps (02307-14)

Module Five
Sizing DWV and Storm Systems (02306-14)

Module Four
Types of Venting
(02305-14)

Module Three
Potable Water Supply Treatment (02303-14)

Module Two
Sizing and Protecting the Water Supply System (02312-14)

Module One
Applied Math
(02301-14)

Plumbing Level Two

Plumbing Level One

Core Curriculum: Introductory Craft Skills

This course map shows all of the modules in *Plumbing Level Three*. The suggested training order begins at the bottom and proceeds up. Skill levels increase as you advance on the course map. The local Training Program Sponsor may adjust the training order.

02301-14

Applied Math

OVERVIEW

Plumbers use math every day, but you don't have to be a mathematician to install a perfectly functioning plumbing system. You may not realize you are using math when you size a drain, level a water heater, or calculate a grade for a run of pipe, but you are. Plumbers use applied mathematics, which is any mathematical process that is used to accomplish a specific task. Applied mathematics enables you to find the length of a run of pipe, the slope of a drain, the area of a floor, the volume of a climb-in interceptor, the temperature of wastewater—anything where you need to find a precise number.

Module One

Trainees with successful module completions may be eligible for credentialing the NCCER Registry. To learn more, go to **www.nccer.org** or contact us at **1.888.622.3720**. Our website has information on the latest product releases and training, as well as online versions of our *Cornerstone* magazine and Pearson's product catalog.

Your feedback is welcome. You may email your comments to **curriculum@nccer.org,** send general comments and inquiries to **info@nccer.org**, or fill in the User Update form at the back of this module.

This information is general in nature and intended for training purposes only. Actual performance of activities described in this manual requires compliance with all applicable operating, service, maintenance, and safety procedures under the direction of qualified personnel. References in this manual to patented or proprietary devices do not constitute a recommendation of their use.

Objectives

When you have completed this module, you will be able to do the following:

1. Identify the weights and measures used in the English and metric systems.
   a. Explain the English system of weights and measures.
   b. Explain the metric system of weights and measures.
   c. Explain how to convert measurements.
2. Describe how to measure area and volume.
   a. Describe how to measure area.
   b. Describe how to measure volume.
3. Describe the practical applications of area and volume in plumbing.
   a. Describe how to perform area calculations in plumbing.
   b. Describe how to perform volume calculations in plumbing.
   c. Describe how to perform load calculations.
4. Explain the concepts of temperature and pressure and how they apply to plumbing installations.
   a. Explain the concept of temperature and how it applies to plumbing installations.
   b. Explain the concepts of pressure and force and how they apply to plumbing installations.
5. Explain the functions and applications of six simple machines: inclined plane, lever, pulley, wedge, screw, and wheel and axle.
   a. Explain how inclined planes work.
   b. Explain how levers work.
   c. Explain how pulleys work.
   d. Explain how wedges work.
   e. Explain how screws work.
   f. Explain how wheels and axles work.

Performance Tasks

This is a knowledge-based module; there is no performance testing.

Trade Terms

Applied mathematics	Fulcrum	Rectangle
Area	Gallon	Right triangle
Bimetallic thermometer	Head	Screw
Celsius scale	Inclined plane	Simple machine
Centigrade scale	Infrared thermometer	Square
Circle	Isosceles triangle	Square foot
Conduction	Kelvin scale	Square meter
Cube	Lever	Temperature
Cubic foot	Liquid thermometer	Thermal expansion
Cubic meter	Liter	Thermometer
Cylinder	Metric system	Volume
Decimal of a foot	Pounds per square inch	Water supply fixture unit (WSFU)
English system	Pressure	Wedge
Equilibrium	Prism	Wheel and axle
Fahrenheit scale	Pulley	Work

Industry-Recognized Credentials

If you're training through an NCCER-accredited sponsor, you may be eligible for credentials from the NCCER Registry. The ID number for this module is 02301-14. Note that this module may have been used in other NCCER curricula and may apply to other level completions. Contact NCCER's Registry at 888.622.3720 or go to **www.nccer.org** for more information.

Code Note

Codes vary among jurisdictions. Because of the variations in code, consult the applicable code whenever regulations are in question. Referring to an incorrect set of codes can cause as much trouble as failing to reference codes altogether. Obtain, review, and familiarize yourself with your local adopted code.

Contents ——————————————

Topics to be presented in this module include:

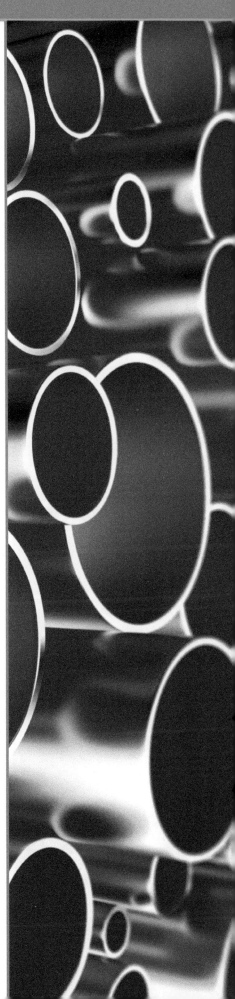

Figures and Tables

1.0.0 WEIGHTS AND MEASURES

Objective

Identify the weights and measures used in the English and metric systems.

    a. Explain the English system of weights and measures.
    b. Explain the metric system of weights and measures.
    c. Explain how to convert measurements.

Trade Terms

Applied mathematics: Any mathematical process used to accomplish a task.

English system: One of the standard systems of weights and measures. The other system is the metric system.

Metric system: A system of measurement in which multiples and fractions of the basic units of measure are expressed as powers of 10.

Systems of measurement allow people to exchange information about all kinds of numbers. Whether it's the angle of a pipe, the amount of water in a tank, the capacity of a pump, or the distance to the sewer line, plumbers are always thinking and working in terms of how much or how little, how large or how small, and how long or how short. Systems of measurement are essential to the plumbing trade. They are an important component of applied mathematics, which is any mathematical process that is used to accomplish a specific task.

Plumbers use two systems of measurement when working with plumbing installations. This section will review the two major systems of weights and measures and how to convert measurements from one system to the other.

1.1.0 Understanding the English System of Weights and Measures

The system of weights and measures currently used in the US is called the English system. It is a version of a system created in England many centuries ago called the British Imperial system. The system was brought over by the European settlers who first established colonies in North America. The system has gradually developed over time to its current form. Some measures and values differ from the original British Imperial system. The US is the only country that primarily uses the English system for many of its weights and measures.

Most of the pipe, tools, fixtures, fittings, and other plumbing and construction equipment in the US are sized according to the English system. Many people in the US are familiar with the weights and measures included in the English system. However, the manner in which the different measures work together (such as how many feet are in a yard) is not always known or understood. *Table 1* lists weights and measures that are commonly used by plumbers.

Table 1 English System—Common Weights and Measures

Linear Measures		
1 foot	=	12 inches
1 yard	=	36 inches
1 yard	=	3 feet
1 mile	=	5,280 feet
1 mile	=	1,760 yards
Area Measures		
1 square foot	=	144 square inches
1 square yard	=	9 square feet
1 acre	=	43,560 square feet
1 square mile	=	640 acres
Volume Measures		
1 cubic foot	=	1,728 cubic inches
1 cubic yard	=	27 cubic feet
1 freight ton	=	40 cubic feet
1 register ton	=	100 cubic feet
Liquid Volume Measures		
1 gallon	=	128 fluid ounces
1 gallon	=	4 quarts
1 barrel	=	31.5 gallons
1 petroleum barrel	=	42 gallons
Weights		
1 pound	=	16 ounces
1 hundred weight	=	100 pounds
1 ton	=	2,000 pounds
Pressure Measures		
1 pound per square inch absolute	=	14.7 pounds at sea level
1 pound per square inch gauge	=	Varies with altitude

1.2.0 Understanding the Metric System of Weights and Measures

The metric system was developed in France in the 18th century. It has since become a worldwide standard in almost all professional and scientific fields. The metric system is also called the Système International d'Unités (SI), or International System of Units.

One of the major advantages of the metric system is that it is rigorously standardized. This makes it easy to memorize the basic system. There are only seven basic units of measurement in the modern metric system (see *Table 2*). All other metric measures result from various combinations of these basic units. Multiples and fractions of the basic units are expressed as powers of 10 (except for seconds, the measure of time, of which there are 60 in a minute). A standard set of prefixes is used to denote these larger or smaller numbers (refer to *Table 2*). That way, you will always know that a kilometer is 1,000 meters just by looking at the prefix. Note that the basic measure of mass, the kilogram, is already a multiple. A kilogram is 1,000 grams. Because a gram is such a small amount of mass, however, a larger base unit seemed more appropriate.

With the growing popularity of international standards, such as the International Plumbing Code®, plumbers will frequently encounter metric weights and measures. The federal government uses metric measurements in its contracts, so plumbers working on a federal government project will need to understand metric weights and measures. And as more and more companies conduct their business internationally, they adopt metric measurements as well.

> **CAUTION**
>
> Always make sure that you express measurements using the correct system. Errors caused by using the wrong system of measurement can cost time and money. Never use a metric tool on English-system pipe and fittings, or vice versa. You could damage both the tool and the fitting.

1.3.0 Converting Measurements

When working with systems of measurement, you may be required to convert decimals to fractions. Converting decimals of a foot to fractions is a two-step process. First, multiply the decimal by 12, which is the number of inches in a foot. Then, multiply that number by the base of the fraction

Table 2 Basic Measures of the Metric System

The unit of...	Is called the...
Length	Meter
Mass	Kilogram
Temperature	Kelvin
Time	Second
Pressure	Pascal
Electric current	Ampere
Light intensity	Candela
Substance amount	Mole

The prefix...	Means...
Micro-	One millionth
Milli-	One thousandth
Centi-	One hundredth
Deci-	One tenth
Deka-	Ten times
Hecto-	One hundred times
Kilo-	One thousand times
Mega-	One million times

you are seeking. For example, multiply by 16 if you are looking for sixteenths of an inch, by 8 if you are looking for eighths of an inch, and so on.

Another important type of conversion is between the English and metric systems. You've probably seen tools marked with lengths in both inches and millimeters, or fixtures with shipping weights in both pounds and kilograms. However, there may be times in the field when you have to perform a conversion calculation yourself. You should become familiar with the basic conversions between the English and metric systems.

Appendix A provides a reference for converting the most common weights and measures. To convert a measurement into its equivalent in the other system, multiply the measurement by the number, called the conversion factor, in the far right column. For example, you want to convert 5 miles into kilometers. Referring to the table, you see that the number must be multiplied by 1.6. The result is that 5 miles equals 8 kilometers.

Sometimes a conversion requires two calculations, one within a system and one between systems. For example, to convert from feet to meters, (where you do not have an exact conversion factor) first use the factor for converting feet to centimeters (30.48) and then divide the result by 100 to convert from centimeters to meters. Practice calculating some conversions of your own, referring to *Table 2* if needed to convert within the metric system.

The Standardization of Measurement Systems

The science of the standardization of measures is called metrology. Beginning in the late 1880s, many countries established laboratories to develop and maintain standards for weights and measures. Two of the most famous of these standards laboratories are the National Institute of Standards and Technology (NIST) in the US and the National Physical Laboratory (NPL) in the United Kingdom. During the twentieth century, these laboratories developed guidelines for an international system of weights and measures. Since then, most countries have adopted these guidelines. The International Bureau of Weights and Measures, located in France, is now responsible for establishing and maintaining international standards. The other standards laboratories help implement these standards in their own countries.

History of the Metric System

The idea of basing weights and measures on a decimal system (numbers that are multiples of 10) was first proposed in the sixtenth century by the Dutch engineer Simon Stevin. In 1790, Thomas Jefferson proposed that the new US develop a decimal weights and measures system. The same year, King Louis XVI of France instructed scientists in his country to develop such a system. Five years later, France adopted the metric system. In 1875, France hosted the Convention of the Metre. Eighteen nations, including the US, signed a treaty that established an international body to govern the development of weights and measures. In 1960, this organization officially changed the name of the metric system to the International System of Units (SI).

The metric system was designed with three goals in mind:

- The system should use units based on unchanging quantities in nature.
- All measurement units should derive from only a few base units.
- The system should use multiples of 10.

The modern metric system still follows these three simple guidelines, except for units of time.

Originally, the meter was measured as one 10-millionth the distance from the North Pole to the equator. Bars of that length were constructed in brass and later in platinum and sent to the standards laboratories of participating countries. Scientists then learned that the actual distance from the pole to the equator was different from the measurement the French scientists used. The length of the meter was recalculated using more precise methods, including the distance traveled by a light beam in a fraction of a second.

In 1994, the US government required that consumer products feature measurements in both metric (SI) and English form. Government-issued construction contracts now use metric measurements. With the advent of the International Plumbing Code® and other international standards, the US may eventually follow the rest of the world and completely adopt the metric system of weights and measures.

Additional Resources

International Plumbing Code®, Latest Edition. Falls Church, VA: International Code Council.
Math to Build On: A Book for Those Who Build. 1993. Johnny and Margaret Hamilton. Clinton, NC: Construction Trades Press.

1.0.0 Section Review

1. The English system of measurement is based on the _____.

   a. decimal system
   b. Colonial system
   c. metric system
   d. British Imperial system

2. The number of basic units of measurement in the modern metric system is _____.

   a. 5
   b. 7
   c. 10
   d. 12

3. The conversion factor for converting feet to centimeters is _____.

   a. 25.4
   b. 0.9144
   c. 30.48
   d. 1.6

SECTION TWO

2.0.0 MEASURING AREA AND VOLUME

Objective

Describe how to measure area and volume.

a. Describe how to measure area.
b. Describe how to measure volume.

Trade Terms

Area: A measure of a surface, expressed in square units.

Circle: A surface consisting of a curve drawn all the way around a point. The curve keeps the same distance from that point.

Cube: A rectangular prism in which the lengths of all sides are equal. When used with another form of measurement, such as cubic meter, the term refers to a measure of volume.

Cubic foot: The basic measure of volume in the English system. There are 7.48 gallons in a cubic foot.

Cubic meter: The basic measure of volume in the metric system. There are 1,000 liters in a cubic meter.

Cylinder: A pipe- or tube-shaped space with a circular cross section.

Decimal of a foot: A decimal fraction where the denominator is either 12 or a power of 12.

Gallon: In the English system, the basic measure of liquid volume. There are 7.48 gallons in a cubic foot.

Isosceles triangle: A triangle in which two of the sides are of equal length.

Liter: In the metric system, the basic measure of liquid volume. There are 1,000 liters in a cubic meter.

Prism: A volume in which two parallel rectangles, squares, or right triangles are connected by rectangles.

Rectangle: A four-sided surface in which all corners are right angles.

Right triangle: A three-sided surface with one angle that equals 90 degrees.

Square: A rectangle in which all four sides are equal lengths. When used with another form of measurement, such as square meter, the term refers to a measure of area.

Square foot: In the English system, the basic measure of area. There are 144 square inches in 1 square foot.

Square meter: In the metric system, the basic measure of area. There are 10,000 square centimeters in a square meter.

Volume: A measure of a total amount of space, measured in cubic units.

Plumbers must know how to measure the size of flat surfaces and the space inside different types of containers. Typical flat surfaces include building foundations, floors, roofs, and parking lots. Pipes are probably the most common container you will work with as a plumber. Swimming pools, interceptors, and lavatories are also types of containers.

In this section, you will learn how to measure flat surfaces and the space inside containers. With practice, these measurements will become as familiar to you as any other tool you use on the job, like a wrench or a plumb bob. *Appendix B* offers a quick reference to the basic formulas for area and volume.

2.1.0 Measuring Area

Area is the measurement of a flat surface (see *Figure 1*). Area is measured using two dimensions: length and width. Floor space is a common type of area that plumbers encounter on the job. Plumbers must take floor space into account when designing plumbing installations. In this section you will learn how to calculate the area of floor spaces and other surfaces.

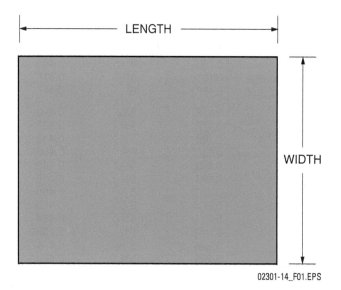

02301-14_F01.EPS

Figure 1 The concept of area.

In the English system, area is expressed in square feet. You will also see this written as sq ft or ft². In the metric system, area is measured in square meters. This can also be written as m².

Length and width are measured in decimals of a foot. You have already learned that a decimal of a foot is a decimal fraction in which the denominator is 12, instead of the usual 10 (see *Table 3*).

You can also convert inches to their decimal equivalent in feet without the table of decimal equivalents. All you need to do is some simple division. First, express the inches as a fraction that has 12 as the denominator. You use 12 because there are 12 inches in a foot. Then reduce the fraction and convert it to a decimal.

Table 3 Inches Converted to Decimals of a Foot

Inches	Decimals of a Foot	Inches	Decimals of a Foot	Inches	Decimals of a Foot	Inches	Decimals of a Foot
1/16	0.005	3 1/16	0.255	6 1/16	0.505	9 1/16	0.755
1/8	0.010	3 1/8	0.260	6 1/8	0.510	9 1/8	0.760
3/16	0.016	3 3/16	0.266	6 3/16	0.516	9 3/16	0.766
1/4	0.021	3 1/4	0.271	6 1/4	0.521	9 1/4	0.771
5/16	0.026	3 5/16	0.276	6 5/16	0.526	9 5/16	0.776
3/8	0.031	3 3/8	0.281	6 3/8	0.531	9 3/8	0.781
7/16	0.036	3 7/16	0.286	6 7/16	0.536	9 7/16	0.786
1/2	0.042	3 1/2	0.292	6 1/2	0.542	9 1/2	0.792
9/16	0.047	3 9/16	0.297	6 9/16	0.547	9 9/16	0.797
5/8	0.052	3 5/8	0.302	6 5/8	0.552	9 5/8	0.802
11/16	0.057	3 11/16	0.307	6 11/16	0.557	9 11/16	0.807
3/4	0.063	3 3/4	0.313	6 3/4	0.563	9 3/4	0.813
13/16	0.068	3 13/16	0.318	6 13/16	0.568	9 13/16	0.818
7/8	0.073	3 7/8	0.323	6 7/8	0.573	9 7/8	0.823
15/16	0.078	3 15/16	0.328	6 15/16	0.578	9 15/16	0.828
1	0.083	4	0.333	7	0.583	10	0.833
1 1/16	0.089	4 1/16	0.339	7 1/16	0.589	10 1/16	0.839
1 1/8	0.094	4 1/8	0.344	7 1/8	0.594	10 1/8	0.844
1 3/16	0.099	4 3/16	0.349	7 3/16	0.599	10 3/16	0.849
1 1/4	0.104	4 1/4	0.354	7 1/4	0.604	10 1/4	0.854
1 5/16	0.109	4 5/16	0.359	7 5/16	0.609	10 5/16	0.859
1 3/8	0.115	4 3/8	0.365	7 3/8	0.615	10 3/8	0.865
1 7/16	0.120	4 7/16	0.370	7 7/16	0.620	10 7/16	0.870
1 1/2	0.125	4 1/2	0.375	7 1/2	0.625	10 1/2	0.875
1 9/16	0.130	4 9/16	0.380	7 9/16	0.630	10 9/16	0.880
1 5/8	0.135	4 5/8	0.385	7 5/8	0.635	10 5/8	0.885
1 11/16	0.141	4 11/16	0.391	7 11/16	0.641	10 11/16	0.891
1 3/4	0.146	4 3/4	0.396	7 3/4	0.646	10 3/4	0.896
1 13/16	0.151	4 13/16	0.401	7 13/16	0.651	10 13/16	0.901
1 7/8	0.156	4 7/8	0.406	7 7/8	0.656	10 7/8	0.906
1 15/16	0.161	4 15/16	0.411	7 15/16	0.661	10 15/16	0.911
2	0.167	5	0.417	8	0.667	11	0.917
2 1/16	0.172	5 1/16	0.422	8 1/16	0.672	11 1/16	0.922
2 1/8	0.177	5 1/8	0.427	8 1/8	0.677	11 1/8	0.927
2 3/16	0.182	5 3/16	0.432	8 3/16	0.682	11 3/16	0.932
2 1/4	0.188	5 1/4	0.438	8 1/4	0.688	11 1/4	0.938
2 5/16	0.193	5 5/16	0.443	8 5/16	0.693	11 5/16	0.943
2 3/8	0.198	5 3/8	0.448	8 3/8	0.698	11 3/8	0.948
2 7/16	0.203	5 7/16	0.453	8 7/16	0.703	11 7/16	0.953
2 1/2	0.208	5 1/2	0.458	8 1/2	0.708	11 1/2	0.958
2 9/16	0.214	5 9/16	0.464	8 9/16	0.714	11 9/16	0.964
2 5/8	0.219	5 5/8	0.469	8 5/8	0.719	11 5/8	0.969
2 11/16	0.224	5 11/16	0.474	8 11/16	0.724	11 11/16	0.974
2 3/4	0.229	5 3/4	0.479	8 3/4	0.729	11 3/4	0.979
2 13/16	0.234	5 13/16	0.484	8 13/16	0.734	11 13/16	0.984
2 7/8	0.240	5 7/8	0.490	8 7/8	0.740	11 7/8	0.990
2 15/16	0.245	5 15/16	0.495	8 15/16	0.745	11 15/16	0.995
3	0.250	6	0.500	9	0.750	7	1.000

Consider the following example. To find the decimal equivalent in feet of 3 inches, follow these three steps:

Step 1 Express 3 inches as a fraction with 12 as the denominator.

$$^3/_{12}$$

Step 2 Reduce the fraction.

$$^3/_{12} = {}^1/_4$$

Step 3 Convert the reduced fraction to a decimal. You can do this by dividing the denominator, 4, into 1.00.

$$1.00 \div 4 = 0.25$$

Thus, 3 inches converts to 0.25 of a foot. For more complicated fractions, you may want to use a calculator.

In plumbing, it will be useful for you to know how to calculate the area of three basic shapes:

- Rectangles
- Right triangles
- Circles

With this knowledge, you will be able to calculate most of the types of areas you will encounter on the job. The following sections review how to calculate the area of rectangles, right triangles, and circles.

2.1.1 Measuring the Area of a Rectangle

A rectangle (*Figure 2A*) is a four-sided figure in which all four corners are right angles. To calculate the area (A) of a rectangle, multiply the length (l) by the width (w). Expressed as a formula, the calculation is:

$$A = lw$$

A square is a type of rectangle in which all four sides are the same length. Use the same formula to determine the area of a square (see *Figure 2B*). You can also calculate the area of a square by squaring the length of one of the sides:

$$A = s^2$$

Try this formula on a rectangle that is 5 feet 1½ inches by 8 feet 10¾ inches:

Step 1 First, convert the inches to decimals of a foot.

$$5 \text{ feet } 1\frac{1}{2} \text{ inches} = 5.1250 \text{ feet}$$

$$8 \text{ feet } 10\frac{3}{4} \text{ inches} = 8.895833 \text{ feet}$$

Step 2 Round your answer to three decimal places, ensuring that roundups are accurate. When rounding numbers, any number followed by 5 through 9 is rounded up, while any number followed by 0 through 4 remains the same.

$$5.1250 \text{ feet} = 5.125 \text{ feet}$$

$$8.895833 \text{ feet} = 8.896 \text{ feet}$$

Note that the number of decimal places you round to depends on the job being done and the desired level of accuracy. When in doubt, ask your supervisor.

Step 3 Multiply the length by the width.

$$A = lw$$

$$A = 5.125 \times 8.896$$

$$A = 45.592 \text{ sq ft}$$

The total area of a rectangle that is 5 feet 1½ inches by 8 feet 10¾ inches, is 45.592 square feet.

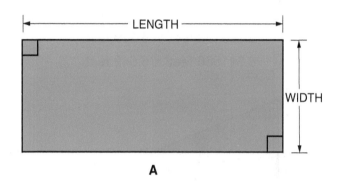

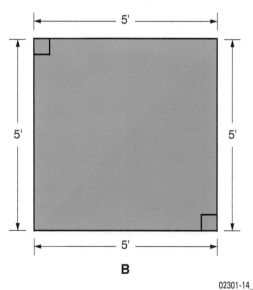

02301-14_F02.EPS

Figure 2 Measuring the area of a rectangle and a square.

2.1.2 Measuring the Area of a Right Triangle

A right triangle is a three-sided figure that has an internal angle that equals 90 degrees (see *Figure 3*). The area of a right triangle is one-half the product of the base (b) and height (h). In other words:

$$A = (\tfrac{1}{2})(bh)$$

An isosceles triangle is a type of triangle that has two sides of equal length. Use the same formula to calculate the area of an isosceles right triangle (see *Figure 3*).

Test this formula on a right triangle where the height is 5 feet 2¾ inches and the base is 6 feet 6½ inches. Calculate the area for this triangle by following these steps:

Step 1 Convert the inches into decimals of a foot.

5 feet 2¾ inches = 5.229166 feet

6 feet 6½ inches = 6.541666 feet

Step 2 Round your answer to three decimal places, ensuring that roundups are accurate.

5.229166 feet = 5.229 feet

6.541666 feet = 6.542 feet

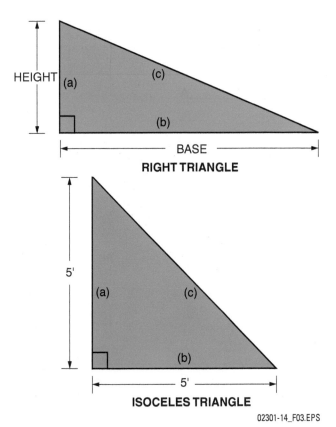

Figure 3 Right triangles.

Step 3 Determine the area of the triangle.

$$A = (\tfrac{1}{2})(bh)$$
$$A = (\tfrac{1}{2})(6.542 \times 5.229)$$
$$A = (\tfrac{1}{2})(34.208)$$
$$A = 17.104 \text{ sq ft}$$

The area of a right triangle with a height of 5 feet 2¾ inches and base of 6 feet 6½ inches long is 17.104 square feet.

Sometimes, you might not know the length of one of the legs in a right triangle. To calculate the missing leg, you can use the Pythagorean theorem. The Pythagorean theorem states the following:

$$a^2 + b^2 = c^2$$

Refer to *Figure 3*. You see that a and b are the two sides that meet at the 90-degree angle. The side marked c is the hypotenuse of the triangle. To find either a or b, use the following calculations:

$$a = \sqrt{(c^2 - b^2)}$$
$$b = \sqrt{(c^2 - a^2)}$$

Use the square root function on your calculator to obtain the square root of a, b, and c. Most calculators have a square root key (see *Figure 4*). It is marked with the $\sqrt{}$ symbol. To use it, first enter the number for which you want to calculate the square root. Then press the square root key. The calculator will display the square root of the number you entered.

If the number you entered represented feet, then the decimals must be converted to inches and fractions of an inch. If the number you entered was in inches, the decimals must be converted to fractions of an inch.

2.1.3 Measuring the Area of a Circle

A circle is a one-sided figure where any point on the side is the same distance from the center. The strainer on a can wash drain is an example of a circular surface you will encounter as a plumber. The area of a circle is a little more complicated to calculate than that of a rectangle or triangle. To calculate the area of a circle, you will use the radius of the circle and pi. The radius of a circle is one-half the diameter (which is formed by a straight line through the center of the circle to its edges). Pi is equal to 3.1416 and is usually represented by the Greek letter π. First, multiply the radius (r) by itself (this is called squaring). Then, multiply the result by pi.

$$A = \pi r^2$$

NCCER – *Plumbing Level Three* 02301-14

To find the area of a circle with a diameter of 4 feet 3½ inches, for example, follow these steps:

Step 1 The radius is one-half the diameter, or 2 feet 1¾ inches. Convert the inches to decimals of a foot.

2 feet 1¾ inches = 2.145833 feet

Step 2 Round your answer to three decimal places, ensuring that roundups are accurate.

2.145833 feet = 2.146 feet

Step 3 Use the formula to calculate the area. Use four decimal places for pi, as given above.

$$A = \pi r^2$$
$$A = (3.1416)(2.146)^2$$
$$A = (3.1416)(4.605)$$
$$A = 14.467 \text{ sq ft}$$

You have determined that a circle with a diameter of 4 feet 3½ inches has an area of 14.467 square feet.

2.2.0 Measuring Volume

Volume is a measure of capacity, the amount of space within something (see *Figure 5*). Volume is measured using three dimensions: length, width, and height. Plumbers must be able to calculate the volumes of various spaces accurately to ensure the efficient operation of plumbing installations. Some of the volumes that plumbers need to be able to calculate include:

- Drain receptors
- Fresh-water pipes
- Drain, waste, and vent (DWV) pipes
- Grease and oil interceptors
- Water heaters
- Septic tanks
- Swimming pools
- Catch basins

In the English system, space is measured in cubic feet (abbreviated as cu ft or ft³). The measure of liquid volume is gallons (abbreviated as gal.). There are 7.48 gallons in 1 cubic foot. In the metric system, the measure of space is cubic meters (abbreviated as m³) and the measure of liquid is in liters (abbreviated as L). There are 1,000 liters in a cubic meter.

You can construct three-dimensional spaces by combining flat surfaces like rectangles, squares, right triangles, and circles. When two parallel rectangles, squares, or right triangles are connected by

SQUARE ROOT KEY

02301-14_F04.EPS

Figure 4 Square root key on a calculator.

rectangles, the space is called a prism. A tube with a circular cross section is called a cylinder.

2.2.1 *Measuring the Volume of a Rectangular Prism*

Rectangular prisms are three-dimensional spaces where rectangles make up the sides, top, and bottom (see *Figure 6*). A 2 × 4 piece of lumber is a rectangular prism, as is the interior of a truck bed. To calculate the volume of a rectangular prism, multiply the length by the width by the height.

For example, assume you have a rectangle with a length of 5 feet 3¾ inches, a width of 2 feet 1½ inches, and a height of 6 feet. Calculate the volume this way:

Step 1 Convert the inches into decimals of a foot.

5 feet 4¾ inches = 5.395833 feet

2 feet 1½ inches = 2.1250 feet

Step 2 Round your answer to three decimal places, ensuring that roundups are accurate.

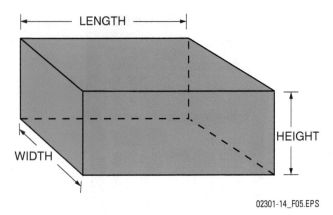

Figure 5 The concept of volume.

5.395833 feet = 5.396 feet
2.1250 feet = 2.125 feet

Step 3 Calculate the volume using the formula.

$$V = lwh$$
$$V = (5.396 \times 2.125)(6)$$
$$V = (11.467)(6)$$
$$V = 68.802 \text{ cu ft}$$

The volume of the given rectangle is 68.802 cubic feet.

A cube is a rectangular prism where the length, width, and height are all the same. To find the volume of a cube, use the formula for finding the volume of a rectangular prism.

2.2.2 Measuring the Volume of a Right Triangular Prism

The volume of a right triangular prism is ½ times the product of the length, width, and height (refer to *Figure 6*). For a right triangle prism with a base and width of 3 feet and a height of 5 feet, the volume would be calculated as follows:

$$V = (\tfrac{1}{2})(bwh)$$
$$V = (\tfrac{1}{2})(3 \times 3)(5)$$
$$V = (\tfrac{1}{2})(9)(5)$$
$$V = (\tfrac{1}{2})(45)$$
$$V = 22.5 \text{ cu ft}$$

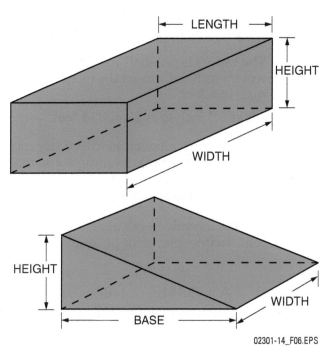

Figure 6 Rectangular prism and right triangular prism.

The volume of the right triangular prism is 22.5 cubic feet.

2.2.3 Measuring the Volume of a Cylinder

A cylinder is a straight tube with a circular cross section (*Figure 7*). It is a common plumbing shape. Pipes are long, narrow cylinders. Many water heaters are also cylindrical. You can calculate the volume of a cylinder in cubic feet by multiplying the area of the circular base (πr^2) by the height (h).

In the following example, a cylinder has an inside diameter of 30 feet and a height of 50 feet 9 inches. Obtain the volume in cubic feet using the following steps:

Step 1 The ends of a cylinder are circles. So first, you must determine the area of the circle. In the example, the radius (half the diameter) is 15 feet.

$$A = \pi r^2$$
$$A = (3.1416)(15^2)$$
$$A = (3.1416)(225)$$
$$A = 706.86 \text{ sq ft}$$

As Easy as π

The value of π represents the ratio of a circle's circumference to its diameter. The circumference is the length of a circle's curved edge. A diameter is a straight line drawn through the middle of a circle, from one side to the other. On the job, you can use 3.1416 to represent π. However, π may actually be an infinite number. Mathematicians are still not sure!

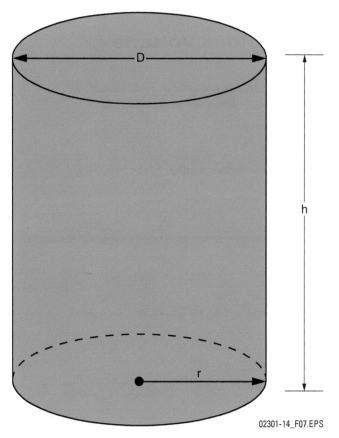

Figure 7 Cylinder.

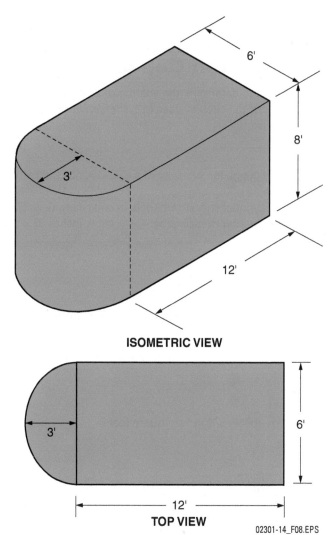

ISOMETRIC VIEW

TOP VIEW

02301-14_F08.EPS

Figure 8 Complex volume.

Step 2 To calculate the volume of the tank in cubic feet, multiply the area of the circle by the height. Remember to convert the measurement into decimals of a foot.

$$V = Ah$$
$$V = (706.86)(50.75)$$
$$V = 35,873.145 \text{ cu ft}$$

The volume of a tank with an inside diameter of 30 feet and a height of 50 feet 9 inches is 35,873.145 cubic feet.

2.2.4 Measuring the Volume of a Complex Shape

You can calculate the volume of a complex shape by calculating the volume of its various parts. For example, divide the shape in *Figure 8* into a half cylinder and a rectangular solid.

The volume of the rectangular portion is found as follows:

$$V = lwh$$
$$V = (12)(6)(8) \text{ feet}$$
$$V = 576 \text{ cu ft}$$

The volume of the half cylinder is found as follows:

$$V = (\pi r^2 h) \div 2$$
$$V = [(3.1416)(3^2)(8)] \div 2$$
$$V = (226.1952) \div 2$$
$$V = 113.098 \text{ cu ft}$$

Now add the two together to get the total volume:

$$V = 576 \text{ cu ft} + 113.098 \text{ cu ft}$$
$$V = 689.098 \text{ cu ft}$$

The total volume of the complex shape is 689.098 cubic feet.

Alternate Method to Calculate Cylinder Volume and Gallon Capacity

A shortcut to determine the volume of a cylinder uses the diameter rather than the radius. First, square the diameter of the circular base (d^2). Then, multiply the result by 0.7846 and, finally, multiply by the length (l). The formula is:

$$V = (d^2)(0.7854)l$$

To find the gallon capacity of a cylinder, use the result of the above calculation in either of the following formulas:

When cylinder measurement is in cubic feet: $G = V(7.48)$
When cylinder measurement is in cubic inches: $G = V \div 231$

Additional Resources

Code Check Plumbing: A Field Guide to Plumbing. 2000. Redwood Kardon, Michael Casey, and Douglas Hansen. Newtown, CT: Taunton Press.

2.0.0 Section Review

1. An abbreviation for square feet is _____.
   a. sf
   b. ft³
   c. ft. sq.
   d. ft²

2. The number of gallons in 1 cubic foot is _____.
   a. 0.7846
   b. 1.315
   c. 7.48
   d. 8.33

SECTION THREE

3.0.0 AREA AND VOLUME IN PLUMBING

Objective

Describe the practical applications of area and volume in plumbing.

   a. Describe how to perform area calculations in plumbing.
   b. Describe how to perform volume calculations in plumbing.
   c. Describe how to perform load calculations.

P lumbers often need to calculate areas and volumes while on the job. You might need to measure a floor in order to calculate drainage. Or you might be asked to determine the space taken up by an appliance to see if it will fit. When that happens, you need to know how to apply the formulas for finding area and volume in order to make accurate measurements and estimates in the field. This section covers some of the ways that the area and volume calculations introduced in the previous section can be applied to real-life situations in the field.

3.1.0 Performing Area Calculations in Plumbing

You have learned how to calculate the area of different shapes. With this knowledge, you can calculate the area of almost any floor plan. Always refer to the scale in a construction drawing to determine the accurate measurements of the floor.

On the job, you may have to calculate the area of a space shaped like a rectangle or a right triangle. Let's say you have to install an area drain in a parking lot that is 30 feet 6½ inches by 50 feet 3 inches. The parking lot's size affects the amount of storm water runoff that the drain will have to handle. This factor determines the size of the area drain you select.

Step 1 First, convert the inches to decimals of a foot.

30 feet 6½ inches = 30.541666 feet
50 feet 3 inches = 50.250 feet

Step 2 Round each answer to three decimal places, ensuring that roundups are accurate.

30.541666 feet = 30.542 feet
50.250 feet = 50.250 feet

Step 3 Multiply the length by the width.

$$A = lw$$
$$A = (30.542)(50.250)$$
$$A = 1{,}534.736 \text{ sq ft}$$

The total area of a parking lot that is 30 feet 6½ inches by 50 feet 3 inches is 1,534.736 square feet.

When calculating the area of a surface that is not a simple rectangle or right triangle, break the floor plan into smaller areas shaped like rectangles and right triangles (see *Figure 9*). Then calculate the area for each of the smaller areas using the appropriate formula.

3.2.0 Performing Volume Calculations in Plumbing

For manufactured items, you can usually find the dimensions and volume in the manufacturer's specifications. For site-built containers, such as large grease interceptors, you will have to calcu-

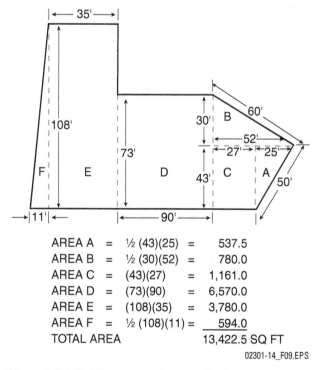

AREA A = ½ (43)(25) = 537.5
AREA B = ½ (30)(52) = 780.0
AREA C = (43)(27) = 1,161.0
AREA D = (73)(90) = 6,570.0
AREA E = (108)(35) = 3,780.0
AREA F = ½ (108)(11) = 594.0
TOTAL AREA   13,422.5 SQ FT

02301-14_F09.EPS

Figure 9 Subdividing an area into simple shapes.

late the volume yourself. Don't forget to calculate the volume of objects inside a container, such as baffles, fittings, or even fixtures. Calculate their volumes and subtract them from the overall volume of the space (see *Figure 10*).

When determining volume, it is also important to consider weight. Take, for example, a swimming pool installed on a patio. The larger the pool, the more water it will hold and the heavier the water inside it will be. Consequently, the deck will require more support to hold the pool. When installing pipes and fixtures, plumbers must be able to calculate the weight of water for a given volume to ensure that the pipe or fixture is adequately supported. Local codes provide requirements on proper support.

Once you have calculated the volume of a container such as a pipe, water tank, slop sink, or interceptor, you must perform one more step to determine its gallon capacity. One gallon of water weighs 8.33 pounds. There are 7.48 gallons in a cubic foot. To find the capacity in gallons of a volume measured in cubic feet, multiply the cubic footage by 7.48. If you multiply the weight of 1 gallon of water by the number of gallons in a cubic foot, you will find that a cubic foot of water weighs 62.31 pounds. Remember that these weights are for

pure water. Wastewater will have other liquids and solids in it. This means the weight of a volume of wastewater may be greater than that of fresh water. To convert measurements in cubic inches to gallons, divide the cubic-inch measurement by 231.

When installing water supply lines, using a smaller pipe size can save the customer both energy and money. You can calculate the difference using the above formula. For example, a ¾-inch-diameter pipe has a cross-sectional area of 0.442 square inches, and a ½-inch-diameter pipe has a cross-sectional area of 0.196 square inches (*Figure 11*). If you perform the calculations, you will find that a ¾-inch pipe will hold 2¼ times the amount of water that a ½-inch pipe will hold. With a larger pipe, a faucet will have to run longer to get hot water to a sink. In addition, the amount of hot water that will be left in the pipe to cool once the faucet is turned off is much greater with a larger pipe. Always consult local codes to determine the water- and energy-saving measures possible for each job.

Plumbers need to be able to calculate the volume and weight of water heaters (*Figure 12*) in order to ensure that there is enough room when installing them, and that the floor can support the weight. Because most tank water heaters are cylindrical in shape, a rough estimate of the volume can be determined by using the measurements provided in the manufacturer's specifications. Obtain the measurements for the diameter of the water heater's base and the water heater's height. Be sure to factor in space for fittings such as valves and controls. This can be done by measuring the height of the fittings and controls. To calculate the area of the circular base of the water heater, use the formula for determining the area of a circle ($A = \pi r^2$). Use chalk or a pencil to draw the resulting circle on the floor in the desired location. Then use a measuring tape to extend a vertical line up from the edge of the circle to the specified height of the water heater, including

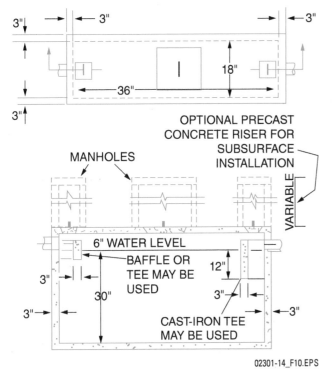

02301-14_F10.EPS

Figure 10 Calculating the volume of a site-built grease interceptor.

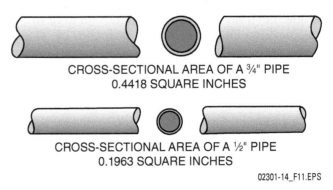

CROSS-SECTIONAL AREA OF A ¾" PIPE
0.4418 SQUARE INCHES

CROSS-SECTIONAL AREA OF A ½" PIPE
0.1963 SQUARE INCHES

02301-14_F11.EPS

Figure 11 Cross-sectional areas of pipe.

the fittings, valves, and controls. Take this measurement from several points along the circumference of the circle, to ensure that the space is clear all around. This will allow you to determine the space to be taken up by the water heater.

To calculate the weight of a full water heater, you will need to know both the weight of the water in the tank and the weight of the water heater itself. First, consult the manufacturer's specifications for the empty weight of the water heater. Then calculate the weight of the water in the tank. Refer to the manufacturer's specifications to determine the gallon or the cubic-foot capacity of the tank. Remember that 1 gallon of water weighs 8.33 pounds and that 1 cubic foot of water weighs 62.31 pounds. Multiply the tank's gallon or cubic-foot capacity by the appropriate weight, and add the weight of the water heater itself to the result. This will give you the total weight of the full water heater.

For example, determine the weight of a water heater that weighs 150 pounds empty and has a tank capacity of 40 gallons. Follow these steps:

Step 1   Multiply the gallon capacity of the tank by the weight of a gallon of water.

40 gallons × 8.33 pounds per gallon = 333.2 pounds

Step 2   Add the dry weight of the water heater to the weight of the water in the tank.

150 pounds + 333.2 pounds = 483.2 pounds

The estimated total weight of the full water heater will be 483.2 pounds.

3.3.0 Performing Load Calculations in Plumbing

When plumbers work on large projects where pipe and fittings are moved by crane, the crane operators need to know the weight of pipe bundles and fittings they have to move. This information allows them to pick up, move, and place their loads safely. Plumbers are responsible for providing this information to crane operators. For safety considerations, it is essential that the information is accurate. If a crane attempts to pick up a load that is heavier or lighter than planned, the crane could topple or the load could swing uncontrolla-

02301-14_F12.EPS

Figure 12 Electric water heater.

bly, causing damage to materials and equipment, and injury or even death.

Fitting and pipe weights can be found in literature provided by pipe manufacturers. Fitting books, also called dimensional catalogs, provide the dimensions and weights of pipe bundles (see *Table 4*) as well as individual pipe lengths and fittings (*Table 5*). Note that weights and dimensions can vary from manufacturer to manufacturer. Always use the fitting book published by the manufacturer of the pipe and fittings that are being installed.

For example, a crane needs to move a load of fifty 8-inch cast-iron quarter-bend fittings. Refer to *Table 5* to find the weight of an individual fitting. Then multiply the weight of the fitting by the number of fittings to be moved:

1 fitting = 23.1 pounds

23.1 pounds × 50 fittings = 1,155 pounds

The estimated total weight of the load to be moved by crane will be 1,155 pounds.

Table 4 Sample Table of Pipe Bundles from a Manufacturer's Fitting Book

Size	Pieces	Weight (in pounds)	Height
1½" × 10'	72	2,052	12"
2" × 10'	54	1,906	11"
3" × 10'	36	1,944	14"
4" × 10'	27	1,922	17"
5" × 10'	24	2,342	20"
6" × 10'	18	2,120	23"
8" × 10'	8	1,367	22"
10" × 10'	6	1,528	26"
12" × 10'	6	1,908	29"
15" × 10'	2	985	19"

Weights are approximate and are for shipping purposes only.

Table 5 Sample Table of Individual Fittings from a Manufacturer's Fitting Book

Size	D	Weight
1½	4¼	1.7
2	4½	2.2
3	5	3.7
4	5½	6.5
5	6½	9.3
6	7	13.4
8	6½	23.1
4 × 3	5½	5.5

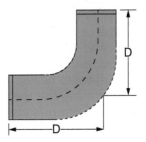

3.0.0 Section Review

1. The area of a 15-foot by 7-foot rectangle is _____.

   a. 2.142 ft²
   b. 8 ft²
   c. 105 ft²
   d. 220 ft²

2. The volume of a length of pipe that is 2 inches in diameter and 6 feet long is _____.

   a. 864 in³
   b. 226.1952 in³
   c. 271.425 in³
   d. 3,542.33 in³

3. A 10-foot length of cast-iron pipe weighs 25 pounds and a cast-iron long bend weighs 6 pounds. A bundle of 20 pipe lengths and 40 fittings weighs _____.

   a. 740 pounds
   b. 500 pounds
   c. 240 pounds
   d. 65 pounds

SECTION FOUR

4.0.0 TEMPERATURE AND PRESSURE

Objective

Explain the concepts of temperature and pressure and how they apply to plumbing installations.

a. Explain the concept of temperature and how it applies to plumbing installations.
b. Explain the concepts of pressure and force and how they apply to plumbing installations.

Trade Terms

Bimetallic thermometer: A thermometer that determines temperature by using the thermal expansion of a coil of metal consisting of two thin strips of metal bonded together.

Celsius scale: A centigrade scale used to measure temperature.

Centigrade scale: A scale divided into 100 degrees. Generally used to refer to the metric scale of temperature measure (see Celsius scale).

Conduction: The transfer of heat energy from a hot object to a cool object.

Equilibrium: A condition in which all objects in a space have an equal temperature.

Fahrenheit scale: The scale of temperature measurement in the English system.

Head: The height of a water column, measured in feet. One foot of head is equal to 0.433 pounds per square inch gauge.

Infrared thermometer: A thermometer that measures temperature by detecting the thermal radiation emitted by an object.

Kelvin scale: The scale of temperature measurement in the metric system.

Liquid thermometer: A thermometer that measures temperature through the expansion of a fluid, such as mercury or alcohol.

Pounds per square inch: In the English system, the basic measure of pressure. Pounds per square inch (psi) is measured in pounds per square inch absolute (psia) and pounds per square inch gauge (psig).

Pressure: The force applied to the walls of a container by the liquid or gas inside.

Temperature: A measure of relative heat as measured by a scale.

Thermal expansion: The expansion of materials in all three dimensions when heated.

Thermometer: A tool used to measure temperature.

Water supply fixture unit (WSFU): The measure of a fixture's load, varying according to the amount of water, the water temperature, and the fixture type.

The previous section showed how to calculate the space inside three-dimensional containers like pipes, interceptors, and swimming pools. Those calculations are an important step when installing a plumbing system. But once a plumbing system is in place, it has to contain and move liquids and gases. Liquids and gases have properties that affect how they behave inside the system. In this section, you will learn about three of the most important of these properties: temperature, pressure, and force.

Temperature, pressure, and force are very important concepts in plumbing. Pipes, fittings, and fixtures are all designed to work within a range of acceptable temperatures and pressures. If those limits are exceeded, the system could be severely damaged, and personal injury could result. A plumbing system that contains materials not appropriate for its operating conditions is also in danger of failing and causing damage and injury. Every plumbing system is designed from the outset with these considerations in mind, so every plumber needs to know the principles that govern temperature and pressure.

4.1.0 Understanding How Temperature Applies to Plumbing Installations

Temperature is a measure of the heat of an object according to a scale. Heat transfers from warmer objects to cooler ones. This is because heat is a form of energy, and energy seeks equilibrium. Equilibrium is a condition in which the temperature is the same throughout the object or space. This is why you get hot when you are working next to a steam pipe; your body is absorbing some of the heat energy of the pipe. The flow of heat energy from a hotter object to a cooler one is called conduction.

Plumbers are very concerned with heat flow when they install a plumbing system. Pipes can lose heat into the surrounding atmosphere, into a nearby wall or ceiling, or into the ground. When the pipe loses too much heat, it freezes. Plumbers are also concerned about how much heat energy enters a plumbing system. Too much heat can cause damage. That is why, for example, hot wastes are allowed to cool before they enter the sanitary system.

4.1.1 Thermometers

Thermometers are tools that measure temperature. They are available in various shapes and are seen in numerous applications. Thermometers are installed wherever there is a need to know a temperature range. Always consult the manufacturer's instructions before installing or using a thermometer.

There are three popular types of thermometers. Liquid thermometers use a glass tube filled with fluid, such as mercury or alcohol. The fluid expands when heated and rises in the tube. Lines on the tube corresponding to the height of the fluid indicate the temperature. Liquid thermometers are commonly used in refrigerators and freezers (see *Figure 13*). Bimetallic thermometers use thermal expansion to show a temperature reading. Inside a bimetallic thermometer is a coil of metal. The coil is made of two thin metal strips that are bonded together. Each of these strips expands at a different rate when heated. This causes the coil to curve when the thermometer is placed near a heat source. An indicator attached to the coil then points to temperature lines marked on a dial (see *Figure 14*). Bimetallic thermometers are found in thermostats and in home and commercial heating/cooling systems. Electrical thermometers (sometimes called digital thermometers) convert electrical resistance or voltages generated by heat into a temperature reading. Handheld infrared thermometers (see *Figure 15*), also called laser thermometers, calculate temperatures by analyzing the heat emitted by an object. This heat is called thermal radiation. Infrared thermometers can sometimes provide false readings when they detect the thermal radiation from other nearby objects. Be sure to follow the manufacturer's in-

02301-14_F13.EPS

Figure 13 Liquid thermometer for refrigerators and freezers.

02301-14_F14.EPS

Figure 14 Bimetallic thermometers.

structions when using infrared thermometers, to ensure you get an accurate measurement.

Thermometers display temperatures using the Fahrenheit scale, the Celsius scale, the Kelvin scale, or sometimes all three (see *Figure 16*). The first two scales are widely used in residential,

Microwave Ovens—The Power to Move Molecules

Microwave ovens are a common kitchen appliance. People use microwaves for everything from thawing frozen food to boiling a cup of water and cooking a steak. Microwave ovens cook food faster and more efficiently than conventional ovens. How do they work? Microwaves are tiny radio waves that are absorbed by water molecules, fats, and sugars. Other materials, such as plastic or ceramic, do not absorb microwaves. When microwaves are absorbed, they convert directly into heat.

When a molecule is heated, scientists say it becomes excited. Heated molecules move fast, bounce against one another, and ricochet off container walls. Though scientists can't actually see individual molecules, they can see the effect of their excited state. When heated molecules bump into a larger particle, the molecules cause the particle to move in a zigzag pattern. This form of motion is called Brownian motion, in honor of Robert Brown, a scientist who mathematically described it in 1827.

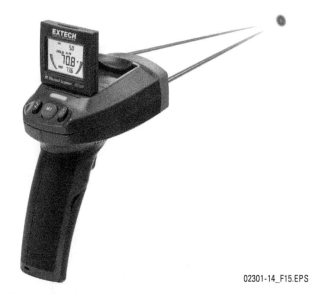

02301-14_F15.EPS

Figure 15 Infrared thermometer.

commercial, and industrial thermometers. On the Fahrenheit (F) scale, 32°F is the freezing point of pure water at sea level. The boiling point of pure water at sea level is 212°F.

Unlike the Fahrenheit scale, the Celsius scale is calibrated to the temperature of pure water at sea-level pressure. On the Celsius (C) scale, therefore, 0°C is the freezing point of pure water and 100°C is the boiling point. The Celsius scale is a centigrade scale because it is divided into 100 degrees (remember that *centi-* means one hundredth).

The Kelvin scale is the basic system of temperature measure in the metric system. Its zero point is the coldest temperature that matter can attain in the universe, according to science. Manufacturers' specifications for fittings and fixtures may be provided in degrees Kelvin.

To convert Fahrenheit measurements into Celsius measurements, first subtract 32 from the Fahrenheit temperature, then multiply the result by 5/9. The formula can be written this way:

$$C = (F - 32) \times (5/9)$$

For example, suppose you want to find the Celsius equivalent of 140°F. Apply the formula as follows:

$$C = (140 - 32) \times (5/9)$$
$$C = [(108) \times 5]/9)$$
$$C = 540/9$$
$$C = 60$$
$$140°F \text{ equals } 60°C$$

To find the Fahrenheit equivalent of a Celsius temperature, first multiply the Celsius number by 9/5. Then add 32. The formula looks like this:

$$F = [C \times (9/5)] + 32$$

Centigrade (Celsius)	Fahrenheit	Kelvin	
C°	F°	K°	
100	212	373	**WATER BOILS**
90	194	363	
80	176	353	
70	158	343	
60	140	333	
50	122	323	
40	104	313	
30	86	303	
20	68	293	
10	50	283	
0	32	273	**WATER FREEZES**
−10	14	263	
−20	−4	253	

02301-14_F16.EPS

Figure 16 Celsius, Fahrenheit, and Kelvin temperature scales.

To test the formula, practice converting 10°C into Fahrenheit:

$$F = [10 \times (9/5)] + 32$$
$$F = (90/5) + 32$$
$$F = 18 + 32$$
$$F = 50$$
$$10°C \text{ equals } 50°F$$

4.1.2 Thermal Expansion

Materials expand when they are heated and contract when they cool. The change in size caused by heat is called thermal expansion. Thermal expansion occurs in all the dimensions of an object: length, width, and height. Thermal expansion is affected by the following factors:

- *Length of the material* – the longer the material, the greater the increase in the length.
- *Temperature change* – the greater the temperature, the greater the increase in size.
- *Type of material* – different materials expand at different rates (see *Figure 17*).

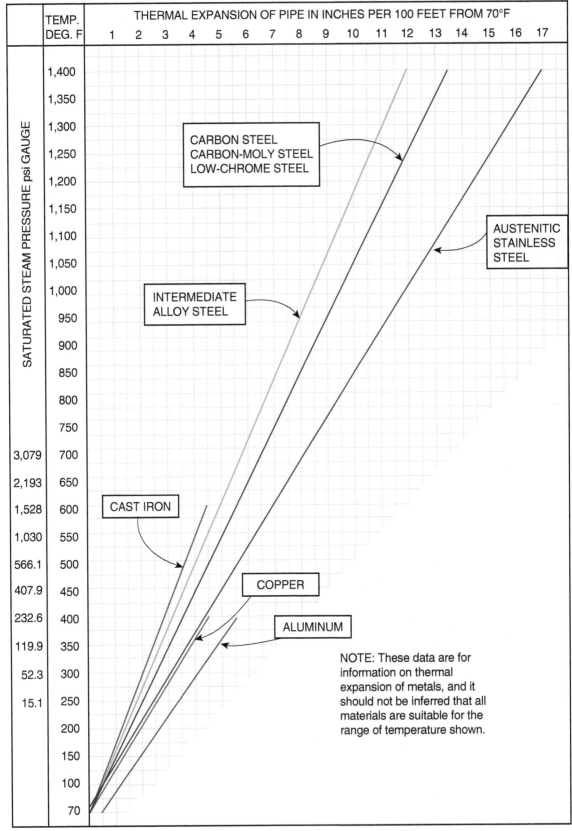

Figure 17 Thermal expansion of different types of pipe materials.

Thermal expansion alters the area and volume of the heated material. For pipe, that means diameter, thread size and shape, joints, bends, and length are all affected. These changes could pose a serious problem for the safe and efficient operation of a plumbing system. Always use materials that are appropriate for the temperature range. Consult your local code or talk to a local code expert about temperature considerations in your area.

The rate of thermal expansion of pipe depends on the temperature of the pipe and the type of material the pipe is made from. Refer to the manufacturer's thermal expansion data (see *Table 6*, for example) for information about the type of pipe you are installing. Your local code will provide information on approved methods for controlling thermal expansion, such as the installation of a pressure-reducing valve. The information will also help you install pipe hangers and supports so that they allow for the correct amount of expansion and contraction.

Using the thermal-expansion information provided by the manufacturer, you can determine pipe expansion (and contraction) at different temperatures. Consider the following example:

You have just installed a 750-foot run of austenitic stainless steel pipe when the air temperature was 50°F. Your supervisor instructs you to calculate the length by which the pipe will expand when steam at 400°F is allowed to flow through the system.

Begin by finding the column for austenitic stainless steel pipe in *Table 6*, and move down the column until you find the expansion coefficient for 400°F (3.88). Now find the expansion coefficient for 50°F (–0.22). Subtract the second expansion coefficient from the first (remember that subtracting negative numbers from positive numbers is the same as adding the numbers):

$$x = 3.88 - (-0.22)$$
$$x = 3.88 + 0.22$$
$$x = 4.1$$

Now multiply the resulting coefficient by the length of the pipe. Notice that the expansion coefficients in *Table 6* are per 100 feet of pipe. This means you must move the decimal point two places to the left for this calculation:

$$7.5 \times 4.1 = 30.75$$

The pipe length will increase by a total of 30.75 inches over its entire length once it has been completely heated by the steam.

4.1.3 Protecting Pipes from Freezing

Plumbers install pipes, fittings, fixtures, and appliances to minimize repeated expansion and contraction. Underground pipes that are at risk of freezing are installed below the frost line. Depending on where in the country you work, the frost-line depth may vary. Consult the local code and building officials. Remember that codes re-

Table 6 Thermal Expansion of Pipe Materials

Degrees Fahrenheit	Expansion (inches per 100 feet from 70° Fahrenheit)			
	Carbon, Carbon/ Molybdenum, and Low Chrome Steel	Austenitic Stainless Steel	Copper	PVC
–50	–0.87	–1.23	–1.32	–4.00
0	–0.51	–0.79	–0.79	–2.33
50	–0.15	–0.22	–0.22	–0.67
70	0.00	0.00	0.00	0.00
100	0.23	0.34	0.34	1.00
150	0.61	0.91	0.91	2.67
200	0.99	1.51	1.51	—
250	1.40	2.08	2.08	—
300	1.82	2.67	2.67	—
350	2.26	3.27	3.27	—
400	2.70	3.88	3.88	—
450	3.16	4.49	4.49	—
500	3.62	5.12	5.12	—
550	4.11	5.74	5.74	—
600	4.60	6.39	6.39	—

quire accessibility to valves on water service pipes and cleanouts on the sanitary sewer drainpipe.

Snow acts like insulation and keeps frost from penetrating the soil too deeply. Therefore, the frost depth on snow-covered ground is significantly less than it is in other areas that are routinely cleared of snow, such as driveways. Therefore, in colder climates, do not locate the service entrance under a sidewalk, driveway, or patio. Note also that frost penetrates wet soils more deeply than it does dry soils. If possible, locate water pipes under lawns or other areas where snow will remain. Inside buildings, use insulation to protect pipes from freezing.

4.2.0 Understanding How Pressure and Force Apply to Plumbing Installations

How can a slug of air in a drainpipe blow out a fixture trap seal? Why does water in a tank high above ground flow with more force than water in a tank at ground level? The answer to both questions has to do with pressure. Pressure is the force created by a liquid or gas inside a container. When a liquid or gas is compressed, it exerts more pressure on its container. If the compressed liquid or gas finds an outlet, it will escape. The speed of the escaping liquid or gas is proportional to the amount of pressure to which it is subjected. In the examples above, the pressure applied by the weight of a water column causes both reactions.

In the English system of measurement, pressure is measured in pounds per square inch (psi). There are two ways to measure psi:

- *Pounds per square inch absolute (psia)* – uses a perfect vacuum as the zero point. The total pressure that exists in the system is called the absolute pressure.

- *Pounds per square inch gauge (psig)* – uses local air pressure as the zero point. Gauge pressure is pressure that can be measured by gauges.

Psig is a relative measure. Absolute pressure is equal to gauge pressure plus the atmospheric pressure. That's why if you are in Miami (at sea level), 0 psig will be 14.7 psia. However, if you are in Denver (elevation 5,280 feet), 0 psig will be 12.1 psia. The relationship between the two forms of measurement is easy to remember:

psia = psig + local atmospheric pressure

Pressure is related to temperature; increasing the pressure in a container will increase the temperature of the liquid or gas inside. Likewise, when you increase the temperature of a liquid or gas in a closed container, the pressure on the container will rise. This is why steam whistles out of a kettle when you boil water on the stove. If the steam did not escape, the pressure created by the steam would eventually blow off the lid of the kettle. In the case of excess temperature and pressure in a 40-gallon water heater, the consequences would be much more severe.

Pressure is a fundamental concept in plumbing. Two of the most important forms of pressure in plumbing are air pressure and water pressure. Atmospheric air is a gas, and therefore it exerts pressure. Air at sea level (an altitude of zero feet) creates 14.7 pounds of pressure per square inch of surface area. Residential and commercial plumbing installations operate at this pressure. Vents maintain air pressure within a plumbing system. Vents allow outside air to enter and replace wastewater flowing out of various parts of the system.

Water pressure is especially important to plumbers. The next sections discuss water hammer, a damaging effect of pressure, and head, a way of measuring water pressure.

Fahrenheit, Celsius, and Kelvin

The Fahrenheit, Celsius, and Kelvin scales are named after pioneering scientists who helped develop modern methods of temperature measure.

Daniel Gabriel Fahrenheit (1686–1736) was a German physicist. In 1709, he invented a liquid thermometer that used alcohol. Five years later, he invented the mercury thermometer. In 1724, he developed a temperature scale that started at the freezing point of salt water. This is the modern Fahrenheit scale.

Anders Celsius (1701–1744) was a Swedish astronomer and professor who developed a centigrade scale in 1742. He set the scale's 0° mark at the freezing point of pure water and the 100° mark at the boiling point of pure water. The modern centigrade temperature scale is the same one that Celsius developed, and it is named in his honor.

William Thomson, Lord Kelvin (1824–1907), was a British physicist and professor. He conducted pioneering research in electricity and magnetism. He also participated in a project to lay the first communications cable across the Atlantic in 1857. Kelvin was also an inventor. In the mid-19th century, he developed a measuring system based on the Celsius scale. The zero point of the scale is the theoretical lowest temperature in the universe.

The Kelvin scale is the metric system's standard temperature scale.

4.2.1 Water Hammer

When liquid flowing through a pipe is suddenly stopped, vibration and pounding noises, called water hammer, result. Water hammer is a destructive form of pressure. The forces generated at the point where the liquid stops act like an explosion. When water hammer occurs, a high-intensity pressure wave travels back through the pipe until it reaches a point of relief. The shock wave pounds back and forth between the point of impact and the point of relief until the energy dissipates. Installing water hammer arresters prevents this type of pressure from reducing the service life of pipes.

The Plumbing & Drainage Institute (PDI) has developed an alphabetical rating system for water hammer arresters. Arresters are rated from A to F by size and shock absorption capability. In the PDI scale, A units are the smallest and F units are the largest. The size of the arrester will depend on the load of the fixtures on the branch. The load factors are represented by a measurement called the water supply fixture unit (WSFU). You will learn how to use WSFUs in the module titled *Sizing and Protecting the Water Supply System*. For now, you only need to know that the greater the number of WSFUs on the branch, the larger the water hammer arrester you should install (see *Table 7*).

Table 7 Sizing a Water Hammer Arrester

PDI Units	A	B	C	D	E	F
WSFUs	1–11	12–32	33–60	61–113	114–154	155–330

4.2.2 Head

Water pressure is measured by the force exerted by a water column. A water column can be a well, a run of pipe, or a water tower. The height of a water column is called head. Head is measured in feet. The pressure exerted by a rectangular column of water 1 inch long, 1 inch wide, and 12 inches (1 foot) high is 0.433 psig. Another way to express the same measurement is to say that 1 foot of head is equal to 0.433 psig. Each foot of head, regardless of elevation, exerts 0.433 psig. It takes 27.648 inches of head to exert 1 psig.

The pressure, however, is determined solely by vertical height. Fittings, angles, and horizontal runs do not affect the pressure exerted by a column of water. You can use a simple formula to calculate psi:

$$\text{psi} = \text{elevation} \times 0.433$$

Use this formula to determine the pressure required to raise water to a specified height and the pressure created by a water column from an elevated supply (see *Figure 18*).

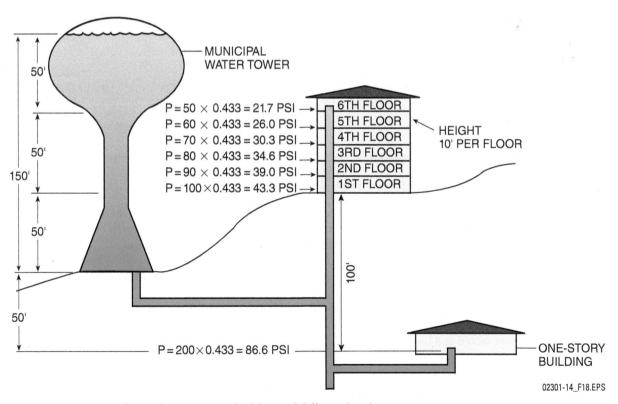

Figure 18 Water storage tank supplying water to buildings of different heights.

If you want to determine the pressure required to raise water to the top of a 50-foot building, perform the following calculation:

$$psi = 50 \times 0.433$$
$$psi = 21.65$$

To raise a column of water 50 feet, therefore, you need to supply 21.65 psi. If you want to determine the pressure created by a 30-foot head of water in a plumbing stack, you would calculate the following:

$$psi = 30 \times 0.433$$
$$psi = 12.99$$

The column of water is creating a force of 12.99 psi on the bottom of the stack.

Consider, for example, a six-floor building with a 150-foot-high tank (refer to *Figure 18*). Note the decrease in pressure as each floor gets higher. You need a force of 43.3 psi on the bottom of the stack, but to raise the column of water to 50 feet, you need to supply 21.65 psi.

Most codes require a minimum of 10 feet of head when testing sanitary systems. Using the formula above, you would find that 10 feet of head exerts a pressure of 4.33 psi.

4.2.3 Calculating Force on Test Plugs

Plumbers are responsible for arranging proper testing of drain, waste, and vent (DWV) systems and water supply piping. Water and air tests are the most common ways to ensure compliance with plans and codes. Prior to inspection, plumbers block pipe openings with test plugs (*Figure 19*) and increase the force in the system to test for leaks. You can calculate the force applied to test plugs using the mathematical formulas you have learned in this module.

Calculating the forces on test plugs will help ensure that the plumbing system can withstand operating forces. Test plugs are rated, so for safety reasons it is important to select a test plug that is suited for the pressures and size of the system. Consider the following example.

To determine the total force on a 4-inch-diameter test plug with a water head of 25 feet, follow these steps:

Step 1   Calculate the force of the water.
$$psi = 25 \times 0.433$$
$$psi = 10.825$$

Step 2   Calculate the area of the test plug.
$$A = \pi \times 22$$
$$A = 3.1416 \times 4$$
$$A = 12.566 \text{ sq in}$$

Step 3   Solve for the force on the plug.
$$psig = (10.825)(12.566)$$
$$psig = 136.027$$

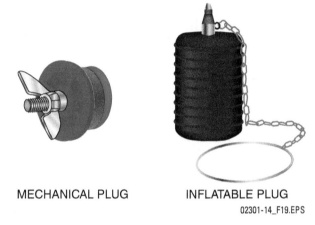

MECHANICAL PLUG   INFLATABLE PLUG

02301-14_F19.EPS

Figure 19 Mechanical and inflatable test plugs.

Water Pressure: Pressure versus Force

If you have two vertical pipes of different diameters, cap the bottom ends of both, and fill them each with water up to the same height (for example, 2 feet of water). The pressure on the cap will be the same in both pipes, but the force of the water will be greater in the wider pipe because there is more water in the pipe.

Pressure is a function of height, not volume, and force is a measure of volume, not height. The reason a higher column of water makes water flow faster through a spigot has nothing to do with the amount of water in the pipe, but has to do with the height of the column.

Think of it this way: take 15 cubic feet of water and put it in a 4-inch pipe. Call the amount of pressure "x" and the amount of force "y." If you took that 15 cubic feet of water and put it in a pipe that was only 2 inches in diameter, the pressure would still be "x" (because the amount of water is the same), but the force would now be "2y" (because the column of water is twice as high).

Plumbers need to understand pressure to make sure the fixtures work the way they are designed to, and force to make sure the pipes they installed on those fixtures don't break.

The force on a 4-inch-diameter test plug from a water head of 25 feet is 136.027 psig.

The proper head is the pressure required by the local code for a pressure test. The larger the diameter of a vertical pipe, the larger the force exerted on the pipe walls, fixtures, and plugs. Although the pressure is the same as for a narrower pipe of the same height, force varies with the volume.

4.2.4 Temperature and Pressure in Water Heaters

Water under pressure boils at a higher temperature than water at normal pressure. At normal (sea level) pressure, water boils at 212°F. However, at 150 pounds per square inch (psi), water boils at 358°F. You are already familiar with special valves that control temperature and pressure in water heaters (see *Figure 20*). Temperature regulator valves operate when the water gets too hot. Pressure regulator valves operate before the pressure of the water becomes too strong for the water heater to withstand. Combination temperature and pressure (T/P) relief valves prevent damage caused by excess temperatures and pressures. Codes require the installation of T/P relief valves on water heaters. Refer to your local code for specific guidelines. Always follow the manufacturer's instructions to ensure that the heater functions within specified operating temperatures and pressures.

02301-14_F20.EPS

Figure 20 Relief valves.

The Concept of Work

The idea of work is common sense, but here is the concept behind it. Work is what happens when force is used to move a load over a distance.

The amount of work (W) can be calculated by multiplying force (F) by distance (D): $W = F \times D$.

Additional Resources

Fundamentals of Temperature, Pressure and Flow Measurements. 1984. Robert P. Benedict. New York: John Wiley & Sons.

4.0.0 Section Review

1. A thermometer that uses the thermal expansion of a coil made of two bonded metal strips is called a(n) _____.
   a. liquid thermometer
   b. galvanic thermometer
   c. infrared thermometer
   d. bimetallic thermometer

2. At normal (sea level) pressure, water boils at _____.
   a. 451°F
   b. 212°F
   c. 12°F
   d. 32°F

5.0.0 THE SIMPLE MACHINES

Objective

Explain the functions and applications of six simple machines: inclined plane, lever, pulley, wedge, screw, and wheel and axle.

  a. Explain how inclined planes work.
  b. Explain how levers work.
  c. Explain how pulleys work.
  d. Explain how wedges work.
  e. Explain how screws work.
  f. Explain how wheels and axles work.

Trade Terms

Fulcrum: In a lever, the pivot or hinge on which a bar, a pole, or other flat surface is free to move.

Inclined plane: A simple machine that has at least one flat surface inclined to the horizontal.

Lever: A simple machine consisting of a bar, a pole, or other flat surface that is free to pivot around a fulcrum.

Pulley: A simple machine consisting of a rope wrapped around a wheel.

Screw: A simple machine consisting of an inclined plane wrapped around a cylinder.

Simple machine: A device that performs work in a single action. There are six types of simple machines: the inclined plane, the lever, the pulley, the wedge, the screw, and the wheel and axle.

Wedge: A simple machine consisting of two back-to-back inclined planes.

Wheel and axle: A simple machine consisting of a circular wheel attached to a central shaft that spins with the wheel.

Work: A measure of the force required to move an object a specified distance.

Machines and people perform work. Whether the work is moving a car on wheels down a road, digging a trench, spinning a drill bit at high speed, or lifting a pallet of materials, machines and people move things by applying some kind of force to them. Muscle power, electricity, air or water pressure, and engines can all supply the force to run machines.

Machines perform work by combining different types of actions or by performing just one action over and over again. A machine that performs a single action is called a simple machine.

There are six types of simple machines:

- The inclined plane
- The lever
- The pulley
- The wedge
- The screw
- The wheel and axle

A complex machine is formed when two or more simple machines are combined.

5.1.0 Understanding How Inclined Planes Work

An inclined plane is a straight, slanted surface (see *Figure 21*). Loads can be raised or lowered along an inclined plane. Ramps are a common type of inclined plane. The longer the inclined plane is, the easier it is to move a load up or down. In other words, with an inclined plane, greater distances require less force.

A pipe with grade is an inclined plane used regularly in plumbing. You have learned that the steeper the grade, the faster the liquid flows through the angled pipe. Plumbers must install pipe with the correct grade. Too much grade means that liquids flow too fast to scour the pipe clean. Too little grade will allow deposits to settle that will eventually block the pipe.

5.2.0 Understanding How Levers Work

A lever is a straight bar, such as a pole or a board, that is free to pivot around a hinge or a support. The hinge or support is called a fulcrum. Operate a lever by applying force to the bar. This causes the bar to rise, carrying a load with it. The longer the lever is (that is, the farther the force is from the fulcrum), the easier it is to move the load. There are three types, or classes, of levers (see *Figure 22*):

- *First-class lever* – The fulcrum is located in the middle of the bar. The force is applied in a downward motion to one side of the fulcrum, and the load is on the opposite side of the ful-

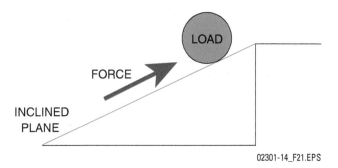

02301-14_F21.EPS

Figure 21 Inclined plane.

crum. A playground seesaw is a familiar type of first-class lever.

- *Second-class lever* – The load is between the fulcrum and the force. The force is applied upward. This type of lever requires much less force to move a load. However, the load does not move as far as when a first-class lever is used. Float arms, used on float-controlled valves in toilets, and wheelbarrows are examples of a second-class lever.
- *Third-class lever* – The force is applied between the fulcrum and the load. This type of lever requires the most force to move the load. However, the distance traveled by the load is also the greatest of the three types of levers. A leaning ladder that is pushed up into place from underneath is an example of a third-class lever. The ladder is the load, the feet are the fulcrum, and the pushing motion is the force.

A pipe dolly (see *Figure 23*) is a second-class lever in which the load actually straddles both sides of the fulcrum. A door is a vertical second-class lever in which the fulcrum is the hinge. Many tools are levers. Wrenches and pliers are good examples. Wrenches are second-class levers because the load being moved is close to the fulcrum. In fact, if you look closely, you will see that the load is the fulcrum (see *Figure 24*). Pliers are actually two levers put together. When you use a claw hammer to pry a nail out of wood, you are using a lever. Can you identify what class of lever a claw hammer is?

5.3.0 Understanding How Pulleys Work

A pulley is a rope wrapped around a wheel (see *Figure 25*). When the rope is pulled or released, the object being lifted also moves. Pulleys and sets of pulleys both work in two different ways:

- They change the direction of the force—for example, pulling down on the rope raises the load.
- They change the amount of force—by combining several pulleys, a lesser force can move a heavier load.

Chain falls and chain come-alongs, are examples of pulleys that plumbers use on the job (see *Figure 26*). A chain fall is a tackle device fitted with an endless chain and is used to hoist heavy loads by hand. It is usually suspended from an overhead track. Pulleys can be useful when

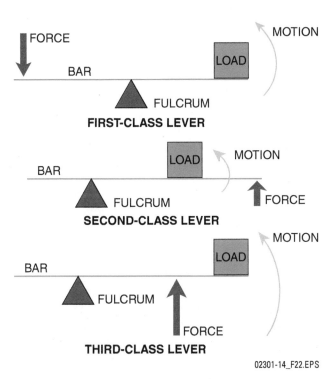

Figure 22 The three classes of levers.

Figure 23 A pipe dolly is a second-class lever.

The Simplicity of Complex Machines

A complex machine is made up of two or more simple machines. Combining different types of simple machines leads to common tools we see everywhere. Wheelbarrows use levers and wheels and axles. Backhoes and cranes both use levers, pulleys, and wheels and axles. Combining tools also allows work to be completed more efficiently. For example, by adding more pulleys, heavier loads can be lifted with less effort. This is called mechanical advantage.

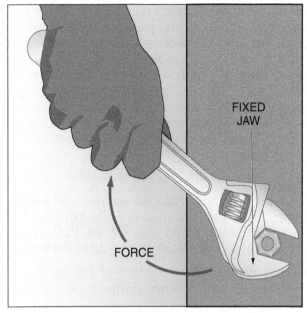

Figure 24 A wrench is a type of lever.

other options for lifting materials, such as elevators, aren't available. So even though they aren't widely used anymore, you should at least be familiar with them in case you need to install one. For example, you could use a chain to turn a gate valve that is high off the ground, thus creating a complex pulley.

5.4.0 Understanding How Wedges Work

A wedge is a modified inclined plane. It is made up of two back-to-back inclined planes (see *Figure 27*). Wedges are used to cut or separate objects, prop something up, or hold something in place. You have probably used a wedge to help level a board,

Figure 26 Chain fall and chain come-along.

hold pipe or batter boards in place, or split wood. A wedge transfers the force along the same line as it is applied. Plumbers use wood chisels and cold chisels, both of which are types of wedges.

5.5.0 Understanding How Screws Work

Like the wedge, the screw is a modification of an inclined plane. A screw is really a very long inclined plane that has been wrapped around a cylindrical shaft. When you apply twisting force to a screw perpendicular to the shaft, the plane transforms that into a motion that is parallel to the shaft. Because the inclined plane on a screw is so long, you need little force to move the screw.

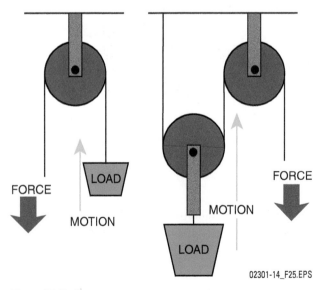

Figure 25 Pulleys.

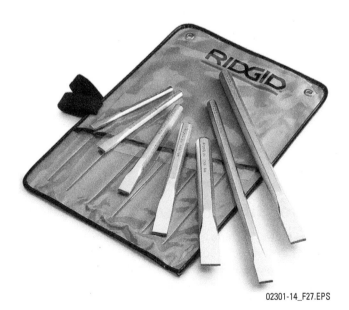

Figure 27 Chisels are a type of wedge.

Plumbers use screws to fasten pipe clamps and hangers to studs, assemble fixtures and appliance components, and drill wells. Adjustable wrenches use screws to widen and narrow the jaws. Drill bits also use the screw principle (see *Figure 28*).

5.6.0 Understanding How Wheels and Axles Work

The wheel and axle is a modification of the basic pulley concept. The wheel is a circle that is attached to a shaft or post, called an axle, that runs through the center. The wheel and axle turn together (see *Figure 29*). The wheel and axle is probably the most common simple machine. Many tools have wheels and axles. Wheelbarrows use them to roll loads along the ground. The handwheel and stem on certain types of valves are another familiar use of the wheel and axle combination.

The Early Wheel

The use of wheels without axles dates back to prehistoric times. These wheels were in the form of rollers in which a succession of logs or pipes would be used to carry a load along a road. As the load rolled along, logs would be brought from the back to the front.

The oldest known wheels—from between 3000 to 2000 BC—were found in Mesopotamian tombs. These were made of wood strips pulled together with cross-struts. A natural knothole acted as the pivot, and the wheel turned on a fixed axle. Wheels with axles were used on Egyptian chariots as early as 1300 BC and on Roman carts dating from AD 70.

02301-14_F28.EPS

Figure 28 Drill bits are a type of screw.

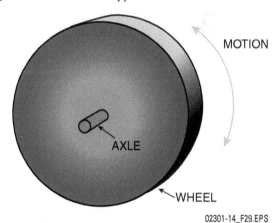

02301-14_F29.EPS

Figure 29 Wheel and axle.

Additional Resources

Mechanics of Materials. 2011. Ferdinand Baer. Science, Engineering & Math Publishing.

5.0.0 Section Review

1. With an inclined plane, the force required to move an object a shorter distance is _____.

   a. less
   b. more
   c. the same
   d. gradually decreased

2. If a lever has its load positioned between the fulcrum and the force, it is a _____.

   a. first-class lever
   b. second-class lever
   c. third-class lever
   d. fourth-class lever

3. The two ways a pulley can work are by _____.

   a. changing the direction and amount of force
   b. changing the speed and direction of motion
   c. slowing and speeding up the load
   d. changing the position and radius of the axle

4. A wedge transfers the force _____.

   a. perpendicular to the line along which it is applied
   b. at an acute angle to the line along which it is applied
   c. along the same line as it is applied
   d. at an obtuse angle to the line along which it is applied

5. Compared to an inclined plane, the force required to move a screw is _____.

   a. proportional
   b. great
   c. the same
   d. little

6. In a simple wheel and axle, the axle _____.

   a. rotates in the opposite direction of the rotation of the wheel
   b. turns together with the wheel
   c. is faster than the rotation of the wheel
   d. remains stationary

Summary

Applied mathematics—math that you use to accomplish a specific task—is an essential part of plumbing. Plumbers use math in almost every step of the plumbing installation process, from system design and sizing to construction and testing. In this module, you learned about the English and metric systems of measurement. Plumbers use these systems to express how high, how heavy, how much, and how many. You learned how to calculate the area of flat surfaces, such as squares, rectangles, and circles. You also reviewed the formulas for calculating the volume of prisms and columns. You were introduced to the concepts of temperature and pressure and how they affect plumbing installations, and you discovered the mechanical principles behind the tools that you use on the job.

Applied mathematics helps plumbers get a job done correctly. It is as important as any other tool that plumbers use. Take the time to master the basic weights, measures, and formulas. Knowledge of applied mathematics is an essential part of your professional development.

1. A mathematical process used to accomplish a specific task is known as _____.

   a. fundamental math
   b. applied mathematics
   c. trade math
   d. practical math

2. Most tools and materials used by plumbers in the US are sized according to the _____.

   a. metric system
   b. British Imperial system
   c. decimal system
   d. English system

3. In the metric system, multiples and fractions of the basic units are expressed as powers of 10 except for the basic unit of _____.

   a. mass
   b. time
   c. light intensity
   d. electric current

4. Of the seven basic metric units, the only one that is also a multiple of another unit of measure is the _____.

   a. kilogram
   b. ampere
   c. second
   d. mole

5. To change a metric quantity to an English quantity (such as centimeters to inches), you multiply by a number called a(n) _____.

   a. conversion constant
   b. algorithm
   c. conversion factor
   d. inverter

6. Length is multiplied by width to find _____.

   a. circumference
   b. area
   c. perimeter
   d. radius

7. A four-sided figure with corners that are all 90-degree angles is a _____.

   a. parallelogram
   b. rhombus
   c. trapezoid
   d. rectangle

8. To ensure accuracy when calculating area or volume, numbers should be rounded to _____.

   a. two decimal places
   b. three decimal places
   c. four decimal places
   d. the nearest whole number

9. To calculate the length of a missing leg in a right triangle, use the _____.

   a. Pythagorean theorem
   b. rule of threes
   c. Phalangean theorem
   d. calculus of angles

10. A circle's radius is one-half its _____.

    a. circumference
    b. area
    c. diameter
    d. volume

11. The abbreviation used for cubic feet is _____.

    a. c^3ft
    b. ft^3
    c. ftc
    d. cf

12. The sides of a prism are _____.

    a. triangles
    b. circles
    c. cylindrical
    d. rectangles

13. A chest freezer that is 3.5 feet high, 4 feet wide, and 2.25 feet deep occupies a space with a volume of _____.

    a. 0.315 cubic feet
    b. 3.15 cubic feet
    c. 31.5 cubic feet
    d. 315 cubic feet

14. When calculating volume for a site-built container, the volume of fittings or other objects inside the container _____.

    a. must be subtracted from the total volume
    b. can be overlooked
    c. must be added to the total volume
    d. is calculated, then doubled

15. To convert volume in cubic inches to volumes in gallons, divide the cubic-inch measurement by _____.

    a. 3.1416
    b. 29
    c. 37.5
    d. 231

16. If the weight of a full tank of water in a water heater is 416.5 pounds, the volume of the tank is _____.

    a. 60 gallons
    b. 50 gallons
    c. 40 gallons
    d. 30 gallons

17. Providing pipe-bundle weight information to the crane operator is the responsibility of the _____.

    a. plumber
    b. pipe manufacturer
    c. project engineer
    d. site supervisor

18. If the temperature is the same throughout a space or an object, the situation is described as a state of _____.

    a. neutrality
    b. equalization
    c. stagnation
    d. equilibrium

19. Alcohol or other fluid that expands in response to an increase in heat is the essential material in a(n) _____.

    a. digital thermometer
    b. infrared thermometer
    c. liquid thermometer
    d. instant-read thermometer

20. On the Celsius scale, the freezing point of pure water is _____.

    a. 32°
    b. 0°
    c. –32°
    d. –100°

21. A temperature of 59° on the Fahrenheit scale is equal to a Celsius temperature of _____.

    a. 43°
    b. 27°
    c. 15°
    d. –11°

22. When materials absorb thermal energy, they change in _____.

    a. chemical composition
    b. size
    c. magnetic orientation
    d. mass

23. Underground pipes that are in danger of freezing must be located _____.

    a. below the frost line
    b. on the south side of the building
    c. in chases
    d. under paved areas

24. Water hammer occurs when liquid flow through a pipe is _____.

    a. increased rapidly
    b. intermittent
    c. restricted
    d. suddenly stopped

25. To test sanitary systems, most codes call for a minimum of 10 feet of _____.

    a. load
    b. head
    c. stack
    d. lead

26. When checking a water supply or DWV system for leaks prior to an inspection, plumbers will install test plugs and _____.

    a. increase the force in the system
    b. fill the system with water
    c. reduce the pressure in the system
    d. fill the system with gas

27. Compared with water at normal pressure, the temperature at which water boils when under pressure is _____.

   a. lower
   b. higher
   c. the same
   d. proportional

28. When two or more simple machines are combined, the result is a _____.

   a. hybrid machine
   b. combination machine
   c. multipurpose machine
   d. complex machine

29. The hinge or support used with a lever is called a(n) _____.

   a. pivot
   b. frog
   c. fulcrum
   d. axis

30. Of the six simple machines, the most common is probably the _____.

   a. wheel and axle
   b. screw
   c. inclined plane
   d. lever

Trade Terms Quiz

Fill in the blank with the correct term that you learned from your study of this module.

1. A(n) _____ is a four-sided surface in which all corners are right angles.

2. The measure of a fixture's load, depending on water quantity, temperature, and fixture type is called _____.

3. A(n) _____ is a thermometer that determines temperature by using the thermal expansion of a coil of metal consisting of two thin strips of metal bonded together.

4. A decimal fraction where the denominator is either 12 or a power of 12 is called a(n) _____.

5. A(n) _____ is a three-sided surface with one angle that equals 90 degrees.

6. The expansion of materials in all three dimensions when heated is called _____.

7. A(n) _____ is a simple machine consisting of two back-to-back inclined planes.

8. The transfer of heat energy from a hot object to a cool object is called _____.

9. A(n) _____ is a thermometer that measures temperature through the expansion of a fluid.

10. The basic measure of volume in the metric system is the _____.

11. A(n) _____ is a volume in which two parallel rectangles, squares, or right triangles are connected by rectangles.

12. A rectangle in which all four sides are equal lengths is called a(n) _____.

13. The _____ and are the standard systems of weights and measures.

14. The measure of the force required to move an object a specified distance is called _____.

15. A(n) _____ is the basic measure of liquid volume in the English system.

16. The height of a water column, measured in feet, is called _____.

17. A(n) _____ is a scale divided into 100 degrees.

18. A straight and slanted surface is called a(n) _____.

19. Any mathematical process used to accomplish a task is called _____.

20. A(n) _____ is a triangle in which two of the sides are of equal length.

21. The condition in which all objects in a space have an equal temperature is called _____.

22. The _____ is the scale of temperature measurement in the metric system.

23. A simple machine consisting of a bar, a pole, or other flat surface that is free to pivot around a fulcrum is called a(n) _____.

24. A measure of a surface, expressed in square units, is called _____.

25. The _____ is the basic measure of liquid volume in the metric system.

26. The basic measure of pressure in the English system is _____.

27. _____ is the force applied to the walls of a container by the liquid or gas inside.

28. The basic measure of volume in the English system is _____.

29. The _____ is a simple machine consisting of a rope wrapped around a wheel.

30. The basic measure of area in the metric system is _____.

31. A(n) _____ is a surface consisting of a curve drawn all the way around a point that keeps the same distance from that point.

32. A thermometer that measures temperature by detecting the thermal radiation emitted by an object is a(n) _____.

33. _____ is a measure of a total amount of space, measured in cubic units.

34. A simple machine consisting of an inclined plane wrapped around a cylinder is called a(n) _____.

35. The _____ is the part of a lever that serves as the pivot or hinge on which a bar, a pole, or other flat surface is free to move.

36. A measure of relative heat as measured by a scale is called _____.

37. A centigrade scale commonly used to measure temperature is the _____.

38. The basic measure of area in the English system is the _____.

39. A(n) _____ is a rectangular prism in which the lengths of all sides are equal.

40. The tool used to measure temperature is called a(n) _____.

41. The _____ is the scale of temperature measurement in the English system.

42. A simple machine consisting of a circle attached to a central shaft that spins with the wheel is called a(n) _____.

43. A(n) _____ is a device that performs work in a single action.

44. A pipe- or tube-shaped space with a circular cross section is called a(n) _____.

Trade Terms

Applied mathematics	English system	Liter	Square meter
Area	Equilibrium	Metric system	Temperature
Bimetallic thermometer	Fahrenheit scale	Pounds per square inch	Thermal expansion
Celsius scale	Fulcrum	Pressure	Thermometer
Centigrade scale	Gallon	Prism	Volume
Circle	Head	Pulley	Water supply fixture unit
Conduction	Inclined plane	Rectangle	(WSFU)
Cube	Infrared thermometer	Right triangle	Wedge
Cubic foot	Isosceles triangle	Screw	Wheel and axle
Cubic meter	Kelvin scale	Simple machine	Work
Cylinder	Lever	Square	
Decimal of a foot	Liquid thermometer	Square foot	

Trade Terms Introduced in This Module

Applied mathematics: Any mathematical process used to accomplish a task.

Area: A measure of a surface, expressed in square units.

Bimetallic thermometer: A thermometer that determines temperature by using the thermal expansion of a coil of metal consisting of two thin strips of metal bonded together.

Celsius scale: A centigrade scale used to measure temperature.

Centigrade scale: A scale divided into 100 degrees. Generally used to refer to the metric scale of temperature measure (see *Celsius scale*).

Circle: A surface consisting of a curve drawn all the way around a point. The curve keeps the same distance from that point.

Conduction: The transfer of heat energy from a hot object to a cool object.

Cube: A rectangular prism in which the lengths of all sides are equal. When used with another form of measurement, such as cubic meter, the term refers to a measure of volume.

Cubic foot: The basic measure of volume in the English system. There are 7.48 gallons in a cubic foot.

Cubic meter: The basic measure of volume in the metric system. There are 1,000 liters in a cubic meter.

Cylinder: A pipe- or tube-shaped space with a circular cross section.

Decimal of a foot: A decimal fraction where the denominator is either 12 or a power of 12.

English system: One of the standard systems of weights and measures. The other system is the metric system.

Equilibrium: A condition in which all objects in a space have an equal temperature.

Fahrenheit scale: The scale of temperature measurement in the English system.

Fulcrum: In a lever, the pivot or hinge on which a bar, a pole, or other flat surface is free to move.

Gallon: In the English system, the basic measure of liquid volume. There are 7.48 gallons in a cubic foot.

Head: The height of a water column, measured in feet. One foot of head is equal to 0.433 pounds per square inch gauge.

Inclined plane: A simple machine that has at least one flat surface inclined to the horizontal.

Infrared thermometer: A thermometer that measures temperature by detecting the thermal radiation emitted by an object.

Isosceles triangle: A triangle in which two of the sides are of equal length.

Kelvin scale: The scale of temperature measurement in the metric system.

Lever: A simple machine consisting of a bar, a pole, or other flat surface that is free to pivot around a fulcrum.

Liquid thermometer: A thermometer that measures temperature through the expansion of a fluid, such as mercury or alcohol.

Liter: In the metric system, the basic measure of liquid volume. There are 1,000 liters in a cubic meter.

Metric system: A system of measurement in which multiples and fractions of the basic units of measure are expressed as powers of 10.

Pounds per square inch: In the English system, the basic measure of pressure. Pounds per square inch (psi) is measured in pounds per square inch absolute (psia) and pounds per square inch gauge (psig).

Pressure: The force applied to the walls of a container by the liquid or gas inside.

Prism: A volume in which two parallel rectangles, squares, or right triangles are connected by rectangles.

Pulley: A simple machine consisting of a rope wrapped around a wheel.

Rectangle: A four-sided surface in which all corners are right angles.

Right triangle: A three-sided surface with one angle that equals 90 degrees.

Screw: A simple machine consisting of an inclined plane wrapped around a cylinder.

Simple machine: A device that performs work in a single action. There are six types of simple machines: the inclined plane, the lever, the pulley, the wedge, the screw, and the wheel and axle.

Square: A rectangle in which all four sides are equal lengths. When used with another form of measurement, such as square meter, the term refers to a measure of area.

Square foot: In the English system, the basic measure of area. There are 144 square inches in 1 square foot.

Square meter: In the metric system, the basic measure of area. There are 10,000 square centimeters in a square meter.

Temperature: A measure of relative heat as measured by a scale.

Thermal expansion: The expansion of materials in all three dimensions when heated.

Thermometer: A tool used to measure temperature.

Volume: A measure of a total amount of space, measured in cubic units.

Water supply fixture unit (WSFU): The measure of a fixture's load, varying according to the amount of water, the water temperature, and the fixture type.

Wedge: A simple machine consisting of two back-to-back inclined planes.

Wheel and axle: A simple machine consisting of a circular wheel attached to a central shaft that spins with the wheel.

Work: A measure of the force required to move an object a specified distance.

CONVERSION TABLES

English to SI Metric Conversions

English to Metric			
	To convert...	**Into...**	**Multiply by...**
LENGTH	Inches	Millimeters	25.4
	Feet	Centimeters	30.48
	Yards	Meters	0.9144
	Miles	Kilometers	1.6
AREA	Square inches	Square centimeters	6.452
	Square feet	Square meters	0.093
	Square yards	Square meters	0.836
	Square miles	Square kilometers	2.589
	Acres	Hectares	0.4047
MASS and WEIGHT	Ounces	Grams	28.35
	Pounds	Kilograms	0.4536
	Short tons	Megagrams (metric tons)	0.907
LIQUID MEASURE	Fluid ounces	Milliliters	29.57
	Pints	Liters	0.473
	Quarts	Liters	0.946
	Gallons	Liters	3.7854
VOLUME	Cubic inches	Cubic centimeters	16.387
	Cubic feet	Cubic meters	0.0283
PRESSURE	Pounds per square inch	Kilopascals	6.89

02301-14_TA01A.EPS

Metric to English			
	To convert...	**Into...**	**Multiply by...**
LENGTH	Millimeters	Inches	0.039
	Centimeters	Feet	0.33
	Meters	Yards	1.0936
	Kilometers	Miles	0.62
AREA	Square centimeters	Square inches	0.155
	Square meters	Square yards	1.196
	Square kilometers	Square miles	0.386
	Hectares	Acres	2.47
MASS and WEIGHT	Grams	Ounces	0.035
	Kilograms	Pounds	2.205
	Megagrams (metric tons)	Short tons	1.1
LIQUID MEASURE	Milliliters	Fluid ounces	0.0338
	Liters	Pints	2.113
	Liters	Quarts	1.057
	Liters	Gallons	0.264
VOLUME	Cubic centimeters	Cubic inches	0.061
	Cubic decimeters	Cubic feet	0.0353
	Cubic meters	Cubic yards	1.308
PRESSURE	Kilopascals	Pounds per square inch	0.145

02301-14_TA01B.EPS

Appendix B

AREA AND VOLUME FORMULAS

Area Formulas		
Rectangle	$A = lw$	Multiply length by width
Right Triangle	$A = \frac{1}{2}(bh)$	One-half the product of the base and height
Circle	$A = \pi r^2$	Multiply pi (3.1416) by the square of the radius
Volume Formulas		
Rectangular Prism	$V = lwh$	Multiply length by width by height
Right Triangular Prism	$V = \frac{1}{2}(bwh)$	Multiply one-half the product of base by width by height
Cylinder	$V = (\pi r^2)h$	Multiply the area of the circle by the height

02301-14_TA02.EPS

Additional Resources

This module presents thorough resources for task training. The following resource material is suggested for further study.

Code Check Plumbing: A Field Guide to Plumbing. 2000. Redwood Kardon, Michael Casey, and Douglas Hansen. Newtown, CT: Taunton Press.

Fundamentals of Temperature, Pressure and Flow Measurements. 1984. Robert P. Benedict. New York: John Wiley & Sons.

International Plumbing Code®, Latest Edition. Falls Church, VA: International Code Council.

Math to Build On: A Book for Those Who Build. 1993. Johnny and Margaret Hamilton. Clinton, NC: Construction Trades Press.

Mechanics of Materials. 2011. Ferdinand Baer. Science, Engineering & Math Publishing.

Figure Credits

Stacy Sabers, CO01

Courtesy of Calculated Industries, Inc., Figure 4

Courtesy of Charlotte Pipe and Foundry Company, Tables 4–5

Courtesy of Taylor Precision Products, Inc., Figure 13

Courtesy of Weiss Instruments, Figure 14

Image courtesy of FLIR Systems, Inc., Figure 15

Courtesy of Watts, Figure 20

Courtesy of Sumner Manufacturing, Co. Inc., Figure 23

Jet Brand of WMH Tool Group, Figure 26

Emerson Tool Company, Ridge Tool Company/RIDGID, Figure 27

Courtesy of IRWIN Tools, Figure 28

Answer	Section Reference	Objective
Section One		
1. d	1.1.0	1a
2. b	1.2.0	1b
3. c	1.3.0	1c
Section Two		
1. d	2.1.0	2a
2. c	2.2.0	2b
Section Three		
1. c*	3.1.0	3a
2. b*	3.2.0	3b
3. a*	3.3.0	3c
Section Four		
1. d	4.1.1	4a
2. b	4.2.4	4b
Section Five		
1. b	5.1.0	5a
2. b	5.2.0	5b
3. a	5.3.0	5c
4. c	5.4.0	5d
5. d	5.5.0	5e
6. b	5.6.0	5f

Section Three Math Calculations

1. $15 \times 7 = 105$

2. $V = \pi r^2 h$ where $r = 1"$ and and $h = (6 \text{ ft})(12 \text{ inches per foot}) = 72"$

   $V = \pi \times 72$

   $3.1416 \times 72 = 226.1952$

3. $25 \times 20 = 500$

   $40 \times 6 = 240$

   $500 + 240 = 740$

NCCER CURRICULA — USER UPDATE

NCCER makes every effort to keep its textbooks up-to-date and free of technical errors. We appreciate your help in this process. If you find an error, a typographical mistake, or an inaccuracy in NCCER's curricula, please fill out this form (or a photocopy), or complete the online form at **www.nccer.org/olf**. Be sure to include the exact module ID number, page number, a detailed description, and your recommended correction. Your input will be brought to the attention of the Authoring Team. Thank you for your assistance.

Instructors – If you have an idea for improving this textbook, or have found that additional materials were necessary to teach this module effectively, please let us know so that we may present your suggestions to the Authoring Team.

NCCER Product Development and Revision
13614 Progress Blvd., Alachua, FL 32615

Email: curriculum@nccer.org
Online: www.nccer.org/olf

❏ Trainee Guide ❏ Lesson Plans ❏ Exam ❏ PowerPoints Other _____

Craft / Level: _____ Copyright Date: _____

Module ID Number / Title: _____

Section Number(s): _____

Description: _____

_____

_____

_____

Recommended Correction: _____

_____

_____

_____

Your Name: _____

Address: _____

_____

Email: _____ Phone: _____

02312-14

Sizing and Protecting the Water Supply System

OVERVIEW

The water supply system is one of the most important systems installed in residential and commercial buildings. Plumbers must size supply systems correctly so that they reliably provide adequate water at the correct pressure. Proper installation of water supply systems requires an understanding of the physical properties of water. Plumbers calculate the correct pipe sizes for the system. Local codes provide tables, graphs, and specifications to help with these calculations.

Module Two

Trainees with successful module completions may be eligible for credentialing through the NCCER Registry. To learn more, go to **www.nccer.org** or contact us at **1.888.622.3720**. Our website has information on the latest product releases and training, as well as online versions of our *Cornerstone* magazine and Pearson's product catalog.

Your feedback is welcome. You may email your comments to **curriculum@nccer.org,** send general comments and inquiries to **info@nccer.org**, or fill in the User Update form at the back of this module.

This information is general in nature and intended for training purposes only. Actual performance of activities described in this manual requires compliance with all applicable operating, service, maintenance, and safety procedures under the direction of qualified personnel. References in this manual to patented or proprietary devices do not constitute a recommendation of their use.

Objectives

When you have completed this module, you will be able to do the following:

1. Determine the factors that affect the sizing of water supply systems.
   a. Determine how temperature and density affect water supply systems.
   b. Determine how flow and friction affect water supply systems.
2. Size a given water supply system for different acceptable flow rates and calculate pressure drops in a given water system.
   a. Determine how to establish system requirements for a given water supply system.
   b. Determine how to calculate demand for a given water supply system.
   c. Determine the correct pipe size based on system and supply pressures in a given water supply system.
   d. Determine how to calculate system losses in a given water supply system.
3. Describe the six basic backflow-prevention devices and the hazards they are designed to prevent.
   a. Describe the principle of backflow due to back siphonage and back pressure.
   b. Describe when and how to install air gaps.
   c. Describe when and how to install atmospheric vacuum breakers.
   d. Describe when and how to install pressure-type vacuum breakers.
   e. Describe when and how to install dual-check valve backflow preventer assemblies.
   f. Describe when and how to install double-check valve assemblies.
   g. Describe when and how to install reduced-pressure zone principle backflow preventer assemblies.
   h. Describe when and how to install specialty backflow preventers.
   i. Describe when and how to troubleshoot and repair backflow preventers.

Performance Tasks

Under the supervision of your instructor, you should be able to do the following:

1. Using design information provided by the instructor, lay out a water supply system and calculate developed lengths of branches.
2. Install common types of backflow preventers.

Trade Terms

Atmospheric vacuum breaker	Demand	Equivalent length	Laminar flow	Reduced-pressure zone principle backflow preventer assembly (RPZ)
Back siphonage	Density	Flow rate	Low hazard	
Backflow	Developed length	Friction	Manufactured air gap	
Backflow preventer with intermediate atmospheric vent	Double-check valve assembly (DCV)	Friction loss	Pressure drop	Test cock
	Dual-check valve backflow preventer assembly (DC)	High hazard	Pressure-type vacuum breaker	Transient flow
		Hose-connection vacuum breaker		Turbulent flow
Cross-connection		In-line vacuum breaker		Viscosity

Industry-Recognized Credentials

If you're training through an NCCER-accredited sponsor, you may be eligible for credentials from NCCER's Registry. The ID number for this module is 02312-14. Note that this module may have been used in other NCCER curricula and may apply to other level completions. Contact NCCER's Registry at 888.622.3720 or go to **www.nccer.org** for more information.

Code Note

Codes vary among jurisdictions. Because of the variations in code, consult the applicable code whenever regulations are in question. Referring to an incorrect set of codes can cause as much trouble as failing to reference codes altogether. Obtain, review, and familiarize yourself with your local adopted code.

Contents ———————————————————

Topics to be presented in this module include:

Figures and Tables

1.0.0 FACTORS AFFECTING WATER SUPPLY SYSTEM SIZING

Objective

Determine the factors that affect the sizing of water supply systems.

    a. Determine how temperature and density affect water supply systems.

    b. Determine how flow and friction affect water supply systems.

Trade Terms

Demand: The measure of the water requirement for the entire water supply system.

Density: The amount of a liquid, gas, or solid in a space, measured in pounds per cubic foot.

Equivalent length: The length of pipe required to create the same amount of friction as a given fitting.

Flow rate: The rate of water flow in gallons per minute that a fixture uses when operating.

Friction: The resistance that results from objects rubbing against one another.

Friction loss: The partial loss of system pressure due to friction. Friction loss is also called pressure loss.

Laminar flow: The parallel flow pattern of a liquid that is flowing slowly. Also called streamline flow or viscous flow.

Pressure drop: In a water supply system, the difference between the pressure at the inlet and the pressure at the farthest outlet.

Transient flow: The erratic flow pattern that occurs when a liquid's flow changes from a laminar flow to a turbulent flow pattern.

Turbulent flow: The random flow pattern of fast-moving water or water moving along a rough surface.

Viscosity: The measure of a liquid's resistance to flow.

To install water supply systems correctly, you must understand the physical properties of water. These properties affect how water behaves inside pipes and fittings. Plumbers must consider the following factors when sizing a water supply system:

- Temperature
- Density
- Flow
- Friction

Each of these factors has an effect on the operation of a water supply system. It is important to install water supply systems so that those effects are controlled. This will help the system last longer and operate more efficiently.

1.1.0 Determining the Effects of Temperature and Density on Water Supply System Sizing

You have already learned about the importance of temperature and of one of its related properties, pressure. Another property of temperature that affects plumbing systems is density. Density is the amount of a substance in a given space. It is measured in pounds per cubic foot. Increasing the temperature of a substance decreases its density (see *Figure 1*). It also increases its volume. Plumbers need to know the temperature of water in a plumbing system to choose the correct pipe size.

Another way to explain density is to say that as a volume of water gets hotter, it gets lighter. At 32°F, a cubic foot of water weighs 62.42 pounds, but at 100°F, the same volume of water weighs 61.99 pounds. At 200°F, it weighs 60.14 pounds.

Overheated water is dangerous. It can split pipes and weaken fittings. It can also cause water heaters to explode. Install temperature/pressure relief valves on hot-water heaters to prevent overheated water from causing damage. Relief valves release water or steam and equalize the pressure

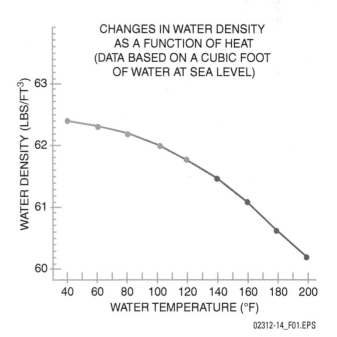

02312-14_F01.EPS

Figure 1 Density of a cubic foot of water at different temperatures.

---

in the tank. Review your local code to determine the proper precautions for hot-water pipes.

1.2.0 Determining the Effects of Flow and Friction on Water Supply System Sizing

In plumbing, flow is the measure of how much liquid moves through a pipe. The measure of how much a liquid resists flow is called its viscosity. You may have encountered this term in reference to motor oil. High-viscosity oil is very thick. In other words, it resists flowing. The concepts of flow and resistance are very important in water supply systems. There are three types of flow:

- Laminar flow
- Transient flow
- Turbulent flow

In cases where the rate of flow is less than 1 foot per second, water flows smoothly. If you could see the molecules in slow-moving water, they would look like they were moving parallel to each other in layers. This is called laminar flow. Laminar flow is also called streamline flow or viscous flow. When water flow changes from one type of flow to another, it becomes erratic and unstable. This type of flow is called transient flow.

Turbulent flow is the random motion of water as it moves along a rough surface, such as the inner surface of a pipe. The faster the water flows along the inner surface, the more turbulent the flow becomes. Turbulent flow is a concern to plumbers, because it can accelerate the wear on pipes.

The different types of flow can be visualized by thinking about how water behaves in a kitchen sink. The water coming out of the faucet moves in a straight line. This is laminar flow. In the basin, it sloshes and spins. This is turbulent flow. The moment at which the water hits the basin and starts to swirl—you may not even be able to see this—is transient flow.

Turbulent flow is caused by friction. Friction is the resistance or slowing down that happens when two things rub against each other. For ex-

ample, run your hand over a smooth surface, like silk or a polished tabletop. Your hand moves easily over the surface. It is harder to run your hand over a bumpy rug or over sandpaper because the uneven surface slows your hand down. This is an example of friction.

Friction makes water supply systems less efficient. It reduces the pressure in the system by slowing water down. This is called friction loss. Friction loss is also known as pressure loss. In water-supply piping, friction is caused by several factors, including:

- The inner surface of a pipe
- Flow-through water meters
- Flow-through fittings such as valves, elbows, and tees
- Increased water velocity

Effective system installation can reduce overall friction and improve system efficiency. Plumbers can reduce friction loss by using pipes with a larger diameter. This is especially true when the water pipe is undersized. When water meters are required, use one that is designed for low friction. Keep the number of valves to a minimum to prevent excessive flow restriction. See the section *Sizing a Water Supply System* in this module to learn how to calculate friction loss in the system.

There is much less friction at the center of a pipe than near its inner surface. This means that the flow rate, or the speed at which water flows, is faster at the center of a pipe and slower near the pipe's inner surface (see *Figure 2*). Flow rate is measured in gallons per minute (gpm). Plumbers install plumbing systems to minimize turbulence and maximize flow. Plumbers can consult published sources to determine the average rate of flow under different conditions. Average rate of flow is determined by a combination of available pressure, demand, system size, and pipe size.

Fittings and valves can also cause friction loss. You can calculate the amount of friction loss caused by fittings and valves by comparing them with lengths of pipe that would cause equal friction loss (see *Table 1*). These are called equivalent lengths. You can use equivalent lengths to calcu-

Facts and Legends about Water

Water is one of the most plentiful substances on Earth. It is common in all living things. It is the internationally recognized standard for temperature and density measurements. However, water behaves very differently from other liquids. Water can't be compressed as much as other liquids. Sometimes, water seems to defy common sense. For example, if water begins to freeze, the volume will decrease until it reaches 39°F, then it will start to expand. Sometimes it takes hot water less time to freeze than cold water!

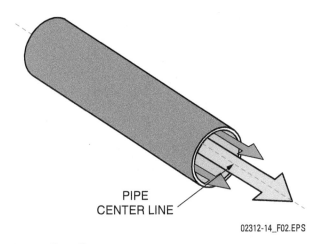

PIPE
CENTER LINE

02312-14_F02.EPS

Figure 2 Pipe flow rates.

late **pressure drop**. This is the difference in pressure between the inlet and the farthest outlet. You will learn more about pressure drop in the section *Determining Pipe Size*.

Burrs on pipe joints are a common cause of turbulence and friction loss in copper tube and polyvinyl chloride (PVC) pipe. Always ream copper tube and PVC pipe after cutting, to ensure that all burrs have been removed. This will ensure smooth flow at pipe joints. You learned how to ream and join copper tube and PVC pipe in the modules, *Copper Tube and Fittings* and *Plastic Pipe and Fittings*.

Refer to *Table 1* to determine the most efficient choice of fittings. For example, say you are going to install two 45-degree elbows to make a right-angle bend. The system uses ⅝-inch tubing. *Table 1* shows that one 45-degree elbow causes the same friction loss as 0.5 feet of pipe. Therefore, the friction losses in two 45-degree elbows equal the loss in 1 foot of pipe. However, one 90-degree elbow causes the same friction loss as 1.5 feet of pipe. Therefore, if the length of pipe between the two 45-degree elbows is less than 0.5 foot, the use of two 45-degree elbows would be the more efficient choice. Always refer to local code for standards in your area.

The table also allows you to compare the friction losses caused by various types of valves. A rule of thumb that many plumbers use is to allow an additional 50 percent of the system's total length for fittings and valves. You will learn how to calculate the total length later in this module. Remember to test the accuracy of your estimate after the pipe sizes have been determined.

Table 1 Sample Equivalent Pipe Lengths Used for Determining Friction Loss in Fittings[a]

| NOMINAL OR STANDARD SIZE (inches) | FITTINGS | | | | | VALVES | | | |
| | Standard Ell | | 90-Degree Tee | | | | | | |
	90 Degree	45 Degree	Side Branch	Straight Run	Coupling	Ball	Gate	Butterfly	Check
⅜	0.5	—	1.5	—	—	—	—	—	1.5
½	1	0.5	2	—	—	—	—	—	2
⅝	1.5	0.5	2	—	—	—	—	—	2.5
¾	2	0.5	3	—	—	—	—	—	3
1	2.5	1	4.5	—	—	0.5	—	—	4.5
1¼	3	1	5.5	0.5	0.5	0.5	—	—	5.5
1½	4	1.5	7	0.5	0.5	0.5	—	—	6.5
2	5.5	2	9	0.5	0.5	0.5	0.5	7.5	9
2½	7	2.5	12	0.5	0.5	—	1	10	11.5
3	9	3.5	15	1	1	—	1.5	15.5	14.5
3½	9	3.5	14	1	1	—	2	—	12.5
4	12.5	5	21	1	1	—	2	16	18.5
5	16	6	27	1.5	1.5	—	3	11.5	23.5
6	19	7	34	2	2	—	3.5	13.5	26.5
8	29	11	50	3	3	—	5	12.5	39

a. Expressed as equivalent length of tube (in feet). Allowances are for streamlined soldered fitting and recessed threaded fitting. For threaded fittings, double the allowances shown in the table The equivalent lengths presented above are based on a C factor of 150 in the Hazen-Williams friction loss formula. The lengths shown are rounded to the nearest half-foot.

Potable Water

Codes are strict about the sources, quantity, and quality of potable water. They also list the sizes and materials allowed for water supply piping. Plumbers can use only pipe and fittings that are approved by national standards organizations. Violators may face strict penalties. It is in your best interest and that of public safety to follow the local code closely.

Additional Resources

Code Check Plumbing & Mechanical: An Illustrated Guide to the Plumbing and Mechanical Codes. 2011. Douglas Hansen, Redwood Kardon, and Paddy Morrisey. Newtown, CT: Taunton Press.
Plumbing Level One. 2012. Alachua, FL: NCCER.

1.0.0 Section Review

1. Density is measured in _____.

   a. pounds per cubic foot
   b. pounds per square inch
   c. inches of mercury
   d. inches of water

2. The random motion of water as it moves along a rough surface is called _____.

   a. chaotic flow
   b. turbulent flow
   c. laminar flow
   d. transient flow

SECTION TWO

2.0.0 SIZING A WATER SUPPLY SYSTEM

Objective

Size a given water supply system for different acceptable flow rates and calculate pressure drops in a given water system.

  a. Determine how to establish system requirements for a given water supply system.
  b. Determine how to calculate demand for a given water supply system.
  c. Determine the correct pipe size based on system and supply pressures in a given water supply system.
  d. Determine how to calculate system losses in a given water supply system.

Performance Task

Using design information provided by the instructor, lay out a water supply system and calculate developed lengths of branches.

Trade Terms

Back siphonage: A form of backflow caused by subatmospheric pressure in the water system, which results in siphoning of water from downstream.

Backflow: An undesirable condition that results when nonpotable liquids enter the potable-water supply by reverse flow through a cross-connection.

Developed length: The length of all piping and fittings from the water supply to a fixture.

Plumbers must consider a number of factors when installing a water supply system. Those factors include the following:

- The system's ability to carry an adequate supply of water.
- The system's ability to operate with a minimum amount of turbulence.
- The system's ability to maintain a reasonable flow rate, which will ensure quiet operation.
- The system's design; it should not be expensively overdesigned.

Careful installation will ensure that the system provides the best possible service. Before you begin, consult the building's plans or drawings. Plumbing drawings are included in the plans for most industrial and commercial buildings. For light residential buildings, consult the floor plan. The drawings will help you understand how the system should be laid out.

2.1.0 Determining System Requirements

Refer to the material takeoff for the types and number of fixtures and outlets used in the system. Determine the flow rate at each fixture and outlet by referring to charts and tables in your local applicable code (see *Table 2*). This information can also be found in the product specifications. Identify both the rate and pressure of flow for each item. Note that the rate and pressure may be less if water conservation devices are used. Determining the flow rate will provide an initial idea of the size of the system.

Table 2 Demand and Flow-Pressure Values for Fixtures

FIXTURE SUPPLY OUTLET SERVING	FLOW RATE (gpm)	FLOW PRESSURE (psi)
Bathtub, balanced-pressure, thermostatic or combination balanced-pressure/thermostatic mixing valve	4	20
Bidet, thermostatic mixing valve	2	20
Combination fixture	4	8
Dishwasher, residential	2.75	8
Drinking fountain	0.75	8
Laundry tray	4	8
Lavatory	2	8
Shower	3	8
Shower, balanced-pressure, thermostatic or combination balanced-pressure/thermostatic mixing valve	3	20
Sillcock, hose bibb	5	8
Sink, residential	2.5	8
Sink, service	3	8
Urinal, valve	12	25
Water closet, blow out, flushometer valve	25	45
Water closet, flushometer tank	1.6	20
Water closet, siphonic, flushometer valve	25	35
Water closet, tank, close coupled	3	20
Water closet, tank, one-piece	6	20

One of the goals of an efficient system installation is to supply adequate water pressure to the farthest point of use during peak demand. If the system can do that, then the points in between are also likely to have adequate service. However, this assumption may not always be true. Pressure and water-volume requirements will vary from fixture to fixture. Keep these issues in mind when installing the water supply system.

Some fixtures may have a recommended pressure that is lower than the system's pressure. To correct this, install a pressure-reducing valve (PRV) on the line before the fixture. PRVs can also reduce the system's water and energy consumption. This is especially true for hot-water lines. In a high-rise building, pressure at one floor may exceed that at another. Install pressure-reducing valves to balance the pressure throughout the system.

Correct pressure must be maintained within the system. Too little pressure will result in poor service. Negative pressure can cause back siphonage, which could contaminate the system. It can also cause the system to emit an objectionable whistling noise. Whistling can also be caused by pipes that are too small.

Some fixtures might require a larger volume of water than others. The installation will have to include a larger pipe to supply that fixture. Remember that water flows more slowly through a wide pipe than a narrow one. Ensure that the rate of flow will not fall below the fixture's requirements.

2.2.0 Calculating Demand

After determining the system's requirements, estimate its demand. The demand is the water requirement for the entire system—pipes, fittings, outlets, and fixtures. Plumbers calculate the rate of flow in gpm according to the number of water supply fixture units (WSFUs) that each fixture requires. WSFUs measure a fixture's load, and vary with the quantity and temperature of the water and with the type of fixture (see *Table 3*). Determine the WSFUs for all fixtures and outlets in the system. Then convert the WSFUs to gpm (see *Table 4*). Add up the results for all fixtures and outlets. This is the total capacity of all the fixtures in the water supply system.

When sizing residential plumbing systems, keep in mind the greater flow rates for newer high-end baths, whirlpools, and showers; and determine whether a fixture or outlet's demand is intermittent or continuous.

Showers and baths have become high-end consumer products. People can now have shower towers, spas, and whirlpool baths installed in their homes (see *Figure 3*). These fixtures demand higher flow rates than normal showers and baths. Consult your local applicable code because it will specify allowable pressures and flow rates. Codes also specify the proper connection to the supply and recirculation systems. Ensure that the supply system can handle the demand. The product's specifications and installation information will provide the required data.

Next, determine whether the water demand will be intermittent or continuous. Outlets such as hose bibbs, air conditioners, sprinkler systems, and irrigation systems are common continuous-demand items. When they are used, the water flow is constant. The system must be able to supply water to the whole structure even when the continuous outlets are operating. Notice that the fixtures listed above are used only at certain times of the year. Lawn sprinklers and irrigation systems are usually used from spring to early fall. Other continuous-demand systems are needed only under special conditions. Fire-suppression sprinklers are an example. Be sure to size fire sprinklers so that they can provide enough water in an emergency. Check the product's instructions and refer to your local applicable code. In contrast, sinks, lavatories, water closets, and similar devices are used for about five minutes or less at a time. They are said to be intermittent-demand fixtures.

Combine the amount of flow from continuous- and intermittent-demand fixtures and outlets. Factor this information into the earlier rate-of-flow estimate. Consult the local code for requirements in your area.

Next, calculate the system's maximum probable flow. This number is an estimate of peak water demand. Each fixture has flow and demand pressure values assigned to it. You will learn how to calculate flow and demand in the *Water Pressure Booster and Recirculation Systems* module. Consult with local plumbing and code experts to learn about the values used in your area.

Finally, determine the developed length of each branch line. The developed length is the total length of piping from the supply to a single fixture. This includes pipes, elbows, valves, tees, and water heaters. Refer to *Table 1* to determine the equivalent lengths for fittings.

> **WARNING!**
>
> Always use water-supply piping that is made from materials specified in the local code. All water-supply piping must be resistant to corrosion. The amount of lead allowed in piping materials is strictly limited. Failure to use approved materials could contaminate the potable-water supply. This could cause illness and even death.

Table 3 Load Values Assigned to Fixtures

FIXTURE	OCCUPANCY	TYPE OF SUPPLY CONTROL	LOAD VALUES, IN WATER SUPPLY FIXTURE UNITS (WSFU)		
			Cold	Hot	Total
Bathroom group	Private	Flush tank	2.7	1.5	3.6
Bathroom group	Private	Flushometer value	6.0	3.0	8.0
Bathtub	Private	Faucet	1.0	1.0	1.4
Bathtub	Public	Faucet	3.0	3.0	4.0
Bidet	Private	Faucet	1.5	1.5	2.0
Combination fixture	Private	Faucet	2.25	2.25	3.0
Dishwashing machine	Private	Automatic	—	1.4	1.4
Drinking fountain	Offices, etc.	⅜" valve	0.25	—	0.25
Kitchen sink	Private	Faucet	1.0	1.0	1.4
Kitchen sink	Hotel, restaurant	Faucet	3.0	3.0	4.0
Laundry tray (1 to 3)	Private	Faucet	1.0	1.0	1.4
Lavatory	Private	Faucet	0.5	0.5	0.7
Lavatory	Public	Faucet	1.5	1.5	2.0
Service sink	Offices, etc.	Faucet	2.25	2.25	3.0
Shower head	Public	Mixing valve	3.0	3.0	4.0
Shower head	Private	Mixing valve	1.0	1.0	1.4
Urinal	Public	1" flushometer valve	10.0	—	10.0
Urinal	Public	¾" flushometer valve	5.0	—	5.0
Urinal	Public	Flush tank	3.0	—	3.0
Washing machine (8 lb)	Private	Automatic	1.0	1.0	1.4
Washing machine (8 lb)	Public	Automatic	2.25	2.25	3.0
Washing machine (15 lb)	Public	Automatic	3.0	3.0	4.0
Water closet	Private	Flushometer valve	6.0	—	6.2
Water closet	Private	Flush tank	2.2	—	2.2
Water closet	Public	Flushometer valve	10.0	—	10.0
Water closet	Public	Flush tank	5.0	—	5.0
Water closet	Public or private	Flushometer valve	2.0	—	2.0

Table 4 Table for Estimating Demand (1 of 2)

SUPPLY SYSTEMS PREDOMINANTLY FOR FLUSH TANKS			SUPPLY SYSTEMS PREDOMINANTLY FOR FLUSHOMETER VALVES		
Load	Demand		Load	Demand	
(Water supply fixture units)	(Gallons per minute)	(Cubic feet per minute)	(Water supply fixture units)	(Gallons per minute)	(Cubic feet per minute)
1	3.0	0.04104	—	—	—
2	5.0	0.0684	—	—	—
3	6.0	0.86892	—	—	—
4	8.0	1.06944	—	—	—
5	9.4	1.256592	5	15.0	2.0052
6	10.7	1.430376	6	17.4	2.326032
7	11.8	1.577424	7	19.8	2.646364
8	12.8	1.711104	8	22.2	2.967696
9	13.7	1.831416	9	24.6	3.288528
10	14.6	1.951728	10	27.0	3.60936
11	15.4	2.058672	11	27.8	3.716304
12	16.0	2.13888	12	28.6	3.823248
13	16.5	2.20572	13	29.4	3.930192
14	17.0	2.27256	14	30.2	4.037136
15	17.5	2.3394	15	31.0	4.14408
16	18.0	2.90624	16	31.8	4.241024
17	18.4	2.459712	17	32.6	4.357968
18	18.8	2.513184	18	33.4	4.464912
19	19.2	2.566656	19	34.2	4.571856
20	19.6	2.620128	20	35.0	4.6788
25	21.5	2.87412	25	38.0	5.07984
30	23.3	3.114744	30	42.0	5.61356
35	24.9	3.328632	35	44.0	5.88192
40	26.3	3.515784	40	46.0	6.14928
45	27.7	3.702936	45	48.0	6.41664
50	29.1	3.890088	50	50.0	6.684
60	32.0	4.27776	60	54.0	7.21872
70	35.0	4.6788	70	58.0	7.72344
80	38.0	5.07984	80	61.2	8.181216
90	41.0	5.48088	90	64.3	8.595624
100	43.5	5.81508	100	67.5	9.0234

Water Pressure Booster Systems

Sometimes the water pressure serving a building is too low for the system's requirements. In these cases, codes require that water pressure booster systems be installed. Fixture outlet pressures vary from code to code. The *International Plumbing Code®* (IPC), for example, requires a flow pressure of 8 pounds per square inch (psi) for most common household fixtures. A temperature-controlled shower requires up to 20 psi, and some urinals require 15 psi. Check your local code's pressure requirements and the manufacturer's instructions before installing the system.

Table 4 Table for Estimating Demand (2 of 2)

| SUPPLY SYSTEMS PREDOMINANTLY FOR FLUSH TANKS | | | SUPPLY SYSTEMS PREDOMINANTLY FOR FLUSHOMETER VALVES | | |
| Load | Demand | | Load | Demand | |
(Water supply fixture units)	(Gallons per minute)	(Cubic feet per minute)	(Water supply fixture units)	(Gallons per minute)	(Cubic feet per minute)
120	48.0	6.41664	120	73.0	9.75864
140	52.5	7.0182	140	77.0	10.29336
160	57.0	7.61976	160	81.0	10.82808
180	61.0	8.15448	180	85.5	11.42964
200	65.0	8.6892	200	90.0	12.0312
225	70.0	9.3576	225	95.5	12.76644
250	75.0	10.026	250	101.0	13.50168
275	80.0	10.6944	275	104.5	13.96956
300	85.0	11.3628	300	108.0	14.43744
400	105.0	14.0364	400	127.0	16.97736
500	124.0	16.57632	500	143.0	19.11624
750	170.0	22.7256	750	177.0	23.66136
1,000	208.0	27.80544	1,000	208.0	27.80544
1,250	239.0	31.94952	1,250	239.0	31.94952
1,500	269.0	35.95992	1,500	269.0	35.95992
1,750	297.0	39.70296	1,750	297.0	39.70296
2,000	325.0	43.446	2,000	325.0	43.446
2,500	380.0	50.7984	2,500	380.0	50.7984
3,000	433.0	87.88344	3,000	433.0	57.88344
4,000	525.0	70.182	4,000	525.0	70.182
5,000	593.0	79.27224	5,000	593.0	79.27224

2.3.0 Determining Pipe Size

Once you have estimated the system's requirements and demands, you can begin to size the system. There is more than one right way to size a water supply system. The method discussed in this section is based on the 2012 *International Plumbing Code®* (IPC). Plumbers in your area may use different engineering practices. Take the time to learn how systems are sized in your area. Whatever method you use, ensure that you follow the highest professional standards. The result will be a water supply system that provides the right amount of water at the right amount of pressure.

Begin with a sketch of the entire system (see *Figure 4*). Then, determine the minimum acceptable pressure to be provided to the highest fixture in the system. Refer to *Table 2* to determine the minimum pressure. If adequate pressure is provided to this fixture, you can assume that there will be enough pressure in the rest of the system. Your local code will specify the minimum pressure for flush valves and flush tanks. The system's pressure requirements must not fall below the minimum pressure or exceed the maximum pressure provided by the water supply.

Create a table, such as the one shown in *Table 5*, to calculate the correct pipe sizes for the system. Divide the drawing of the system (refer to *Figure 4*) into sections. The sections should occur where there are branches or changes in elevation.

Install Fittings for Future Access

Install valves, regulators, and strainers so that they are accessible, because you may need to repair or replace these items periodically. Valves and strainers should be installed so that they can be repaired without removing them.

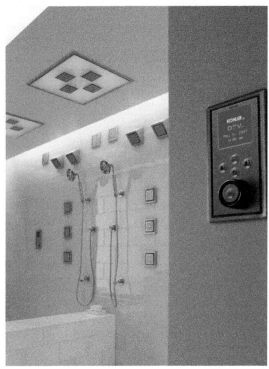

Figure 3 Luxury shower tower installation.

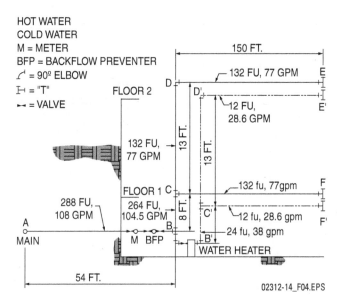

HOT WATER
COLD WATER
M = METER
BFP = BACKFLOW PREVENTER
⌐ = 90º ELBOW
⊢ = "T"
►◄ = VALVE

Figure 4 Water supply system drawing.

Add each section to the table. If the system design warrants it, consider calculating the hot- and cold-water requirements separately. Enter the flow in gpm for each section.

Find the minimum pressure available from the supply. Then calculate the highest pressure required in the system. Using your local code, calculate the pressure drop in the system due to the various fittings. Pressure drop is a loss of water pressure caused by:

- Water meters
- Water-main taps
- Water filters and softeners
- Backflow preventers
- Pressure regulators
- Valves and fittings (refer to *Table 1*)
- Pipe friction (see *Figure 5*)

Your local code will provide tables and graphs to help you calculate pressure drop. Note that the charts and tables will vary with the pipe material.

> **NOTE**
>
> Always save copies of the isometric sketches because you will provide them to the owner when you are finished.

2.4.0 Calculating System Losses

Calculate the static head loss, which is the difference in elevation between the supply line and the system's highest fixture. Enter all this information into the table on the appropriate lines and add them together. The result is the total overall losses and requirements.

Subtract the total losses from the minimum pressure available. This is the system pressure that is available to overcome friction loss. You can use this information to select the appropriate pipe size for each section of the system. Note that in some systems, the main is above the highest fixture. In this case, the main pressure must be added to the system, instead of subtracted from it.

Add the lengths of all the sections and select a trial pipe size using the following equation:

$$\text{psi} = (\text{available pressure}) \times 100 \div (\text{total pipe length})$$

Compare this number to the table on friction loss in fittings (refer to *Table 1*). The table indicates the equivalent lengths for trial pipe size of fittings and valves. Add the section lengths and equivalent lengths to find the total equivalent length of the system.

Finally, refer to your local code to determine the friction loss per 100 feet of pipe (refer to *Figure 5*). Multiply this number by the total equivalent length of each section. The result is the friction loss for each section. Add the friction losses together, and subtract the available pressure from the total pipe-friction losses. The result is the excess pressure after friction losses. It should be a small positive number. If it is, that means that the trial pipe size is correct for the system. If the result is a large positive number, then the pipe size is too large and can be reduced. Perform a new set of calculations using a new table and the above steps. Always ensure that the final pipe sizes are not less than the minimum size specified in the local code.

Table 5 Sample Table for Calculating Pipe Sizes

Line	COLUMN 1 Description	2 Lb per square inch (psi)	3 Gal.per min through section	4 Length of section (feet)	5 Trial pipe size (inches)	6 Equivalent length of fittings and valves (feet)	7 Total equivalent length col. 4 and col. 6 (100 feet)	8 Friction loss per 100 feet of trial size pipe (psi)	9 Friction loss in equivalent length col. 8 × col. 7 (psi)	10 Excess pressure over friction losses (psi)
A	Service and cold water distribution piping[b] — [b]Minimum pressure available at main 55.00	55.00								
B	Highest pressure required at a fixture (Table 604.3) 15.00	15.00								
C	Meter loss 2" meter 11.00	11.00								
D	Tap in main loss 2" tap [Table E103.3(4)] 1.61	1.61								
E	Static head loss 21 × 43 psi 9.03	9.03								
F	Special fixture loss backflow preventer 9.00	9.00								
G	Special fixture loss—Filter 0.00	0.00								
H	Special fixture loss—Other 0.00	0.00								
I	Total overall losses and requirements (Sum of Lines B through H) 45.64	45.64								
J	Pressure available to overcome pipe friction (Line A minus Lines B to H) 9.36	9.36								
	DESIGNATION	FU								
	Pipe section (from diagram) AB	288	108.0	54	2½	15.00	0.69	3.2	2.21	—
	Cold water BC	264	104.5	8	2½	0.5	0.85	3.1	0.26	—
	Distribution CD	132	77.0	13	2½	7.00	0.20	1.9	0.38	—
	piping CF[c]	132	77.0	150	2½	12.00	1.62	1.9	3.08	—
	DE[c]	132	77.0	150	2½	12.00	1.62	1.9	3.08	—
K	Total pipe friction losses (cold)								5.93	
L	Difference (Line J minus Line K)									3.43
	Pipe section (from diagram) A'B'	288	108.0	54	2½	12.00	0.69	3.3	2.21	—
	Diagram B'C'	24	38.0	8	2	7.5	0.16	1.4	0.22	—
	Hot water C'D'	12	28.6	13	1½	4.0	0.17	3.2	0.54	—
	Distribution C'F'[c]	12	28.6	150	1½	7.00	1.57	3.2	5.02	—
	Piping D'E'[c]	12	28.6	150	1½	7.00	1.57	3.2	5.02	—
K	Total pipe friction losses (hot)								7.99	
L	Difference (Line J minus Line K)									1.37

a. Based on information shown in Figure 4
b. To be considered as pressure gain for fixtures below main (to consider separately, omit from "I" and add to "J").
c. To consider separately, in K use C-F only if greater loss than above.

Standards Organizations in the IPC

Plumbers using codes based on the *International Plumbing Code®* must use pipes and fittings that conform to strict standards. These standards have been developed by several organizations. Two of the most frequently cited are located in the United States, another is in Canada.

Take the time to obtain and read these standards carefully. Talk to local plumbing experts and learn how the standards are used in your area. This information will help you broaden your professional knowledge. It will also be reflected in the quality of your work.

American Society for Testing and Materials International (ASTM)
100 Barr Harbor Drive, West Conshohocken, PA 19428-2959 **www.astm.org**
ASTM develops voluntary consensus standards and related technical information. It provides public health and safety guidelines, reliability standards for materials and services, and standards for commerce.

Canadian Standards Association (CAN/CSA)
178 Rexdale Boulevard, Toronto, Ontario M9W 1R3 Canada **www.csa.ca**
CSA develops standards for public safety and health, the environment, and trade. It also provides education and training for people who use their standards.

American Water Works Association (AWWA)
6666 W. Quincy Ave., Denver, CO 80235 **www.awwa.org**
AWWA is a not-for-profit organization dedicated to improving the quality and supply of drinking water. It specializes in scientific research and educational outreach. AWWA is the world's largest organization of water-supply professionals.

History of Pipes and Pipe Sizing in the United States

In the eighteenth and nineteenth centuries, pipe sizing was an art, not a science. Early plumbers used hollowed-out logs. They were limited to the thickness of the logs they could find, generally 9- to 10-inch-wide elm and hemlock trees. Wooden pipes were not ideal because they often sagged, and insects bred in the stagnant puddles that formed at the low points. Plus, log pipes gave water an unappetizing "woody" flavor.

Vents were among the first pipes to be sized. Until the late nineteenth century, vents were designed too small and frequently became clogged with ice or debris. In 1874, an unknown plumber invented a vent that balanced system pressure with the outside air pressure. The system used ½-inch pipe, which was wider than pipes used previously. The plumber also extended the pipe outside the building. It worked for awhile, but then plugged up. Through constant experimentation, plumbers eventually discovered the proper size of pipe to use in vents. Plumbers were able to apply this knowledge to sizing supply and drainpipes.

Today's standards for pipe sizing have come a long way since the early days of trial and error.

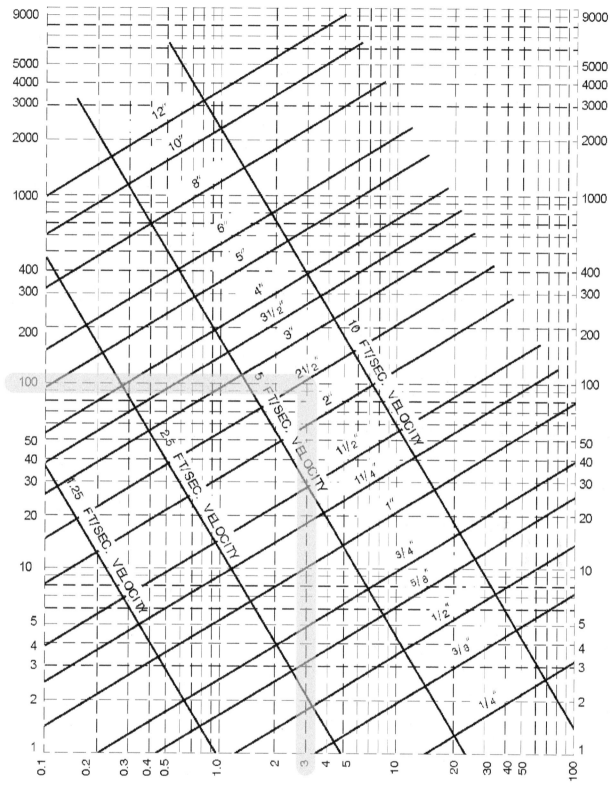

Note: Fluid velocities in excess of 5 to 8 feet/second are not usually recommended.

02312-14_F05.EPS

Figure 5 Sample table for calculating friction loss in smooth pipe.

Additional Resources

International Plumbing Code®, Latest Edition. Falls Church, VA: International Code Council.
Plumbers and Pipefitters Handbook. 1996. William J. Hornung. Englewood Cliffs, NJ: Prentice Hall College Division.
Standard Plumbing Engineering Design, Second Edition. 1982. Louis S. Neilsen. New York: McGraw-Hill.
Pipefitters Handbook. 1967. Forrest R. Lindsey. New York: Industrial Press Inc.

2.0.0 Section Review

1. A factor that plumbers do *not* have to consider when installing a water supply system is the system's ability to _____.
   a. operate with a minimum amount of turbulence
   b. carry an adequate supply of water
   c. prevent transient flow near fittings
   d. maintain a reasonable flow rate

2. Outlets such as sinks, lavatories, and water closets are considered _____.
   a. intermittent-demand items
   b. high use
   c. continuous-demand items
   d. low flow

3. When preparing an isometric sketch of a water supply system to help determine pipe size, divide the sketch into sections _____.
   a. where continuous-demand fittings are located
   b. by elevation within the system
   c. where branches or changes in elevation occur
   d. by fixture output in gallons per minute

4. The equation for selecting a trial pipe size is _____.
   a. gpm = (available pressure) ÷ 100 × (total pipe length)
   b. psi = (available pressure) ÷ 100 × (total pipe length)
   c. gpm= (available pressure) × 100 ÷ (total pipe length)
   d. psi = (available pressure) × 100 ÷ (total pipe length)

SECTION THREE

3.0.0 BACKFLOW PREVENTION

Objective

Describe the six basic backflow-prevention devices and the hazards they are designed to prevent.

a. Describe the principle of backflow due to back siphonage and back pressure.
b. Describe when and how to install air gaps.
c. Describe when and how to install atmospheric vacuum breakers.
d. Describe when and how to install pressure-type vacuum breakers.
e. Describe when and how to install dual-check valve backflow preventer assemblies.
f. Describe when and how to install double-check valve assemblies.
g. Describe when and how to install reduced-pressure zone principle backflow preventer assemblies.
h. Describe when and how to install specialty backflow preventers.
i. Describe when and how to troubleshoot and repair backflow preventers.

Performance Task

Install common types of backflow preventers.

Trade Terms

Atmospheric vacuum breaker: A backflow preventer designed to prevent back siphonage by allowing air pressure to force a silicon disc into its seat and block the flow.

Backflow preventer with intermediate atmospheric vent: A specialty backflow preventer used on low-flow and small supply lines to prevent back siphonage and back pressure. It is a form of double-check valve.

Cross-connection: A direct link between the potable-water supply and water of questionable quality.

Double-check valve assembly (DCV): A backflow preventer that prevents back pressure and back siphonage through the use of two spring-loaded check valves that seal tightly in the event of backflow. DCVs are larger than dual-check valve backflow preventers and are used for heavy-duty protection. DCVs have test cocks.

Dual-check valve backflow preventer assembly (DC): A backflow preventer that uses two spring-loaded check valves to prevent back siphonage and back pressure. DCs are smaller than double-check valve assemblies and are used in residential installations. DCs do not have test cocks.

High hazard: A classification denoting the potential for an impairment of the quality of the potable water that creates an actual hazard to the public health through poisoning or through the spread of disease by sewage, industrial fluids, or waste. This type of water-quality impairment is called contamination.

Hose-connection vacuum breaker: A backflow preventer designed to be used outdoors on hose bibs to prevent back siphonage.

In-line vacuum breaker: A specialty backflow preventer that uses a disc to block flow when subjected to back siphonage. It works the same as an atmospheric vacuum breaker.

Low hazard: A classification denoting the potential for an impairment in the quality of the potable water to a degree that does not create a hazard to the public health but does adversely and unreasonably affect the aesthetic qualities of such potable water for domestic use. This type of water-quality impairment is called pollution.

Manufactured air gap: An air gap that can be installed on a reduced-pressure zone principle backflow preventer to prevent back pressure and back siphonage.

Pressure-type vacuum breaker: A backflow preventer installed in a water line that uses a spring-loaded check valve and a spring-loaded air inlet valve to prevent back siphonage.

Reduced-pressure zone principle backflow preventer assembly (RPZ): A backflow preventer that uses two spring-loaded check valves plus a hydraulic, spring-loaded pressure-differential relief valve to prevent back pressure and back siphonage.

Test cock: A valve in a backflow preventer that permits the testing of individual pressure zones.

Once a water supply system has been sized and installed, it needs to be protected. Normally, water supply systems are separated from drain, waste, and vent (DWV) systems that remove wastewater. However, sometimes problems or malfunctions can force water to flow backward through a system. When this happens, wastewater and other liquids can be siphoned into the fresh-water supply. This can cause contamination, sickness, and even death.

Codes, health departments, and water purveyors all require plumbing systems to provide protection against reverse flow. Protection can be in the form of a gap or a barrier between the two sources. Check valves can be effective barriers and can be arranged to provide increasing protection against contamination. In this section, you will learn how to install devices that protect potable water from pollution and contamination.

3.1.0 Understanding Backflow Caused by Back Siphonage and Back Pressure

The reverse flow of nonpotable liquids into the potable-water supply is called backflow. Backflow can contaminate a fixture, a building, or a community. There are two types of backflow: back pressure and back siphonage. Back pressure is any elevation of pressure in the downstream direction of a potable-water system that increases the downstream pressure to the point where reversal of flow direction occurs. Back siphonage is a form of backflow that, due to a reduction in the system pressure upstream, causes a subatmospheric pressure to exist at any site in the water system and results in siphoning of water from downstream.

Backflow cannot happen unless there is a direct connection between the potable-water supply and another source. This condition is called a cross-connection. Cross-connections are not hazards all by themselves (see *Figure 6*). In fact, sometimes they are required. Cross-connections are sometimes hard for the public to spot. Many people might not know that a hose left in a basin of wastewater could cause a major health hazard by creating a backflow. Consider the cross-connection in *Figure 6*. If someone opens the faucet while the washing machine is running, the wastewater in the sink would be forced by the higher pressure back into the potable-water supply, contaminating the building supply and possibly even the municipal water supply system.

When something creates a backflow, a cross-connection becomes dangerous. Several things can cause backflow, which, in turn, can force wastes through a cross-connection:

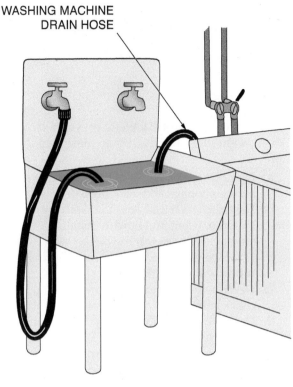

WASHING MACHINE DRAIN HOSE

02312-14_F06.EPS

Figure 6 Simple cross-connection.

- Cuts or breaks in the water main (see *Figure 7*)
- Failure of a pump
- Injection of air into the system
- Accidental connection to a high-pressure source

Plumbers use backflow preventers to protect against cross-connections. They provide a gap or a barrier to keep backflow from entering the water supply. Many local codes now require certain types of backflow preventers for most fixtures in a new structure, while an air gap is required for others. Tank trucks filled with sewage, pesticides, or other dangerous substances must also use backflow preventers. Plumbers are responsible for installing backflow preventers in plumbing systems. Only certified people, who may or may not be plumbers, are allowed to service them.

Backflow Blotter: Seattle, Washington, 1979

A pump used in the scrubbing cycle at a car wash broke down. To keep the scrubber working, the staff ran a hose from the scrubber cycle to the rinse cycle. The rinse cycle was connected to the potable-water system. Two days later, a plumber fixed the pump. However, no one remembered to remove the hose that connected the scrubbers to the potable-water system. The new high-pressure scrubber pump created back pressure. This forced wastewater into the fresh-water system. An eight-block area was contaminated, and two people got sick. The city ordered the car wash owner to install a reduced-pressure zone principle backflow preventer, which he did the following day.

Source: US Environmental Protection Agency

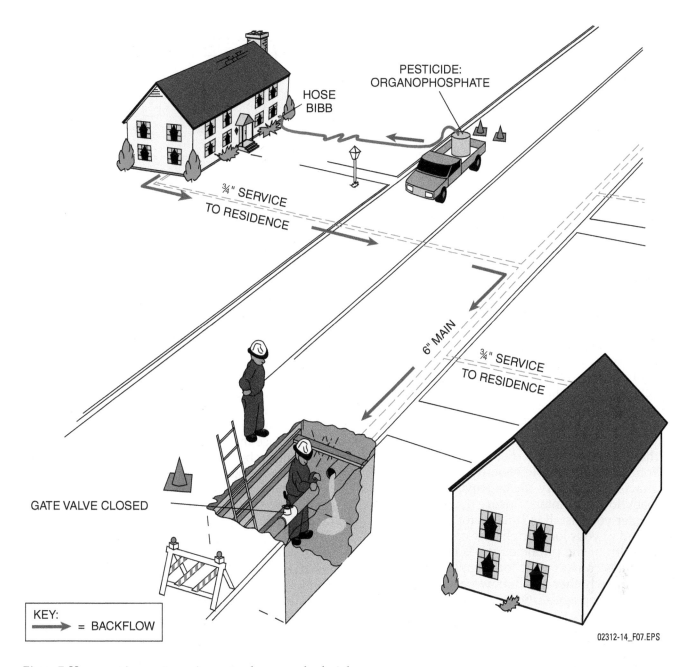

PESTICIDE:
ORGANOPHOSPHATE

HOSE
BIBB

¾" SERVICE
TO RESIDENCE

6" MAIN

¾" SERVICE
TO RESIDENCE

GATE VALVE CLOSED

KEY:
= BACKFLOW

02312-14_F07.EPS

Figure 7 How a cut in a water main creates dangerous back siphonage.

Backflow preventers help keep drinking water safe from contamination caused by back siphonage and back pressure. They are an important part of a plumbing system because they help maintain public health. There are six basic types of backflow preventers, each designed to work under different conditions. Some protect against both back pressure and back siphonage, while others can handle

Standards for Backflow Preventers

Backflow preventers permitted for use in water supply systems in the United States are designed according to one or more of the national standards developed by the following organizations:

- American Water Works Association (AWWA)
- American Society of Sanitary Engineers (ASSE)
- University of California Foundation for Cross-Connection Control and Hydraulic Research

only one type of backflow. Before installing a back-flow preventer, always ensure that it is appropriate for the type of installation. Consider several factors:

- Risk of cross-connection
- Type of backflow
- Health risk posed by contaminants

Preventers must be installed correctly. Select the proper preventer for the application, and review the manufacturer's instructions before installing a preventer. Never attempt to install a backflow preventer in a system for which it is not designed because it may fail, resulting in backflow. Refer to your local code for specific guidance.

Backflow preventers are classified according to whether they are used in low hazard or high hazard applications. Both "low hazard" and "high hazard" refer to the possibility of a threat to the quality of the potable-water supply. The difference is that a low-hazard application does not threaten to create a public-health hazard although the pollutants would adversely and unreasonably affect the appearance, taste or other qualities of the water. In a high-hazard application, the potential for water quality impairment would create an actual hazard to public health through poisoning or through the spread of disease.

Your local applicable code will include tables or charts that specify the approved applications for backflow preventers. Ensure that the backflow preventer you install is suitable for the hazard as specified in your local applicable code. The safety and health of the public depends on it. Remember, always consult your local applicable code to determine whether a backflow preventer is suitable for a given application.

> **NOTE**
>
> Remember to calculate the pressure drop across backflow preventers when sizing water supply systems.

3.2.0 Installing Air Gaps

An air gap is a physical separation between a potable-water supply line and the flood-level rim of a

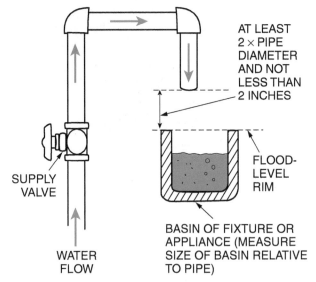

Figure 8 Air gap for supply pipe greater than 1" in diameter.

fixture (see *Figure 8*). It is the simplest form of backflow preventer. Air gaps can be used in low-hazard intermittent pressure applications with a risk of back siphonage. Ensure that the supply pipe terminates above the flood-level rim at a distance that is at least twice the diameter of the pipe. The air gap should never be less than 1 inch. If the supply pipe diameter is greater than 1 inch, the air gap must be at least 2 inches. Some codes have specific requirements for fixtures with openings that are less than 1 inch (see *Table 6*).

All fixture faucets must incorporate an air gap. Air gaps can be used to prevent back pressure, back siphonage, or both. Never submerge hoses or other devices connected to the potable-water supply in a basin that may contain contaminated liquid. Even seemingly innocent activities such as filling up a wading pool with a garden hose could be a problem if, elsewhere, workers digging with a backhoe accidentally cut through the water main.

Because air gaps permit a loss of system pressure, they are most often used at the end of a service line where they empty into storage tanks or reservoirs. Never attempt to eliminate splashing during drainage by moving the end of the line into

Backflow Blotter: Unidentified Town, Connecticut, 1982

Workers repairing a tank at a propane storage facility purged the tank with water. They used a hose connected to a fire hydrant. However, pressure in the tank was about 20 psi higher than that of the water system. About 2,000 cubic feet of propane gas—enough to fill 1 mile of an 8-inch-diameter water main—backflowed into the water system. The city was alerted by reports of gas hissing out of fixtures. Hundreds of homes and businesses had to be evacuated. Two homes caught fire, and one of them was completely gutted. At another house, a washing machine exploded.

Source: US Environmental Protection Agency

Table 6 Minimum Air Gaps from a Model Code

FIXTURE	MINIMUM AIR GAP	
	Away from a wall (inches)	Close to a wall (inches)
Lavatories and other fixtures with effective opening not greater than ½ inch in diameter	1	1½
Sink, laundry trays, gooseneck back faucets and other fixtures with effective openings not greater than ¾ inch in diameter	1½	2½
Over rim bath fillers and other fixtures with effective openings not greater than 1 inch in diameter	2	3
Drinking water fountains, single orifice not greater than 7/16 inch in diameter or multiple orifices with a total area of 0.150 square inch (area of circle 7/16 inch in diameter)	1	1½
Effective openings greater than 1 inch	Two times the diameter of the effective opening	Three times the diameter of the effective opening

the receptacle. This defeats the purpose of the air gap and creates a cross-connection situation.

3.3.0 Installing Atmospheric Vacuum Breakers

If it is not possible to install an air gap where there is a risk of back siphonage, consider using an atmospheric vacuum breaker (see *Figure 9*). Atmospheric vacuum breakers (AVBs) use a silicon disc float to control water flow. In normal operation, the float rises to allow potable water to enter. If back siphonage causes low pressure, air enters the breaker from vents and forces the disc into its seat. This shuts down the flow of fresh water (see *Figure 10*).

Use AVBs for intermittent-use, low-hazard nonpotable-water systems where there is a risk of back siphonage, such as the following:

- Commercial dishwashers
- Sprinkler systems
- Outdoor faucets

Some codes allow AVBs to be used where a pipe or hose terminates below the flood-level rim of a basin, as long as no conditions for back pressure are present (see *Figure 11*). Because AVBs are meant to be used only in low-hazard applications, AVBs do not offer sufficient protection for health care or food-processing applications. Other pro-

02312-14_F09.EPS

Figure 9 Atmospheric vacuum breaker.

tection upstream of the AVB would be required in such applications. Refer to your local applicable code before installing AVBs to ensure that they are permitted for such applications.

AVBs are designed to operate at normal air pressure. Do not use them where there is a risk of back-pressure backflow. Check the manufac-

Freeze Protection for Backflow Preventers

Codes require backflow preventers to be installed outside. Depending on the climate, codes may specify taking steps to protect backflow preventers from freezing. Freeze protection is a good idea even if the codes don't require it. Install insulated protective cases around exposed backflow preventers and wrap pipes with heat tape. Always refer to the manufacturer's instructions first to ensure that you are using appropriate materials and methods.

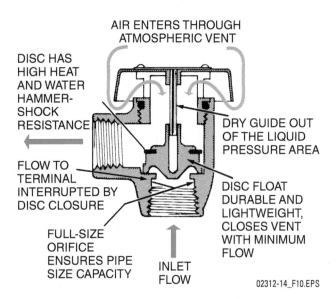

DISC HAS HIGH HEAT AND WATER HAMMER-SHOCK RESISTANCE

AIR ENTERS THROUGH ATMOSPHERIC VENT

DRY GUIDE OUT OF THE LIQUID PRESSURE AREA

FLOW TO TERMINAL INTERRUPTED BY DISC CLOSURE

DISC FLOAT DURABLE AND LIGHTWEIGHT, CLOSES VENT WITH MINIMUM FLOW

FULL-SIZE ORIFICE ENSURES PIPE SIZE CAPACITY

INLET FLOW

02312-14_F10.EPS

Figure 10 Cut-away of atmospheric vacuum breaker in the closed position.

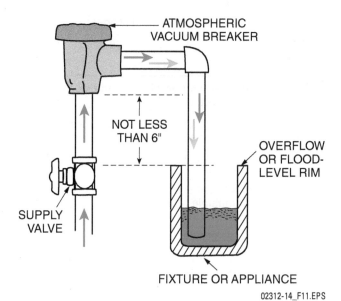

ATMOSPHERIC VACUUM BREAKER

NOT LESS THAN 6"

OVERFLOW OR FLOOD-LEVEL RIM

SUPPLY VALVE

FIXTURE OR APPLIANCE

02312-14_F11.EPS

Figure 11 Atmospheric vacuum breaker installation.

turer's specifications for the proper operating temperatures and pressures. If conditions exceed these limits, consult the manufacturer first. Unlike other types of backflow preventers, AVBs are not designed to be routinely tested.

Install an AVB after the last control valve, and ensure that it is at least 6 inches above the fixture's flood-level rim. Install the AVB so that the fresh-water intake is at the bottom, and ensure that the water supply is flowing in the same direction as the arrow on the breaker. Use AVBs on applications with intermittent supply pressure. Do not install shutoff valves downstream from an AVB.

Hose-connection vacuum breakers (*Figure 12*) are designed for use where there is a hose connection. Some codes require them when there is a low-hazard risk of back siphonage in hoses attached to any of the following:

- Sill cocks (hose bibbs)
- Laundry tubs
- Service sinks
- Photograph-developing sinks
- Dairy barns
- Wash racks

Hose-connection vacuum breakers work the same way and under the same conditions as AVBs. Note that hose connections do not allow water to drain from an outdoor faucet that is not protected

02312-14_F12.EPS

Figure 12 Hose-connection vacuum breaker.

from freezing. If there is a danger of freezing, install a nonfreeze faucet with a built-in vacuum breaker. This will allow the faucet to drain. Both types of AVBs can be equipped with strainers.

3.4.0 Installing Pressure-Type Vacuum Breakers

Install pressure-type vacuum breakers (see *Figure 13*) to prevent back siphonage only, in either low- or high-hazard applications. They should be used in water lines that supply the following:

Containment and Isolation Backflow Preventers

Backflow preventers installed at the meter or curb protect the public water supply system. They are called containment backflow preventers. Backflow preventers on a house system protect the occupants from external contaminants. They are called isolation backflow preventers.

- Cooling towers
- Swimming pools
- Heat exchangers
- Degreasers
- Lawn sprinkler systems with downstream zone valves

A pressure-type vacuum breaker (PVB) has a spring-loaded check valve and a spring-loaded air inlet valve. The PVB operates when line pressure drops to atmospheric pressure or less. The air inlet valve opens the vent to the outside air. The check valve closes to prevent back siphonage.

PVBs are suitable for applications with constant inlet water pressure. PVBs are equipped with test cocks, which are valves that allow testing of the individual pressure zones within the device. They can also be equipped with strainers if required. Shutoff valves can be installed downstream from PVBs.

Vacuum breakers are also installed on hose bibbs (see *Figure 14*). As you learned in the *Installing and Testing Water Supply Piping* module, codes typically require at least one hose bibb in a residential structure, although additional hose bibbs may be installed according to the wishes of the designer or the customer. Hose-bibb outlets are usually ¾ inch in diameter, although ½-inch piping may be used to access the outlet.

3.5.0 Installing Dual-Check Valve Backflow Preventer Assemblies

Homes are full of potential sources of contamination. The most common include the following:

- Hose-attached garden spray bottles
- Lawn sprinkler systems
- Bathtub whirlpool adapters
- Hot tubs
- Water-closet bowl deodorizers
- Wells or backup water systems
- Photography darkrooms
- Misapplied pest extermination chemicals

To keep residential backflow from reaching the water main, use a dual-check valve backflow preventer assembly (DC) (see *Figure 15*). These preventers feature two spring-loaded check valves (see *Figure 16*). They work similarly to double-check valve assemblies (DCVs), which are discussed in the section *Installing Double-Check Valve Assemblies*. Note that two check valves in series are not automatically a backflow preventer. A DC is a specially designed and sized single unit.

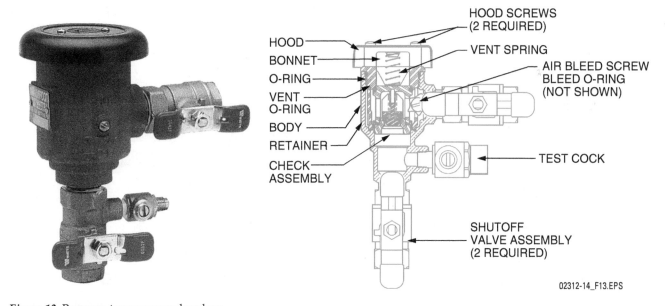

Figure 13 Pressure-type vacuum breaker.

Backflow Preventers in Restaurants

Many fixtures and appliances in restaurants depend on clean, fresh water. Steamers, dishwashers, coffee machines, icemakers, and soda dispensers all connect to the fresh-water supply. They all require backflow preventers. Public health inspectors regularly test these systems to ensure that they are working properly. A restaurant in violation of the local health code can be shut down. An accidental cross-connection, therefore, could not only cause sickness and death but could also damage a restaurant's reputation and threaten its economic survival.

Figure 14 Hose bibb with vacuum breaker.

02312-14_F15.EPS

Figure 15 Dual-check valve backflow preventer assembly.

DCs protect against back pressure and back siphonage in low-hazard applications. Because DCs are small, they are commonly used in residential installations. These preventers tend to be less reli-

able than DCVs. Unlike DCVs, they are normally not fitted with shutoff valves or test cocks. This makes it difficult for plumbers to remove, test, or replace them. Check the local code before installing DCs. Install DCs on the customer side of a residential water meter. They can be installed either horizontally or vertically, depending on the available space and the existing piping arrangements.

3.6.0 Installing Double-Check Valve Assemblies

Where heavy-duty protection is needed, install a double-check valve assembly to protect against back siphonage, back pressure, or both in low-hazard, continuous-pressure applications (see *Figure 17*). A DCV has two spring-loaded check

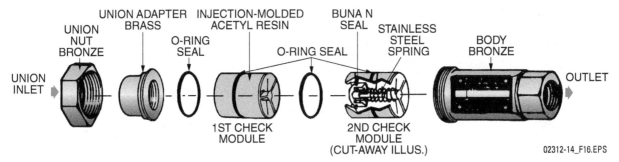

Figure 16 Cut-away of dual-check valve backflow preventer assembly showing check valves.

Backflow Blotter: Lacy's Chapel, Alabama, 1986

The driver of a tanker truck flushed his truck's tank with water. The tank had been filled with the chemical sodium hydroxide. The driver filled the tank from a connection at the bottom of the truck. There was no backflow preventer at the site where the driver filled his truck. At the same time, the town's water main broke. Because of the sudden drop in pressure, the contents of the tanker truck siphoned into the system. The city water service did not have a cross-connection program. As a result, the wastewater entered the potable-water supply. People reported burned mouths and throats after drinking the water. One man suffered red blisters all over his body after showering in the water. He and several others were treated at a nearby emergency room. Sixty homes were contaminated. The mains had to be flushed, and health officials had to inspect all homes in the affected area.

Source: US Environmental Protection Agency

valves and includes a test cock. They are commonly used in food-processing steam kettles and apartment buildings. During normal operation, both check valves are open; however, backflow causes the valves to seal tightly.

> **WARNING!**
>
> Never install brass DCVs in soda carbonation systems. The carbon dioxide (CO_2) gas used to add bubbles to soda will chemically react with brass and create carbonic acid, a liquid that is lethal if ingested. Use stainless steel DCVs instead.

DCVs are available in a wide range of sizes from ¾ inch up to 10 inches. DCVs have a gate or ball shutoff valve at each end. If required, DCV inlets can be equipped with strainers.

3.7.0 Installing Reduced-Pressure Zone Principle Backflow Preventer Assemblies

Reduced-pressure zone principle backflow preventer assemblies (RPZs) protect the water supply from contamination by dangerous liquids (see *Figure 18*). They may be used in both low- and high-hazard applications where there is a risk of back siphonage and back pressure.

RPZs are the most sophisticated type of backflow preventer. They feature two spring-loaded check valves plus a hydraulic, spring-loaded pressure-differential relief valve. The relief valve is located underneath the first check valve. The device creates a low pressure between the two check valves. If the pressure differential between the inlet and the space between the check valves drops below a specified level, the relief valve opens. This allows water to drain out of the space between the check valves (see *Figure 19*). The relief valve will also operate if one or both check valves fail. This offers added protection against backflow. RPZs also have shutoff valves at each end. Sometimes they are also fitted with strainers.

Use an RPZ if a direct connection is subject to back siphonage, back pressure, or both. A wide variety of industrial and agricultural operations require RPZs, including:

- Car washes
- Poultry farms
- Dairies
- Medical and dental facilities
- Wastewater treatment plants
- Manufacturing plants

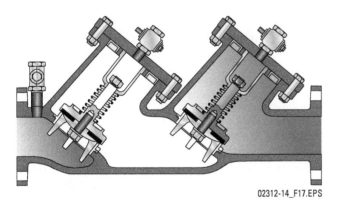

02312-14_F17.EPS

Figure 17 Double-check valve assembly.

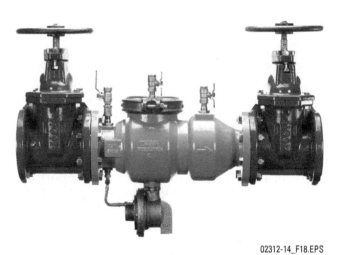

02312-14_F18.EPS

Figure 18 Reduced-pressure zone principle backflow preventer assembly.

Backflow Blotter: Woodsboro, Massachusetts, 1983

Someone working at an agricultural farm left a gate valve to a holding tank open, creating a cross-connection. The tank was filled with a powerful herbicide. About the same time, a pump failed in the town's water system, reducing the system's pressure. The herbicide from the tank siphoned into the potable-water supply. People reported that yellow water was coming from their fixtures. Firefighters warned people to avoid drinking, cooking, or bathing with the town's water. The town had to flush the entire water system. Tanker trucks had to be brought in to supply fresh water, and sampling and testing had to be carried out for several weeks.

Source: US Environmental Protection Agency

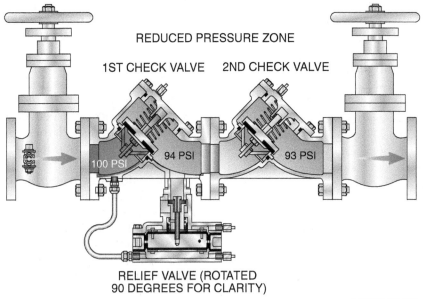

REDUCED PRESSURE ZONE

1ST CHECK VALVE   2ND CHECK VALVE

100 PSI   94 PSI   93 PSI

RELIEF VALVE (ROTATED
90 DEGREES FOR CLARITY)

02312-14_F19.EPS

Figure 19 Operation of a reduced-pressure zone principle backflow preventer assembly with check valves parallel to flow.

WARNING!

Codes prohibit the installation of any type of bypass around a backflow preventer. Accidental backflow through the bypass will pollute or contaminate the system, and can result in a serious health hazard.

Some codes require RPZs on additional types of installations. RPZs are available for all operating temperatures and pressures. Refer to your local code to see where RPZs are required. Many RPZs come with union or flanged connections, which allow fast and easy removal for servicing. Install RPZs so that their lowest point is at least 12 inches above grade, and provide support for the RPZ (see *Figure 20*). Otherwise, sagging could damage the RPZ. Select a safe location that will help prevent damage to and vandalism of the RPZ. Install a manufactured air gap on all RPZ installations (see *Figure 21*). It acts as a drain receptor to prevent siphonage.

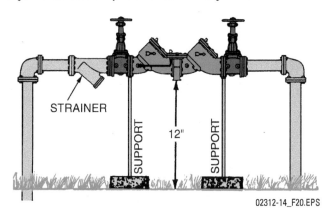

STRAINER   SUPPORT   12"   SUPPORT

02312-14_F20.EPS

Figure 20 Reduced-pressure zone principle backflow preventer assemblies installed with supports.

Some codes permit two RPZs to be installed in a parallel fashion (see *Figure 22*). This allows the system to continue working if an RPZ malfunctions or is being serviced. If two RPZs are used in parallel, a smaller size can be used than if only one RPZ were installed. Locate a strainer in the pipe before both backflow preventers. If there

Cross-Connection and the Spread of Bacteria

Escherichia coliform, also known as E. coli, and Legionnaires' disease, are two potentially deadly bacteria that can be spread by cross-connection. E. coli is a common type of bacteria. It can be found in animal and human intestines, where it provides natural vitamins, but E. coli in animal wastes can be dangerous. It can enter the fresh-water supply through untreated runoff and cross-connections. E. coli turns safe potable water into a disease delivery system and can contaminate meat and other foods. Because plumbers are responsible for installing and maintaining safe plumbing systems, E. coli is a real concern for them, too.

DIRECTION OF FLOW

DIRECTION OF FLOW

02312-14_F21.EPS

Figure 21 Manufactured air gap.

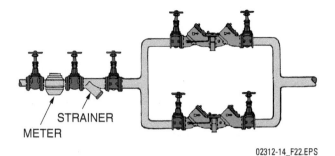

METER

STRAINER

02312-14_F22.EPS

Figure 22 Reduced-pressure zone principle backflow preventer assemblies installed in parallel.

is sediment in the system, it may clog a single strainer. In that case, install separate strainers in advance of each RPZ. Consult with the project engineer and the local plumbing code official before installing parallel backflow preventers.

Health officials regularly test for outbreaks of the bacteria *Escherichia coliform*, also known as E. coli, in the water supply. When an outbreak occurs, it can cause chaos and severe illness. Residents have to use bottled water or tanker trucks have to bring in water for drinking and cooking. Contaminated water must be boiled before people can use it to even brush their teeth. Testing, sam-

pling, and cleanups can take weeks. Sometimes months go by before the health department declares a system completely clean.

Bacteria cause other health problems, including Legionnaires' disease. The first major reported outbreak of this disease occurred at an American Legion conference in 1976. Scientists named the bacteria *Legionella*. It reproduces in warm water. *Legionella* can grow in hot-water tanks and some plumbing pipes. People inhale the bacteria through contaminated water mist from large air conditioning installations, whirlpool spas, and showers. It is not contagious, but between 5 and 30 percent of people who contract Legionnaires' disease die from it. Antibiotics do not kill *Legionella*, but they can keep the bacteria from multiplying.

Plumbers can help block the spread of disease caused by waterborne bacteria. Design plumbing systems so that cross-connections can't happen. Install interceptors and pumps to keep out harmful waste and to prevent water from stagnating. Plumbers play a vital role in maintaining public health.

Backflow Blotter: Groveton, Pennsylvania, 1981

A building was being treated for termites with two toxic pesticides: chlordane and heptachlor. The pesticide contractor diluted the mixture by running water from a hose bibb to his chemical tanks. The end of the hose was submerged in the chemical solution, which set up the link for the cross-connection. At the same time, a plumber cut into a 6-inch main line to install a gate valve. As the line was cut and the water drained, a siphoning effect was created. This drew the toxic chemicals into the water main. When the water main was recharged, these chemicals were dispersed throughout the potable-water system beyond the newly installed gate valve. As a result, the potable-water system in 75 housing units was contaminated. Continued flushing of the water lines failed to remove the chemicals from the water supply. All of the water-supply piping had to be replaced, which cost $300,000. A simple hose-connection vacuum breaker could have prevented this costly mistake.

Source: US Environmental Protection Agency

3.8.0 Installing Specialty Backflow Preventers

Often, low-flow and small supply lines require backflow preventers. Use a backflow preventer with intermediate atmospheric vent for this kind of line (see *Figure 23*). The intermediate atmospheric vent is a form of DCV that also protects against back pressure and back siphonage in low-hazard applications. This type of breaker is therefore suitable for use on the following installations:

- Most laboratory equipment
- Processing tanks
- Sterilizers
- Residential boiler feeds

Your local code or health department will identify low-hazard and high-hazard applications and specify the types of backflow preventers that can be used in such cases. Because plumbers are responsible for purchasing and installing backflow preventers, you should take the time to become familiar with the brands approved for use in your area.

Laboratories and hospitals use chemicals that can cause illness or death. Special backflow preventers help protect the fresh-water supply in such applications. Laboratory faucets with hose attachments require in-line vacuum breakers (see *Figure 24*). In-line vacuum breakers work the same way as atmospheric vacuum breakers. A disc allows fresh water to flow to the faucet. Negative pressure causes the disc to close the water inlet (see *Figure 25*). In-line vacuum breakers protect against back siphonage in high-hazard applications. Install them on new and existing faucets. Changes to the plumbing are not required.

> **WARNING!**
> Never install a backflow preventer in a pit. Refer to your local code for approved installation options.

3.9.0 Testing and Repairing Backflow Preventers

Backflow preventers can break down. When they do, they need to be repaired or replaced. Many state codes require regular testing of preventers to identify whether repair or replacement is needed. Once the backflow preventer has been tested, they can be serviced using repair kits provided by the manufacturer. Always follow the manufactur-

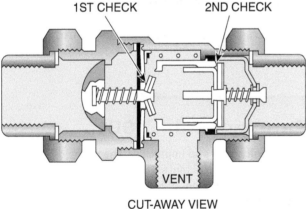

1ST CHECK   2ND CHECK

VENT

CUT-AWAY VIEW

02312-14_F23.EPS

Figure 23 Intermediate-atmospheric-vent vacuum breaker.

02312-14_F24.EPS

Figure 24 In-line vacuum breaker.

Selecting Backflow Preventers

When determining which type of backflow preventer to install, you must always consider its application. Because backflow preventers help to maintain public health, liability risks affect how potential hazards are defined. The type of installation varies depending on the degree of hazard presented by the potential contaminants. Therefore, always check your local code before installing a backflow preventer.

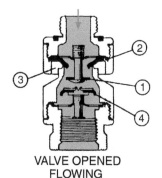

With flow-through valve, primary check (1) opens away from diaphragm seal (2). Atmospheric port (3) remains closed by deflection of diaphragm seal. Secondary check (4) opens away from downstream seat, permitting flow of water through valve.

VALVE OPENED
FLOWING
UNDER PRESSURE

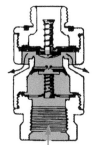

With a back-siphonage condition, secondary check seals tightly against downstream seat. Primary check seals against diaphragm. Atmospheric port is open, permitting air to enter air break chamber. In the event of fouling of downstream check valve, leakage would be vented to the atmosphere through the vent port.

VALVE CLOSED BY
BACK-SIPHONAGE
IN SYSTEM

02312-14_F25.EPS

Figure 25 Operation of an in-line vacuum breaker.

er's instructions when testing or repairing backflow preventers. Your local code will specify the requirements for testing and servicing backflow preventers in your jurisdiction.

3.9.1 Testing Backflow Preventers

Backflow preventers are required to be tested when installed, whenever they are serviced or relocated, and also annually. Only certified technicians can test backflow preventers. Testers are certified in each state. Testers who are not plumbers are not allowed to repair or replace backflow preventers. However, a plumber may work on a backflow preventer and then have a certified tester test the device. The tester will then send a report of the repair work to the authority having jurisdiction.

> **CAUTION**
>
> Under no circumstances should an uncertified person attempt to test backflow preventers. Likewise, a person who is certified to test backflow preventers is not necessarily certified to also repair water supply systems. Plumbers can use a certified tester's report to determine repair needs, however.

Testers must take written and practical exams offered at a site authorized by the local chapter of the American Water Works Association (AWWA). Applicants must have a high school diploma or General Educational Development (GED) equivalent, be at least 18 years old, and speak and understand English. In many states, testers do not have to be licensed plumbers. The certification is good for three years.

Many backflow preventers have test cocks, which are used for troubleshooting (see *Figure 26*). Each type of backflow preventer will have its own testing procedure. Once the test kit has been configured according to the manufacturer's instructions, flush out the test cocks and bleed the air from the test kit.

Proceed from valve to valve, checking for tightness against reverse flow. Refer to the gauge on the test kit to ensure that the pressure differential is within the range specified by the manufacturer. Once the valves have been checked, test the pressure-differential relief valve to ensure that it is maintaining the reduced pressure zone between the two check valves. The manufacturer's instructions will specify the required pressure differential.

> **NOTE**
>
> Backflow preventer test equipment must be calibrated yearly. Refer to the manufacturer's instructions for the calibration schedule and method.

3.9.2 Repairing Backflow Preventers

Only certified technicians are allowed to service backflow preventers. As with backflow preventer testers, service personnel are certified in each state. If you are not certified to repair backflow preventers in your jurisdiction, do not attempt to do so.

Foreign matter blocking the valves can cause a malfunction; as a result, many backflow preventers have strainers to trap dirt and other matter.

Certified Backflow Preventer Assembly Testers

In many states, plumbers can install backflow preventers. However, preventer testing must be left to certified testers. Codes provide stiff penalties for a noncertified plumber who carries out an inspection. Refer to the local code.

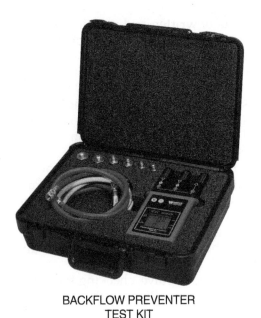

BACKFLOW PREVENTER
TEST KIT

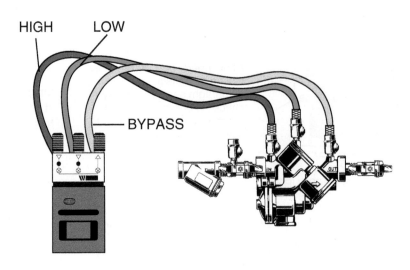

TEST KIT CONNECTED TO
TEST COCKS ON RPZ

02312-14_F26.EPS

Figure 26 Testing the pressure zones of an RPZ.

Strainers can be cleaned without having to disconnect the preventer. If sediment interferes with the operation of a backflow preventer, the system needs to be flushed.

Manufacturers provide repair kits and specifications for backflow preventers that allow them to be serviced in the field without having to replace the entire device, which can be expensive. Note that some codes require malfunctioning or damaged backflow preventers to be replaced rather than repaired. Always consult your local applicable code before attempting to repair a backflow preventer.

The steps for servicing a backflow preventer will vary depending on the type of device and the manufacturer. However, there are some general requirements that are common for many types of backflow prevention devices. The general steps for servicing an RPZ, for example, are covered in the following sections. Always refer to the manufacturer's specifications for the particular device.

To repair or replace the relief valve in an RPZ (see *Figure 27*), follow these steps:

Step 1 Turn off the water supply to the RPZ and bleed off the pressure.

Step 2 Loosen and then remove the bolts holding the valve cover. Hold the cover in place while doing this.

Step 3 Remove the cover, stem, and diaphragm assembly. Note that these may be a single unit depending on the model of RPZ.

Step 4 Remove the relief valve seat. Clean or replace the valve disc as required. Be careful not to damage the rubber seal on the disc during the process.

Step 5 Reassemble the relief valve by pressing the seat firmly into place and ensuring that the spring is centered on the seat. Replace the stem, diaphragm assembly, and valve cover. Ensure that the valve cover is properly aligned before inserting and tightening the bolts to hold the cover in place.

Step 6 Turn on the water supply to the RPZ and check for leaks. If a leak is detected, repair it according to the manufacturer's instructions.

To repair or replace the primary and secondary check valves in an RPZ (see *Figure 28*), follow these steps:

Step 1 Turn off the water supply to the RPZ and bleed off the pressure.

Step 2 Remove the valve retainer from the RPZ body bore and remove the check-valve modules by hand or by using a screwdriver.

Step 3 Remove the check-valve seat by holding the check cage in one hand while with the other hand pushing and rotating the seat clockwise. This will separate the check cage and spring from the disc assembly.

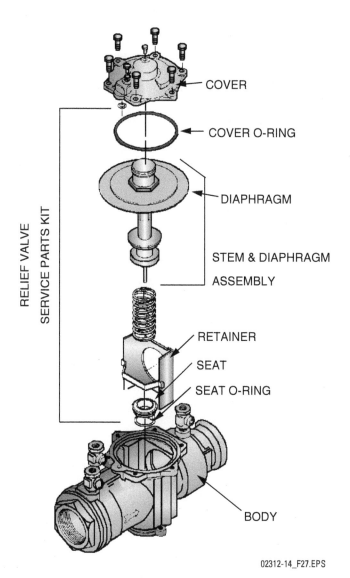

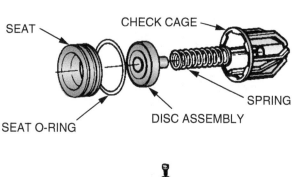

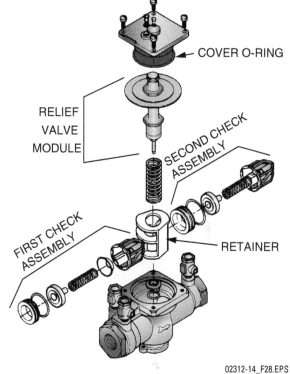

02312-14_F27.EPS

Figure 27 Exploded diagram of a relief valve in an RPZ.

02312-14_F28.EPS

Figure 28 Exploded diagram of a check valve in an RPZ.

Step 4 Clean or replace the valve disc assembly as required. Be careful not to damage the rubber O-ring seal on the disc during the process.

Step 5 Reassemble the check-valve modules, being careful to insert each check cage, spring, and valve assembly into its corresponding valve module. These components are not interchangeable. Install the modules into the body of the RPZ, ensuring that their seats face the valve inlet. Replace the valve retainer.

Step 6 Turn on the water supply to the RPZ and check for leaks. If a leak is detected, repair it according to the manufacturer's instructions.

> **CAUTION**
>
> The seats and springs of the two check valves in an RPZ are not interchangeable. Be sure to insert the correct seat and spring into each valve when repairing or replacing check-valve components.

3.0.0 Section Review

1. For backflow to occur, there must be _____.
   a. lower pressure in the potable-water supply
   b. a direct connection between the potable-water supply and another source
   c. a leak in one or more pipes in the DWV system
   d. contamination of the potable-water supply system

2. To ensure a proper air gap, the supply pipe must terminate above the flood-level rim at a distance that is _____.
   a. at least five times the diameter of the pipe
   b. at least four times the diameter of the pipe
   c. at least three times the diameter of the pipe
   d. at least twice the diameter of the pipe

3. Atmospheric vacuum breakers are intended for use on nonpotable-water systems that are _____.
   a. intermittent use, low hazard
   b. continuous use, low hazard
   c. intermittent use, high hazard
   d. continuous use, high hazard

4. Pressure-type vacuum breakers are suitable for applications with _____.
   a. constant inlet water pressure
   b. variable inlet water pressure
   c. constant inlet water temperature
   d. variable inlet water temperature

5. Dual-check valve backflow preventer assemblies are commonly used in residential installations because they are _____.
   a. more reliable than double-check valve assemblies
   b. smaller than other types of backflow preventer assemblies
   c. designed for use in high-hazard applications
   d. fitted with shutoff valves and test cocks

6. During normal operation of a double-check valve assembly, _____.
   a. both check valves are open
   b. both check valves are closed
   c. the primary check valve is open and the secondary check valve is closed
   d. the primary check valve is closed and the secondary check valve is open

7. The most sophisticated type of backflow preventer is the _____.
   a. dual-check valve backflow preventer assembly
   b. double-check valve assembly
   c. reduced-pressure zone principle backflow preventer assembly
   d. pressure-type vacuum breaker

8. In-line vacuum breakers protect against _____.
   a. back pressure in low-hazard applications
   b. back pressure in high-hazard applications
   c. back siphonage in low-hazard applications
   d. back siphonage in high-hazard applications

9. Backflow preventer testers are required to take written and practical exams offered by a local chapter of the _____.
   a. American Backflow Prevention Association
   b. International Code Council
   c. American Water Works Association
   d. American Society for Testing and Materials International

Summary

A properly sized water supply system ensures that customers have adequate water at the right pressure when they need it. Plumbers install water supply systems tailored for each building. There are several steps and rules that plumbers must follow when installing a supply system. These apply to all systems, no matter how large or small.

To install a system correctly, plumbers must know about the physical properties of water. These properties include temperature, density, flow, and friction. Before installing the system, plumbers also must find the water requirements for all fixtures and outlets. Then they can estimate the total length of all pipes and fittings. They also calculate how much water the system can handle.

All of this information needs to be collected before the pipes can be sized. Pipe sizing involves several steps. The first is to identify the WSFUs for each branch in the system. Next, the plumber selects the right size of pipe for the branches. The plumber then calculates friction loss in the whole system. With that information, the plumber can select the correctly sized supply line. By carefully following all of these steps, the plumber will be able to install an efficient water supply system.

Problems and malfunctions in a plumbing system can force water to flow backward through a system. Backward flow can draw contaminants into a plumbing system through a cross-connection. This is called backflow. Back pressure and back siphonage are the two kinds of backflow. Backflow preventers protect the potable-water supply from contamination. There are six main types of backflow preventers: air gap, atmospheric vacuum breaker, pressure-type vacuum breaker, double-check valve, dual-check valve assembly, and reduced-pressure zone principle backflow preventer assembly. Each of these is designed for specific applications, and some work on both types of backflow.

Codes require certified technicians to test and service backflow preventers. Do not attempt to service or test a preventer without certification. Regular tests ensure that backflow preventers continue to operate correctly. Always consult with the project engineer and local code officials before selecting and installing parallel backflow preventers. Public safety depends on the reliable operation of backflow preventers, and codes require it.

1. To prevent overheated water from damaging a water heater, plumbers install _____.
   a. safety switches
   b. steam valves
   c. overflow drains
   d. temperature/pressure relief valves

2. The resistance of a liquid to flow is called _____.
   a. reluctance
   b. viscosity
   c. inertia
   d. turbidity

3. The term *viscous flow* is another way of describing _____.
   a. turbulent flow
   b. parallel flow
   c. laminar flow
   d. transient flow

4. Friction loss can be reduced by _____.
   a. increasing pipe diameter
   b. arranging piping in shorter runs
   c. decreasing the slope (grade) of the pipe run
   d. decreasing pipe diameter

5. When planning a piping system, a rule of thumb that plumbers use to account for friction loss from fittings and valves is to calculate the total system length plus an additional allowance of _____.
   a. 15 percent
   b. 30 percent
   c. 45 percent
   d. 50 percent

6. To get an initial estimate of water system requirements, the plumber should determine each fixture's and outlet's _____.
   a. volume
   b. flow rate
   c. friction loss
   d. connection size

7. If the recommended water pressure for a fixture is lower than the system pressure, adjust the pressure by installing a(n) _____.
   a. pressure-reducing valve
   b. feedback loop
   c. throttle valve
   d. equalizer

8. Water supply fixture units are _____.
   a. used to calculate system water temperature
   b. the total load requirements of the system
   c. load values assigned to individual fixtures
   d. used only for commercial systems

9. An example of a common continuous-demand component in a residential water system is a _____.
   a. kitchen sink
   b. lawn sprinkler
   c. water closet
   d. bathtub

10. When estimating system requirements, the total length of piping from the supply to a single fixture is considered the _____.
    a. integrated length
    b. optimum length
    c. effective length
    d. developed length

11. An important step in determining pipe size requirements is to determine the minimum acceptable pressure for the system's _____.
    a. highest fixture
    b. smallest fixture
    c. largest fixture
    d. lowest fixture

12. After calculating total system losses, subtract them from the minimum pressure available to determine the system pressure available to overcome _____.
    a. elevation changes
    b. friction loss
    c. turbulence
    d. static head loss

13. For backflow to occur, there must be a(n) _____.
    a. indirect connection to the potable-water supply
    b. backwater valve on the supply line
    c. cross-connection
    d. air gap

14. People who service backflow preventers must be _____.
    a. utility employees
    b. certified
    c. trained by the manufacturer
    d. inspectors

15. Applications for backflow preventers that could affect public health are classified as _____.
    a. critical
    b. potentially dangerous
    c. life threatening
    d. high hazard

16. An air gap used to prevent back siphonage cannot be less than _____.
    a. ½ inch
    b. 1 inch
    c. 1½ inches
    d. 2 inches

17. To control water flow, an atmospheric vacuum breaker uses a(n) _____.
    a. silicon disc float
    b. polycarbonate insert
    c. ceramic plunger
    d. pneumatic valve

18. In a pressure-type vacuum breaker installation, a line pressure drop to atmospheric pressure or less will cause _____.
    a. both the air inlet valve and the check valve to close
    b. the air inlet valve to close and the check valve to open
    c. the air inlet valve to open and the check valve to close
    d. both the air inlet valve and the check valve to open

19. Heavy-duty protection in low-hazard, continuous-pressure applications is provided by installing a(n) _____.
    a. dual-check backflow preventer assembly
    b. pressure-type vacuum breaker
    c. in-line vacuum breaker
    d. double-check valve assembly

20. The type of backflow preventer required for a particular application is specified by _____.
    a. the *International Plumbing Code*®
    b. OSHA
    c. the local code or health department
    d. the contractor or building owner

Trade Terms Quiz

Fill in the blank with the correct term that you learned from your study of this module.

1. A direct link between the potable-water supply and water of questionable quality is called a(n) _____.

2. _____ is the amount of a liquid, gas, or solid in a space, measured in pounds per cubic foot.

3. An undesirable condition that results when nonpotable liquids enter the potable-water supply by reverse flow through a cross-connection is called _____.

4. A(n) _____ is a backflow preventer that uses two spring-loaded check valves to prevent back siphonage and back pressure.

5. The measure of a liquid's resistance to flow is called _____.

6. A(n) _____ is a backflow preventer that uses two spring-loaded check valves plus a hydraulic, spring-loaded pressure-differential relief valve to prevent back pressure and back siphonage.

7. An air gap that can be installed on a reduced-pressure zone principle backflow preventer to prevent back pressure and back siphonage is called a(n) _____.

8. _____ is the difference between the pressure at the inlet and the pressure at the farthest outlet in a water supply system.

9. A backflow preventer that prevents back pressure and back siphonage through the use of two spring-loaded check valves that seal tightly in the event of backflow is called a(n) _____.

10. _____ is a classification denoting the potential for an impairment of the quality of the potable water that creates an actual hazard to the public health through poisoning or through the spread of disease by sewage, industrial fluids, or waste.

11. The classification denoting the potential for an impairment in the quality of the potable water to a degree that does not create a hazard to the public health but does adversely and unreasonably affect the aesthetic qualities of such potable water for domestic use is called _____.

12. _____ is the rate of water flow in gallons per minute that a fixture uses when operating.

13. The length of pipe required to create the same amount of friction as a given fitting is called the _____.

14. A(n) _____ is a specialty backflow preventer used on low-flow and small supply lines to prevent back siphonage and back pressure.

15. The partial loss of system pressure due to friction is called _____.

16. _____ is the erratic flow pattern that occurs when a liquid's flow changes from a laminar flow to a turbulent flow pattern.

17. The measure of the water requirement for the entire water supply system is called _____.

18. A(n) _____ is a specialty backflow preventer that uses a disc to block flow when subjected to back siphonage.

19. The resistance that results from objects rubbing against one another is called _____.

20. _____ is the random flow pattern of fast-moving water or water moving along a rough surface.

21. A(n) _____ is a backflow preventer installed in a water line that uses a spring-loaded check valve and a spring-loaded air inlet valve to prevent back siphonage.

22. The form of backflow caused by subatmospheric pressure in the water system, which results in siphoning of water from downstream, is called _____.

23. The parallel flow pattern of a liquid that is flowing slowly is called _____.

24. A(n) _____ is a backflow preventer designed to prevent back siphonage by allowing air pressure to force a silicon disc into its seat and block the flow.

25. A valve in a backflow preventer that permits the testing of individual pressure zones is called a(n) _____.

26. The _____ is the length of all piping and fittings from the water supply to a fixture.

27. A backflow preventer designed to be used outdoors on hose bibbs to prevent back siphonage is called a(n) _____.

Trade Terms

Atmospheric vacuum breaker
Back siphonage
Backflow
Backflow preventer with intermediate atmospheric vent
Cross-connection
Demand
Density
Developed length
Double-check valve assembly (DCV)

Dual-check valve backflow preventer assembly (DC)
Equivalent length
Flow rate
Friction
Friction loss
High hazard
Hose-connection vacuum breaker
In-line vacuum breaker
Laminar flow
Low hazard

Manufactured air gap
Pressure drop
Pressure-type vacuum breaker
Reduced-pressure zone principle backflow preventer assembly (RPZ)
Test cock
Transient flow
Turbulent flow
Viscosity

Trade Terms Introduced in This Module

Atmospheric vacuum breaker: A backflow preventer designed to prevent back siphonage by allowing air pressure to force a silicon disc into its seat and block the flow.

Back siphonage: A form of backflow caused by subatmospheric pressure in the water system, which results in siphoning of water from downstream.

Backflow: An undesirable condition that results when nonpotable liquids enter the potable-water supply by reverse flow through a cross-connection.

Backflow preventer with intermediate atmospheric vent: A specialty backflow preventer used on low-flow and small supply lines to prevent back siphonage and back pressure. It is a form of double-check valve.

Cross-connection: A direct link between the potable-water supply and water of questionable quality.

Demand: The measure of the water requirement for the entire water supply system.

Density: The amount of a liquid, gas, or solid in a space, measured in pounds per cubic foot.

Developed length: The length of all piping and fittings from the water supply to a fixture.

Double-check valve assembly (DCV): A backflow preventer that prevents back pressure and back siphonage through the use of two spring-loaded check valves that seal tightly in the event of backflow. DCVs are larger than dual-check valve backflow preventers and are used for heavy-duty protection. DCVs have test cocks.

Dual-check valve backflow preventer assembly (DC): A backflow preventer that uses two spring-loaded check valves to prevent back siphonage and back pressure. DCs are smaller than double-check valve assemblies and are used in residential installations. DCs do not have test cocks.

Equivalent length: The length of pipe required to create the same amount of friction as a given fitting.

Flow rate: The rate of water flow in gallons per minute that a fixture uses when operating.

Friction: The resistance that results from objects rubbing against one another.

Friction loss: The partial loss of system pressure due to friction. Friction loss is also called pressure loss.

High hazard: A classification denoting the potential for an impairment of the quality of the potable water that creates an actual hazard to the public health through poisoning or through the spread of disease by sewage, industrial fluids, or waste. This type of water-quality impairment is called contamination.

Hose-connection vacuum breaker: A backflow preventer designed to be used outdoors on hose bibbs to prevent back siphonage.

In-line vacuum breaker: A specialty backflow preventer that uses a disc to block flow when subjected to back siphonage. It works the same as an atmospheric vacuum breaker.

Laminar flow: The parallel flow pattern of a liquid that is flowing slowly. Also called streamline flow or viscous flow.

Low hazard: A classification denoting the potential for an impairment in the quality of the potable water to a degree that does not create a hazard to the public health but does adversely and unreasonably affect the aesthetic qualities of such potable water for domestic use. This type of water-quality impairment is called pollution.

Manufactured air gap: An air gap that can be installed on a reduced-pressure zone principle backflow preventer to prevent back pressure and back siphonage.

Pressure drop: In a water supply system, the difference between the pressure at the inlet and the pressure at the farthest outlet.

Pressure-type vacuum breaker: A backflow preventer installed in a water line that uses a spring-loaded check valve and a spring-loaded air inlet valve to prevent back siphonage.

Reduced-pressure zone principle backflow preventer assembly (RPZ): A backflow preventer that uses two spring-loaded check valves plus a hydraulic, spring-loaded pressure-differential relief valve to prevent back pressure and back siphonage.

Test cock: A valve in a backflow preventer that permits the testing of individual pressure zones.

Transient flow: The erratic flow pattern that occurs when a liquid's flow changes from a laminar flow to a turbulent flow pattern.

Turbulent flow: The random flow pattern of fast-moving water or water moving along a rough surface.

Viscosity: The measure of a liquid's resistance to flow.

Additional Resources

This module presents thorough resources for task training. The following resource material is suggested for further study.

Backflow Prevention: Theory and Practice. 1990. Robin L. Ritland. Dubuque, IA: Kendall/Hunt Publishing Company.

Code Check Plumbing & Mechanical: An Illustrated Guide to the Plumbing and Mechanical Codes. 2011. Douglas Hansen, Redwood Kardon, and Paddy Morrisey. Newtown, CT: Taunton Press.

International Plumbing Code®, Latest Edition. Falls Church, VA: International Code Council.

Manual M14, *Recommended Practice for Backflow Prevention and Cross-Connection Control*, Third Edition. 2004. Denver, CO: American Water Works Association.

Manual of Cross-Connection Control, Tenth Edition. 2009. Foundation for Cross-Connection Control and Hydraulic Research. Los Angeles: University of Southern California.

Plumbers and Pipefitters Handbook. 1996. William J. Hornung. Englewood Cliffs, NJ: Prentice Hall College Division.

Plumbing Level One. 2012. Alachua, FL: NCCER.

Standard Plumbing Engineering Design, Second Edition. 1982. Louis S. Neilsen. New York: McGraw-Hill.

Pipefitters Handbook. 1967. Forrest R. Lindsey. New York: Industrial Press Inc.

Figure Credits

Answer	Section Reference	Objective
Section One		
1. a	1.1.0	1a
2. b	1.2.0	1b
Section Two		
1. c	2.0.0	2
2. a	2.2.0	2b
3. c	2.3.0	2c
4. d	2.4.0	2d
Section Three		
1. b	3.1.0	3a
2. d	3.2.0	3b
3. a	3.3.0	3c
4. a	3.4.0	3d
5. b	3.5.0	3e
6. a	3.6.0	3f
7. c	3.7.0	3g
8. d	3.8.0	3h
9. c	3.9.1	3i

NCCER CURRICULA — USER UPDATE

NCCER makes every effort to keep its textbooks up-to-date and free of technical errors. We appreciate your help in this process. If you find an error, a typographical mistake, or an inaccuracy in NCCER's curricula, please fill out this form (or a photocopy), or complete the online form at **www.nccer.org/olf**. Be sure to include the exact module ID number, page number, a detailed description, and your recommended correction. Your input will be brought to the attention of the Authoring Team. Thank you for your assistance.

Instructors – If you have an idea for improving this textbook, or have found that additional materials were necessary to teach this module effectively, please let us know so that we may present your suggestions to the Authoring Team.

NCCER Product Development and Revision

13614 Progress Blvd., Alachua, FL 32615

Email: curriculum@nccer.org
Online: www.nccer.org/olf

❏ Trainee Guide ❏ Lesson Plans ❏ Exam ❏ PowerPoints Other _____

Craft / Level: _____ Copyright Date: _____

Module ID Number / Title: _____

Section Number(s): _____

Description: _____

_____

_____

_____

_____

Recommended Correction: _____

_____

_____

_____

_____

Your Name: _____

Address: _____

_____

Email: _____ Phone: _____

02303-14

Potable Water Supply Treatment

OVERVIEW

Water is made safe through disinfection, filtration, and softening. Although municipal water utilities are required to disinfect their public water supply systems, private water supply systems require their own disinfection devices. The most commonly encountered problems in water supply systems are hardness, discoloration, acidity, bad odors and flavors, and turbidity. Plumbers should always refer to the local code before attempting to correct a water treatment problem.

Module Three

Trainees with successful module completions may be eligible for credentialing through the NCCER Registry. To learn more, go to **www.nccer.org** or contact us at **1.888.622.3720**. Our website has information on the latest product releases and training, as well as online versions of our *Cornerstone* magazine and Pearson's product catalog.

Your feedback is welcome. You may email your comments to **curriculum@nccer.org,** send general comments and inquiries to **info@nccer.org**, or fill in the User Update form at the back of this module.

This information is general in nature and intended for training purposes only. Actual performance of activities described in this manual requires compliance with all applicable operating, service, maintenance, and safety procedures under the direction of qualified personnel. References in this manual to patented or proprietary devices do not constitute a recommendation of their use.

Objectives

When you have completed this module, you will be able to do the following:

1. Identify the methods for disinfecting the water supply and determine the sources of contamination they address.
   a. Determine when and how to install chlorinators.
   b. Determine when and how to install pasteurization systems.
   c. Determine when and how to install ultraviolet-light systems.
2. Identify the methods for filtering and softening the water supply and determine the sources of contamination they address.
   a. Explain how municipal water treatment systems work.
   b. Determine when and how to install ion-exchange systems.
   c. Determine when and how to install filtration systems.
   d. Determine when and how to install precipitation systems.
   e. Determine when and how to install reverse-osmosis systems.
   f. Determine when and how to install distillation systems.
3. Determine how to troubleshoot water supply problems caused by contamination.
   a. Determine how to troubleshoot problems caused by hardness.
   b. Determine how to troubleshoot problems caused by discoloration.
   c. Determine how to troubleshoot problems caused by acidity.
   d. Determine how to troubleshoot problems caused by foul odors and flavors.
   e. Determine how to troubleshoot problems caused by turbidity.

Performance Tasks

Under the supervision of your instructor, you should be able to do the following:

1. Flush out visible contaminants from a plumbing system.
2. Install a reverse-osmosis system.
3. Identify the basic equipment necessary to solve specific water quality problems.

Trade Terms

Adsorption	Coagulation	Floc	Pasteurization	Precipitation
Alum	Contact time	Grains per gallon	Point-of-entry	Sodium
Backwashing	Contamination	Ion	(POE) unit	hypochlorite
Calcium	Diatomaceous	Ion exchange	Point-of-use	Solenoid valve
hypochlorite	earth	K value	(POU) unit	Turbidity unit
Chlorination	Disinfection	Osmosis	Pollution	Ultraviolet light
Chlorinator	Distillation	Oxidizing agent	Precipitate	Water conditioner

Industry-Recognized Credentials

If you're training through an NCCER-accredited sponsor, you may be eligible for credentials from NCCER's Registry. The ID number for this module is 02303-14. Note that this module may have been used in other NCCER curricula and may apply to other level completions. Contact NCCER's Registry at 888.622.3720 or go to **www.nccer.org** for more information.

Code Note

Codes vary among jurisdictions. Because of the variations in code, consult the applicable code whenever regulations are in question. Referring to an incorrect set of codes can cause as much trouble as failing to reference codes altogether. Obtain, review, and familiarize yourself with your local adopted code.

Contents

Topics to be presented in this module include:

Figures and Tables

SECTION ONE

1.0.0 DISINFECTING THE WATER SUPPLY SYSTEM

Objective

Identify the methods for disinfecting the water supply and determine the sources of contamination they address.

   a. Determine when and how to install chlorinators.

   b. Determine when and how to install pasteurization systems.

   c. Determine when and how to install ultraviolet-light systems.

Performance Task

Flush out visible contaminants from a plumbing system.

Trade Terms

Calcium hypochlorite: The solid form of chlorine. It comes in powder and tablet form.

Chlorination: The use of chlorine to disinfect water.

Chlorinator: A device that distributes chlorine in a water supply system.

Contact time: The amount of time that a chlorine solution must remain in a water supply system in order to disinfect it, as determined by the type and amount of contamination in the system, the concentration of chlorine used, the pH (acidity or alkalinity) level of the water, and the temperature of the water.

Contamination: An impairment of potable water quality that creates a hazard to public health.

Disinfection: The destruction of harmful organisms in water.

K value: A mathematical variable used to determine contact time based on the highest pH level and lowest temperature of the water in a water supply system.

Pasteurization: Heating water to kill harmful organisms in it.

Point-of-entry (POE) unit: A water conditioner that is located near the entry point of a water supply.

Point-of-use (POU) unit: A water conditioner that is located at an individual fixture.

Pollution: An impairment of the potable water quality that does not create a hazard to public health but affects the water's taste and appearance.

Sodium hypochlorite: The liquid form of chlorine, commonly found in laundry bleach.

Solenoid valve: An electronically operated plunger used to divert the flow of water in a water conditioner.

Ultraviolet light: Type of disinfection in which a special lamp heats water as it flows through a chamber.

Water conditioner: Device used to remove harmful organisms and materials from water.

If you could look at a drop of untreated water under a microscope, you would see hundreds of tiny living creatures. Most of these creatures are harmless to humans but others can cause disease and illness if ingested or inhaled. Certain types of bacteria cause cholera, dysentery, and typhoid fever. Viruses can cause polio, hepatitis, and meningitis. These are just some of the potential dangers caused by harmful organisms in water.

Water is made safe through disinfection, filtration, and softening. To disinfect means to destroy harmful organisms in the water. Filtration is the process of scouring water to remove particles and chemicals. Softening removes magnesium and calcium salts that cause scale on the inside of pipes and fittings. Water utility companies disinfect, filter, and sometimes soften public water supplies. Some public water requires additional treatment. All well water should be disinfected, filtered, and if necessary, softened.

You can take steps to prevent contamination while installing a water supply system. Ensure that pipes are stored and handled properly. Do not store pipes in dirty or wet locations. Such places are ideal breeding grounds for harmful organisms. Cap all installed pipe at the end of each workday. When the water supply system has been completely installed, disinfect the system using the following procedure:

Step 1 Flush the completed pipe system with potable water until no more dirty water can be seen coming from any outlet.

Step 2 Fill the entire system with a water/chlorine solution. Valve off the system. Use a solution of at least 50 parts per million (ppm) of chlorine and let the system stand for 24 hours. Alternatively, use a solution of at least 200 ppm of chlorine and let the system stand for 3 hours. Refer to your local code for specific procedures.

Step 3 Flush the system with potable water until the system is completely purged of chlorine. Test the water to determine whether the system is free of chlorine.

Step 4 Using materials and procedures as specified in the local applicable code, collect water samples and submit them to your local health department or a testing lab to test for harmful bacteria in the water.

Step 5 If the test results indicate that harmful bacteria remain in the water, repeat the previous steps until the system is free of the bacteria.

Municipal water utilities are required to disinfect their public water supply systems. Private water supply systems are required to have their own disinfection devices. Plumbers can choose one of several methods to disinfect a private system. The most common methods are chlorination, pasteurization, and ultraviolet light. These methods are discussed in more detail in the following sections. Take the time to review each method carefully. Discuss them with an experienced plumber. Review your local code to learn when to use these methods in your area.

One important distinction to remember is the difference between pollution and contamination. Pollution is an impairment in potable water quality that adversely and unreasonably affects the water's aesthetic qualities (making the water unsightly or murky, and/or with an unpleasant odor) but is not hazardous to public health. Contamination is an impairment in potable water quality that creates an actual hazard to public health via poisoning or the spread of disease from sewage, industrial fluids, or waste.

In this module, you will learn about the different types of water problems and how to treat them. Devices that treat water are called water conditioners. Plumbers can install water conditioners in two locations. Conditioners located near where the water supply enters a building are called point-of-entry (POE) units. Conditioners installed at individual fixtures are called point-of-use (POU) units. The type of unit to install depends on the water problem and the amount of

water to be treated. Often, more than one type of conditioning is required.

Regardless of the type of water conditioning system, ensure that it is well maintained. Otherwise, pollutants will collect in the device and the results could be worse than the original problem. Install only certified water conditioners. Check the label on the unit. It should say that the unit meets the standards of the National Sanitation Foundation (NSF) or the Water Quality Association (WQA). If it doesn't say that, don't let your customers use it. Don't let your customers risk their health just to save a few dollars.

The plumber's responsibility for installing water conditioners will vary depending on the local code requirements. Review your local code before you recommend or install a water conditioner. Customers have a right to expect clean, healthful water. It is your responsibility to ensure that their needs are met.

1.1.0 Determining When and How to Install Chlorinators

People who own swimming pools know how important it is to add chlorine to the water regularly. Chlorine kills harmful organisms in the water. The process of using chlorine to disinfect water is called chlorination. Even a small amount of chlorine makes a good disinfectant. A gallon of chlorine bleach at a solution of only 1 ppm will treat about 10,000 gallons of water. The same amount of bleach at a 5-ppm solution will treat 50,000 gallons. Chlorine kills bacteria and viruses and also eliminates many of the causes of odor and foul taste. Chlorine can be used in a water system of any size.

Chlorine is distributed through a water supply system by a device called a chlorinator. Chlorine is available in liquid and solid forms. The liquid form is called sodium hypochlorite. It is commonly found in laundry bleach. A gallon of domestic laundry bleach contains 5.25 percent sodium hypochlorite. Commercial laundry bleach has 19 percent sodium hypochlorite. Because of the higher concentration, commercial bleach is more economical to use for larger water supply systems. The solid form of chlorine is called cal-

Sterilizing a Water Supply System

Treated water still has living organisms in it. Some businesses need water that is sterile. Sterile water has no living organisms in it at all. Sterilization requires that an entire water supply system be heated to the boiling point. This is difficult to accomplish in most systems. Home and office systems rarely need to be sterilized. Refer to your local code for guidelines if a customer requires a sterilized water supply system.

cium hypochlorite. It is available in powder and tablet form and contains anywhere from 30 percent to 75 percent chlorine.

Chlorine removes iron and sulfur from the water. If a system has iron- or sulfur-rich water, monitor and adjust the system's chlorine use regularly. High levels of alkaline in the water may slow the disinfection process. Cold water will also slow the disinfection process. Suspended particles in the water can shield some bacteria from the chlorine. Install charcoal filters to remove any residual chlorine taste from the water. Calcium hypochlorite may form deposits in pipes. Always review the manufacturer's instructions before you use any form of chlorine.

Plumbers calculate the amount of time that a chlorine solution must remain in a water supply system in order to disinfect it. This is called the solution's contact time. The contact time is determined by several factors:

- The type and amount of contamination in the system
- The concentration of chlorine used
- The pH (acidity or alkalinity) level of the water
- The temperature of the water

To calculate contact time, follow these steps:

Step 1 Determine the highest pH level expected in the system. This information can be obtained from information provided by the municipal water utility.

Step 2 Determine the lowest water temperature expected in the system.

Step 3 Use these figures to determine the K value for the system. The K value is a mathematical variable that can be used to determine the contact time based on the highest pH level and lowest temperature of the water in the system. Use *Table 1* to calculate the K value for the system.

Step 4 Divide the resulting K value by the concentration of chlorine in the chlorine/water solution:

Time = K value/concentration in ppm

Table 1 K Values for Chlorine Contact Time

Highest pH	Lowest Water Temperature (°F)		
	>50	45	<40
6.5	4	5	6
7.0	8	10	12
7.5	12	15	18
8.0	16	20	24
8.5	20	25	30
9.0	24	30	36

Step 5 The resulting number is the amount of time, in minutes, required for the water/chlorine solution to remain in the water supply system to ensure the system has been disinfected.

For example, say that you want to disinfect a water supply system in which the highest estimated pH level is 8 and the lowest estimated temperature is 50°F. You are using a water/chlorine solution of 0.5 ppm. Referring to *Table 1*, you see that the corresponding K value for the system is 16. Divide the K value by the concentration of the chlorine solution to determine the contact time:

$$Time = 16/0.5$$

$$Time = 32$$

The required contact time for the system is 32 minutes.

When the water/chlorine solution has remained in the water supply system for the specified contact time, flush the system thoroughly with potable water until the chlorine has been completely purged from the water supply system. If tests show that the system is still contaminated, repeat the chlorination procedure and test again.

> **WARNING!**
> Soft water contains sodium chloride (salt). Softened water may pose a health risk to people on strict low-sodium diets. Ensure that the customer's health will not be affected by the installation of a soft-water treatment system.

Clean Water: Then and Now

Until the middle of the nineteenth century, people in Europe and the United States regularly got sick from drinking water. Rural rivers and streams contain sand, clay, and other particles. City water contained germs and discarded chemicals. People who drank such water risked illness and even death. Contaminated water even caused several devastating plagues. The lack of clean water is still a major concern in areas of the world with high poverty, natural disasters, and war. Without clean water, the quality and span of life are drastically reduced.

1.2.0 Determining When and How to Install Pasteurization Systems

The next time you are in the supermarket, look at the label on a milk jug or carton. It may state that the milk is pasteurized. This means that it was heated to kill harmful organisms. Pasteurization can be used to disinfect water, too (see *Figure 1*). Untreated water passes through a heat exchanger, where it is heated to about 150°F. A pump draws the water into an electric heating chamber. There, the temperature of the water is raised to 161°F for at least 15 seconds. This exposure is enough to kill organisms in the water.

After heating, the water flows past a thermostat connected to a solenoid valve. A solenoid valve is an electronically operated plunger. If the water temperature is less than 161°F, the valve routes the water back to the heating chamber. If the water is still at 161°F, it is sent to the heat exchanger. There, it heats more incoming untreated water. This method heats only a small amount of water at a time. Treated water is stored in a pressure tank and pumped to the supply system by another pump.

Pasteurization systems are not suitable for all applications. The process is commonly used to disinfect water used in the manufacture of beer and other beverages that require water at an industrial scale. Pasteurization kills harmful organisms much faster than chlorine does and kills both bacteria and viruses. The process is not affected by alkaline levels and does not affect the water's taste. However, most systems can disinfect only 20 to 24 gallons per hour. The heat may cause minerals in the water to settle on the pipe walls as scale. Once water enters the pressure tank, it is not protected from further contamination. Be sure to review the customer's needs before you install this type of system.

1.3.0 Determining When and How to Install Ultraviolet-Light Systems

Natural sunlight exposes harmful organisms in streams and rivers to ultraviolet light. This kills the organisms the same way that heat does. Ultraviolet light is a higher frequency than humans can see. In an ultraviolet-light system (*Figure 2*), a solenoid valve directs water from a storage tank into a disinfecting chamber. Inside the chamber is an ultraviolet lamp surrounded by a quartz sleeve. The sleeve prevents the water from cooling the lamp. The ultraviolet light heats the water as it flows through the chamber. The heat kills all bacteria and many viruses. Then the disinfected water flows to the outlet. Ultraviolet-light units are available in a range of sizes. Small units can disinfect 20 gallons per hour. The largest units can disinfect up to 20,000 gallons per hour. Ultraviolet-light systems are commonly used in water reclamation plants, which treat wastewater before returning it to the natural environment or re-using it in a nonpotable water system.

02303-14_F01.EPS

Figure 1 Typical pasteurization system installation.

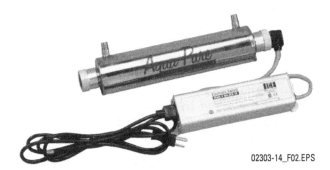

02303-14_F02.EPS

Figure 2 An ultraviolet-light system.

Bottled Water—What Are You Drinking?

The bottled-water industry has grown rapidly in the past decade. You can buy bottled water in grocery stores and convenience stores. Companies lease and sell water dispensers to homes and offices. Many people now prefer bottled water to tap water. However, in many cases, the source of the water is the same for both. Bottlers are not required to identify where the water comes from.

The US Food and Drug Administration (FDA) regulates bottled water. The FDA has instituted fair-labeling rules for bottled water. It has also developed standards for terms such as *mineral water*, *artesian water*, and *distilled water*. Bottled water samples must be tested regularly to help protect consumers against harm.

Install ultraviolet units on the outlet side of the storage tank. They can also be installed near the fixture outlet. Most ultraviolet units are equipped with a photoelectric cell. This device, also called an electric eye, measures the light given off by the lamp. When an old lamp begins to fade, the eye shuts off the water supply. The old lamp must then be replaced. Install a time-delay switch to allow the lamp to reach full intensity before water flows into the chamber. In systems without electric eyes or time-delay switches, replace lamps regularly according to the manufacturer's instructions and your local code.

Like pasteurization, ultraviolet light kills organisms quickly and does not affect taste. Suspended particles make the lamp less efficient. Minerals and alkalines can coat the lamp sleeve. Cold water also reduces the lamp's efficiency. Ensure that filters are installed in the water supply system and that the water temperature is within the correct range, in accordance with the manufacturer's specifications.

Common Types of Chlorinators

The three most common types of chlorinators are:

- Diaphragm-pump chlorinators
- Injector chlorinators
- Tablet chlorinators

Diaphragm-pump chlorinators use a diaphragm pump to introduce chlorine into a pressurized water supply system. The diaphragm pump draws chlorine from a storage tank by drawing back the diaphragm. As the diaphragm is pushed out, the chlorine is pumped into a pressurized storage tank. After each cycle, the chlorine mixes with the potable water.

Injector chlorinators use a water pump to draw chlorine into the system. When the pump turns on, it creates low pressure in the suction line that draws water from the pressure tank, where it flows through a device called an injector. As the water flows through the injector, the flow siphons chlorine from a storage tank and mixes it with water. The chlorinated water is then pumped into the supply system.

In a tablet chlorinator system, water is pumped through a container filled with calcium hypochlorite tablets. A restricting valve on the pump outlet line creates a slight pressure increase on the pump side of the valve and a slight decrease in pressure on the tank side of the valve. The pressure increase diverts some water into the tablet tank, where it is chlorinated. The low pressure then draws the chlorinated water into the pressure tank. The chlorine is diluted in the pressure tank, resulting in the proper solution of chlorine in the water.

The type of chlorinator that you use will depend on the local applicable code, the project specifications, and cost.

Louis Pasteur (1822–1895)

In 1864, Louis Pasteur, a French biologist, discovered that heat kills organisms in water. This is one of his most famous discoveries. The process is even named after him: *pasteurization.* Pasteur also made many other contributions to science. He championed clean and sanitary hospitals as a way to curb the spread of germs. He discovered how to prevent disease through vaccination. His research into the cause of rabies revealed the existence of viruses. For his work, he became world famous. He summed up the secret of his success: "Fortune favors only the prepared mind."

Additional Resources

International Plumbing Code®, Latest Edition. Falls Church, VA: International Code Council.
Pipefitters Handbook. 1967. Forrest R. Lindsey. New York: Industrial Press Inc.
Principles of Water Treatment. 2012. George Tchobanoglous, Kerry J. Howe, David W. Hand, and John C. Crittenden. New York: Wiley.
Water Treatment: Principles and Design, Second Edition. 2005. New York: Wiley.

1.0.0 Section Review

1. When disinfecting a water supply system once the system has been completely installed, a solution of at least 200 ppm of chlorine can be used if the system is allowed to stand for _____.

   a. 1 hour
   b. 3 hours
   c. 12 hours
   d. 24 hours

2. In a pasteurization system, in order to kill organisms in the water, the electric heating chamber raises the temperature of the water to _____.

   a. 150°F for at least 60 seconds
   b. 161°F for at least 60 seconds
   c. 150°F for at least 15 seconds
   d. 161°F for at least 15 seconds

3. In an ultraviolet-light system, the lamp's efficiency will be reduced by _____.

   a. high chlorine levels
   b. cold water
   c. organic matter in the water
   d. charcoal filters

SECTION TWO

2.0.0 FILTERING AND SOFTENING THE WATER SUPPLY SYSTEM

Objective

Identify the methods for filtering and softening the water supply and determine the sources of contamination they address.

   a. Explain how municipal water treatment systems work.
   b. Determine when and how to install ion-exchange systems.
   c. Determine when and how to install filtration systems.
   d. Determine when and how to install precipitation systems.
   e. Determine when and how to install reverse-osmosis systems.
   f. Determine when and how to install distillation systems.

Performance Tasks

Install a reverse-osmosis system.

Trade Terms

Adsorption: A type of absorption that happens only on the surface of an object.

Alum: A chemical used for coagulation. It consists of aluminum and sulfur.

Backwashing: The cleaning process of a fully automatic water softener.

Coagulation: The bonding that occurs between contaminants in the water and chemicals introduced to eliminate them.

Diatomaceous earth: Soil consisting of the microscopic skeletons of plants that are all locked together.

Distillation: The process of removing impurities from water by boiling the water into steam.

Floc: Another term for precipitate.

Ion: An atom with a positive or negative electrical charge.

Ion exchange: A technique used in water softeners to remove hardness by neutralizing the electrical charge of the mineral atoms.

Osmosis: The unequal back-and-forth flow of two different solutions as they seek equilibrium.

Oxidizing agent: A chemical that is used for oxidation and is made up largely of oxygen.

Precipitate: A particle created by coagulation that settles out of the water.

Precipitation: The process of removing contaminants from water by coagulation.

Unfiltered water contains particles such as sand, mud, and silt. This is called turbidity. Turbidity clogs pipes and causes excessive wear on faucets, valves, and appliances. Water can also contain chemicals such as iron, hydrogen sulfide, acids, fluoride, and vegetable tannins. Chemicals discolor water, stain fixtures and clothes, and make the water smell and taste bad. Hard water contains large amounts of mineral salts. Mineral salts coat pipes, fittings, and water heaters with scale. Unfiltered water can reduce the efficiency of a water supply system. It can also shorten the life of appliances and cause health hazards.

Water filters and water softeners separate out undesirable materials. The result is clean, safe water. Filters pass water through a porous material such as a membrane or charcoal. The porous material separates the particles and chemicals from the water. In POU systems like sand filters, the particles and chemicals are then flushed away. Other systems use cartridge filters that trap particles and chemicals in a disposable cartridge. Water softeners use chemicals to remove mineral salts from the water.

A drain and an electrical outlet should be in close proximity to the location where you install water filters and softeners. This will reduce the amount of water pipe required for the installation. The building layout will affect the placement of filters and softener piping and equipment. In basements, install the unit near the water heater. Install shutoff valves on the inlet and outlet lines. In crawl spaces, the piping can be run under the floor. In a slab-on-grade foundation, both the main supply and the drain will be in the floor (see *Figure 3*).

When roughing-in the piping for a water supply system, leave room for filtering and softening equipment. During the installation, try to keep contaminants out of the system. Despite your best efforts, dirt, sand, flux, pipe joint compound, metal filings, and oil will probably get inside the pipes during installation. Follow these steps to remove contaminants from a new water supply system:

Step 1   Remove the aerators and filters from all faucets in the system.

Step 2   Open all valves and faucets as wide as possible.

Step 3 Charge the system with clean water. Flush the system until the water runs clean from all faucets and valves. Disinfect the system (use the procedure given in the section *Disinfecting the Water Supply System*).

Step 4 Reinstall all aerators and filters.

Always test the water before you install a filter or softener. The test will identify any problems with the water quality. This information will allow you to select the appropriate filter or softener. Professionals usually perform these tests. Municipal water utilities also have staff that performs water quality tests. Ask an experienced plumber to recommend a tester in your area.

You have already been introduced to the concepts of water filtering and softening. In this section, you will learn about the different ways to filter and soften water. You will also learn how to install water filters and water softeners. Refer to your local code when installing filtering and softening systems. Always take the time to diagnose the problem before you recommend a filter or softener. Both your reputation and the customer's health are at stake.

2.1.0 How Municipal Water Treatment Systems Work

Public water utilities treat water to ensure that it is potable. Utilities are required to follow local codes and federal water-quality guidelines. Most systems can treat thousands of gallons at a time. The process begins when water is pumped to the treatment facility from a lake, river, well, or reservoir. Chemicals are added to separate out turbidity and kill harmful organisms. The water is then pumped to a settling basin, where solids sink and are filtered out. The water is then pumped through sand filters and treated with lime to remove acidity. It is then stored until it is needed.

Treated water is often stored in open reservoirs. There, it is exposed to contaminants in the atmosphere. Additional treatment is required before this water can be allowed to enter the public water supply system. The water is filtered and disinfected in a storage tank. Diatomaceous earth is a commonly used filter. Diatomaceous earth consists of the fossilized remains of microscopic shells created by one-celled plants, called diatoms. The fossilized remains are locked together, creating a natural filter that traps particles in the

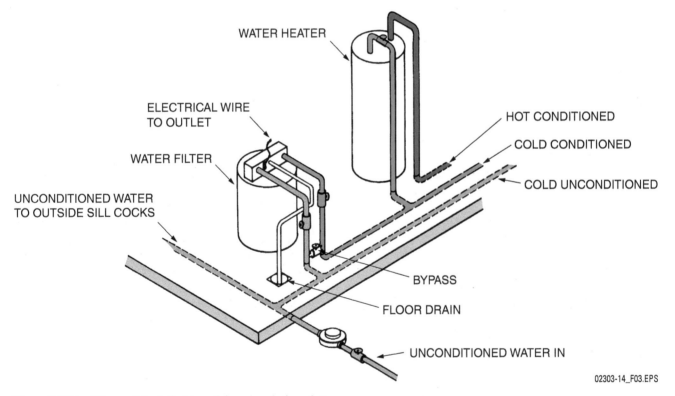

02303-14_F03.EPS

Figure 3 Water filter unit installed in a slab-on-grade foundation.

water. A chlorine-and-ammonia solution is added to the water to disinfect it. When the water has been filtered and disinfected, it is safe for use.

Plumbers install additional water treatment systems as needed. Buildings supplied by a private well must have these systems. The most common water treatment systems are described below. Refer to your local code to learn which systems are permitted in your area.

2.2.0 Determining When and How to Install Ion-Exchange Systems

Look at the inside of an old coffeepot or teakettle. The bottom may be coated with a rough, grayish material, deposits of minerals such as calcium, magnesium, and sodium. Heat causes the minerals to settle out of the water and stick to the metal, resulting in scale. This settling is called precipitation. Scale can block pipes and reduce the efficiency of water heaters and other appliances. Water that contains large amounts of minerals is called hard water.

Heat causes scale by actually changing the structure of calcium, magnesium, and sodium atoms. Heat provides the atom with extra energy. The energy makes the atom shed or absorb electrons, giving the atom an electrical charge. The charge can be either positive or negative. An atom with electrical charge is called an ion. The electrical charge allows the ion to bond with a metal surface. When enough atoms bond to the surface, they become visible as scale.

Water softeners use a chemical process called ion exchange to remove mineral ions from hard water. Ion exchange is the process of bonding ions with other atoms. Once the ions bond with another atom, they lose their electrical charge. They then settle out of the water.

A typical residential ion-exchange system consists of two tanks, a resin (or mineral) tank and a brine tank (see *Figure 4*). Inside the resin tank are pellets of a resinous substance saturated with sodium. Hard water enters the resin tank, where the sodium bonds with the mineral ions. The resulting atoms are washed out of the tank's drain. The softened water then flows into the water supply system. The brine tank supplies sodium-rich water to replenish the resin in the resin tank.

Softeners come in a variety of designs. For fully automatic units, the owner regularly refills the brine tank with salt tablets. The unit recharges the resin tank by itself. Fully automatic units also clean themselves; this cleaning process is called backwashing. A timer controls the recharging and backwashing. Semiautomatic and manual units were used in the past, and are now rare.

Use portable water softeners to treat water when the water supply system is under construction and not yet tied into the municipal system. Portable softeners can be rented from local plumbing suppliers. Connect the unit to the water supply using temporary fittings. The dealer will recharge the system. The customer has no equipment costs or maintenance problems. When the hard-water problem clears up, unhook the unit and return it to the dealer. Ensure that the water supply system's connections are sealed properly. Ask an experienced plumber to recommend a reputable dealer in your area.

Ensure that the water softener is installed level. The softener will operate less efficiently if it is not level. Installation of a bypass valve on the supply line (see *Figure 5*) allows the softener to be removed for maintenance or replacement without disrupting the water supply. Use either a three-way combination valve or three separate valves. Refer to your local code.

Do not drain a water softener into the main waste system. Softener wastewater can contaminate a waste system. Codes require that a softener drain into a laundry tray or floor or other trapped drain. Ensure that there is an air gap of at least 1½ inches between the softener drainpipe and the receiving drain. Some codes allow softeners to drain into dry wells. Always refer to your local code before you install a softener drain.

> **WARNING!**
> Water heaters that are used on the same water supply line as an ion-exchange system must be isolated from the ion-exchange system according to your local code requirements. Otherwise, chlorine can contaminate the water heater and cause sickness and even death.

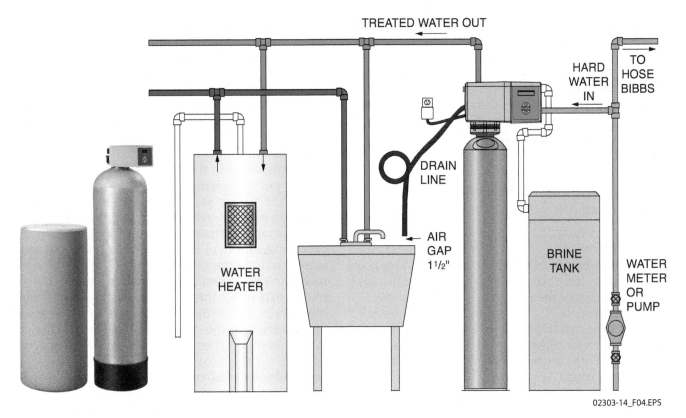

Figure 4 Ion-exchange water softener.

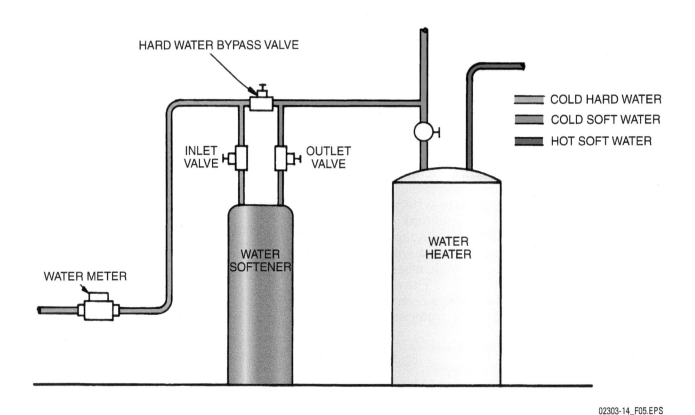

Figure 5 Installation of a bypass valve on a water softener.

After you install a water softener, disinfect it by adding chlorine to the resin tank. The chlorine will flush out any dirt and harmful organisms that may have collected in the tanks. The amount of chlorine depends on both the softener's capacity and the concentration of the chlorine. For example, you need to use 1.2 fluid ounces of domestic bleach per cubic foot of resin to disinfect an ion-exchange system. Use the following steps to disinfect a water softener:

Step 1   Add chlorine to the softener according to the softener's operating instructions. Allow the system to stand for 20 minutes.

Step 2   Drain the softener completely. If the drained water smells like chlorine, the system has been disinfected. If the water does not smell like chlorine, repeat Step 1.

Step 3   When the system has been disinfected, run fresh water through the softener until chlorine odor is gone.

Step 4   Charge the softener according to the operating instructions.

> **CAUTION**
> Always consult the manufacturer's recommendations for the chlorine to be used in an ion-exchange system. Chlorine can affect the resin bed material.

2.3.0 Determining When and How to Install Filtration Systems

There are many different types of filters. Each is designed to treat a specific water quality problem. The installation requirements for filters vary depending on type and manufacturer. Always follow the manufacturer's instructions when installing water filters. Refer to your local code for guidelines on particular types of filters.

Most filters work because of a principle called adsorption. When you stick a sponge in water, the sponge expands as it fills with water. This, as you know, is called absorption. Adsorption is absorption that happens only on the surface of an object.

If you had an adsorbent sponge, it would not expand when placed in water. It would pick up only as much water as could stick to the outside of the sponge. While that behavior makes for a very inefficient sponge, it makes for a very efficient filter.

Activated carbon (AC) is a highly effective adsorbent. AC is very porous. That means it has many crevices in which to trap molecules. In fact, 1 pound of AC has between 60 and 150 acres of surface area. Thus, AC is ideal for adsorption. As water flows through an AC filter, molecules become trapped in the carbon. AC eliminates the organic chemicals that cause foul odor, taste, and color. It also removes chlorine and harmful chemicals, such as pesticides and solvents. AC filters do not remove bacteria and viruses, sodium, hardness, fluorides, nitrates, or heavy metals.

Install AC filters at individual fixtures or on the main supply line. Low-volume units are effective for removing chemicals that cause foul odor, taste, and smell. They do not trap other materials. High-volume units have more AC and can remove harmful chemicals more effectively. Low-volume units can be installed on a faucet. Larger units should be located under a sink.

Change AC filters regularly to prevent them from clogging. Refer to the manufacturer's instructions. Many companies suggest that filters be changed after a certain amount of water has flowed through the unit. Install a sediment filter upstream of the AC filter. This will prevent the AC from clogging with silt. Some types of bacteria can grow in AC filters. They are generally harmless. Many AC filters feature inhibitors designed to prevent bacteria growth.

Other common types of filters include the following:

- Mechanical filters
- Neutralizing filters
- Oxidizing filters

Mechanical filters trap large particles, such as sand, silt, and clay. They work by passing water through sand or other grainy materials. They are not effective against fine particles. Install me-

Lime and Granulated-Anthracite Water Softening Systems

In addition to the systems discussed in this section, lime and granulated anthracite are also used as softening agents in water softening systems. Lime is a mineral that is rich in calcium. When combined with water to create limewater (calcium hydroxide), lime is very effective at removing organic matter from water. The waste sludge produced by large-scale lime softening can be used to treat acidic soil and can even be used in low-strength concrete. Anthracite is a hardened form of coal with high carbon content and very few impurities. In granular (particle) form, anthracite is effective at removing turbidity and chemicals that can cause hard water.

chanical filters along with other water treatment systems to ensure total filtration. Mechanical filters are also called microfilters.

Neutralizing filters remove acidity from water. They work by trapping the acid in material rich in calcium. Common sources of granular calcium are marble, lime, and calcium carbonate.

Oxidizing filters use a chemically neutral base material treated with magnesium. As the iron-rich water passes through the filter, the magnesium turns the iron into rust, which can then be trapped inside the filter. Oxidizing filters are also effective against manganese and some sulfides.

Always follow the manufacturer's instructions when installing these systems. Test the water to determine the type of contamination. This will ensure that you select the correct filtration system.

2.4.0 Determining When and How to Install Precipitation Systems

Extremely fine particles and certain chemicals may be able to pass through filters. These contaminants can be removed by adding precipitation as an extra step in the filtration process. Chemicals are introduced into the water to bond with the contaminants. The action of bonding is called coagulation. The coagulated matter is large and heavy enough to settle out of the wa-

ter. The separated matter, called a precipitate, can then be filtered. Precipitates are also sometimes called floc.

Iron must be converted into much-larger rust particles to be a precipitate. To turn the iron into rust, add an oxidizing agent to the treatment system. An oxidizing agent is a chemical that is made up largely of oxygen. It bonds with chemicals and materials that react to air. You know that exposure to air and water causes iron to rust. When the oxidizing agent bonds with iron, it turns the iron into rust. This process is called oxidation.

To filter out fine solid matter, add alum to the water. Alum is a chemical combination of aluminum and sulfur. The small particles coagulate onto the particles of alum. The result is larger particles that can be filtered out. For acidic water, add a soda-ash solution to the water. These items are available from your local plumbing supplier. Refer to your local code for approved methods and chemicals in your area.

Many coagulants need to be in the water for a while before they can treat the water completely. To accomplish this, install a chemical feeder and retention tank (see *Figure 6*). Chemical feeders inject small amounts of coagulants and chemicals into the water at regular intervals. The feeder draws the chemicals from a storage tank. A timer controls the feeder. Retention tanks allow the chemicals

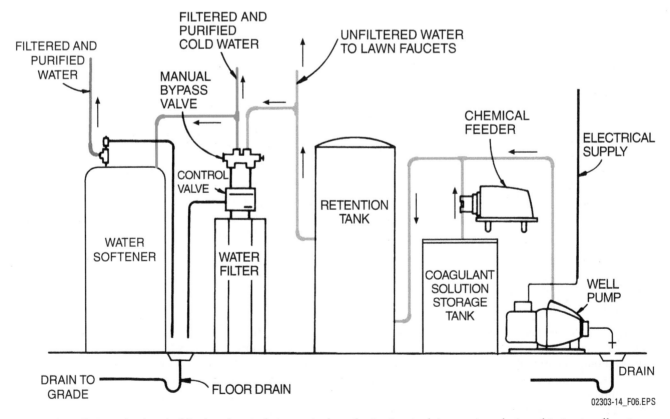

Figure 6 Installation of a chemical feeder, chemical storage tank, and retention tank in a system designed to treat well water.

time to circulate through the water. A local water-treatment specialist can suggest suitable tanks and feeders. Follow the manufacturer's directions when installing the equipment.

2.5.0 Determining When and How to Install Reverse-Osmosis Systems

Imagine that you have a container full of water divided vertically in half by a fine mesh. Add a small amount of salt to the water in one half (side A) and a large amount of salt to the other (side B). You will have two separate saltwater solutions. What happens next is illustrated in *Figure 7*. Solutions tend to seek equilibrium, or balance. The water in side A will pass through the mesh to balance the stronger solution in side B with more water. At the same time, the salt water in side B will try to flow through the mesh to the other side to give it more salt. The flow from side A to side B is stronger because there are many more salt particles in side B to get caught in the mesh as they try to pass through. Eventually, there will be more water in side B than side A. This unequal back-and-forth flow is called osmosis. The living cells in plants and animals use osmosis to obtain nutrients.

Some water treatment systems use a variation of the principle of osmosis to filter out harmful material. Using the example above, imagine that side B is subjected to an increase in pressure. The stronger solution is forced to pass through the membrane. The pressure overcomes the flow of the weaker solution from side A. The membrane traps the salt as the water flows through. This flow is called reverse osmosis (see *Figure 8*).

In a reverse-osmosis system, a pump forces water through a cellulose acetate membrane at a pressure of 200 to 400 pounds per square inch (psi). The holes in the membrane are small enough to let only water and a few chemicals through. The other contaminants are trapped on the membrane. A directed flow of water scours the contaminants off the membrane. The contaminants are then washed away as wastewater. Residential reverse-osmosis systems are small enough to be installed on or under a sink (*Figure 9*). Commercial systems are larger because they are designed to treat larger quantities of water (*Figure 10*). Reverse osmosis is widely used for water treatment in hospitals, science and medical laboratories, computer facilities, and in manufacturing facilities.

Reverse osmosis is a very effective way to eliminate most harmful organisms and contaminants. However, residential systems can treat only small amounts of water at a time. The process also uses a lot of water to clean the membrane. This cleaning water is directly washed out as wastewater. Depending on the installation, this can be wasteful. Codes require that all reverse-osmosis systems drain into the waste system through an air gap.

CELLULOSE
ACETATE
MEMBRANE

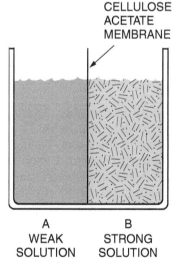

A
WEAK
SOLUTION

B
STRONG
SOLUTION

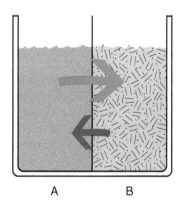

A   B

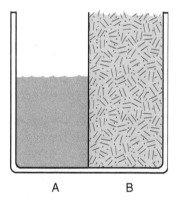

A   B

02303-14_F07.EPS

Figure 7 How osmosis works.

Licensed Water-Conditioner Installers

In most states, plumbers can get a license to install water conditioners. Getting a license may not be a requirement, but it can be a big step up in your professional development. To get a license, you usually need to have experience installing and servicing conditioners. Six months' experience is a common standard. You then have to pass an exam. The exams vary from state to state. Contact your local health department to apply for an exam in your area.

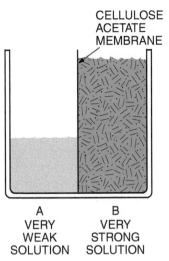
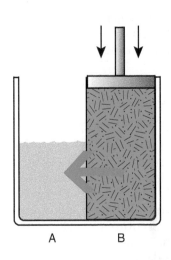
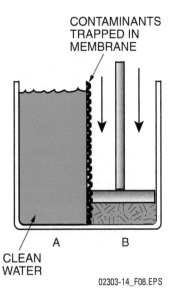

CELLULOSE
ACETATE
MEMBRANE

A
VERY
WEAK
SOLUTION

B
VERY
STRONG
SOLUTION

A   B

CONTAMINANTS
TRAPPED IN
MEMBRANE

A   B

CLEAN
WATER

02303-14_F08.EPS

Figure 8 How reverse osmosis works.

02303-14_F09.EPS

Figure 9 Residential reverse-osmosis system.

2.6.0 Determining When and How to Install Distillation Systems

Converting water to steam and back again is a very effective way to remove contaminants. As water flashes to steam, it leaves behind most impurities. This process is called distillation. It is one of the oldest and simplest forms of water purification. It does not require sodium, chlorine, or other chemicals. Distillation removes more than 99 percent of impurities from water, including the following:

• Bacteria
• Hardness
• Heavy metals
• Nitrates
• Organic compounds
• Sodium
• Turbidity

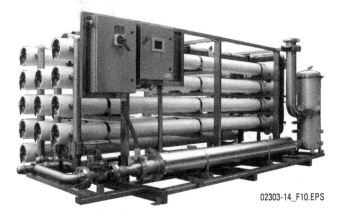

02303-14_F10.EPS

Figure 10 Commercial reverse-osmosis system.

The body of a typical distiller is divided into a boiling chamber and a storage chamber (see *Figure 11*). Cold, untreated water is pumped through a condensing coil into the boiling chamber. There, it is heated by an electric heating element. The heat kills bacteria and viruses in the water. As the heated water turns to steam, it rises and leaves most impurities behind. The purified steam then flows past the cold condensing coil. Contact with the coil turns the steam back into a liquid. The water collects in the storage chamber, where it is fed into a storage tank for use. The contaminated water that remains in the boiling chamber is flushed into a drain.

Distillers are also called stills. They can usually process only small amounts of water at a time. Typical home units process 3 to 11 gallons per day. They use more energy than other treatment systems. Also, some chemicals, such as benzene and toluene, have a lower boiling point than water. They are called volatile organic compounds (VOCs). VOCs can be trapped in steam. They can remain trapped in the water after it cools. Most distillers have filters or vents to allow VOCs to escape. Refer to your local code for guidelines on distillers.

Distillers may be installed at individual fixtures. Ensure that the distiller is properly vented and that all filters are clean. Bacteria can collect around the cooling coils when the distiller is not in use. Ensure that the coils are clean before each use. Scale may form around the heating element. Clean the heating element regularly. Vinegar is effective for cleaning heating elements. Follow the manufacturer's cleaning instructions. Some harsh cleaners can damage the distiller's casing. With proper care, a home distiller can last up to 15 years.

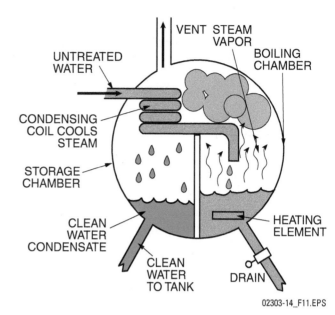

02303-14_F11.EPS

Figure 11 Diagram showing how a distiller works.

Additional Resources

"Distillation for Home Water Treatment." Purdue University Cooperative Extension Service. **http://www.extension.purdue.edu/**
Practical Principles of Ion Exchange Water Treatment. 1985. Dean L. Owens. Littleton, CO: Tall Oaks Publishing.
"Treatment Systems for Household Water Supplies: Activated Carbon Filtration." University of Minnesota Extension. **http://www.extension.umn.edu**
"Treatment Systems for Household Water Supplies: Distillation." University of Minnesota Extension. **http://www.extension.umn.edu**

2.0.0 Section Review

1. A commonly used filter in municipal water treatment systems is _____.

   a. activated charcoal
   b. diatomaceous earth
   c. granulated anthracite
   d. sodium chloride

2. The two tanks in a typical residential ion-exchange system are _____.

   a. the resin tank and the brine tank
   b. the hot-water tank and the cold-water tank
   c. the storage tank and the filter tank
   d. the storage tank and the adsorption tank

3. In an oxidizing filter, magnesium is used to convert iron in the water into _____.

   a. gas
   b. rust
   c. anthracite
   d. acid

4. In a precipitation system, the action of chemicals bonding with contaminants in the water is called _____.

   a. adsorption
   b. osmosis
   c. coagulation
   d. floc

5. Codes require that reverse-osmosis systems drain into the waste system through a(n) _____.

   a. air gap
   b. check valve
   c. reduced pressure-zone principle backflow preventer
   d. intermediate filter

6. Distillation systems are effective at eliminating _____.

   a. alum
   b. precipitates
   c. coagulants
   d. hardness

3.0.0 TROUBLESHOOTING PROBLEMS CAUSED BY CONTAMINATION

Objective

Determine how to troubleshoot water supply problems caused by contamination.

a. Determine how to troubleshoot problems caused by hardness.
b. Determine how to troubleshoot problems caused by discoloration.
c. Determine how to troubleshoot problems caused by acidity.
d. Determine how to troubleshoot problems caused by foul odors and flavors.
e. Determine how to troubleshoot problems caused by turbidity.

Performance Task

Identify the basic equipment necessary to solve specific water quality problems.

Trade Terms

Grains per gallon: A measure of the hardness of water. One grain equals 17.1 ppm of hardness.

Turbidity unit: A measurement of the percentage of light that can pass through a sample of water.

The following are the most commonly encountered problems in water supply systems:

- Hardness
- Discoloration
- Acidity
- Foul odors and flavors
- Turbidity

In this section, you will learn how to identify and correct each of these problems. You will learn how to identify the causes and symptoms of the different problems and how to test the water to confirm your diagnosis. The measurements and treatments discussed may vary depending on the local code.

Always refer to your local code before you attempt to correct a water treatment problem. Consult with expert plumbers about common water supply problems in your area, especially while you are still learning on the job. Later, as you become more experienced, you can return the favor to apprentices.

3.1.0 Determining How to Troubleshoot Problems Caused by Hardness

Hardness is caused by calcium and magnesium salts in the water. The salts may be in the form of bicarbonates, sulfates, or chlorides. Iron salts can also contribute to hardness. When soap is added to hard water, a sticky curd will form on the surface of the water. Suds do not form readily in hard water, and your skin feels rough after you wash in it. Hard water will cause glassware to appear streaked and murky. It also forms hard, scaly deposits on the inside of metal pipes.

Water hardness is measured in grains per gallon. One grain equals 17.1 ppm of hardness. To determine the degree of water hardness, draw a 2-ounce sample of the water. Add one drop of a soap solution to the water and shake it to form lasting suds. If no suds form, continue to add one drop at a time and shake until lasting suds form. The number of drops used is approximately equal to the hardness in grains per gallon. Consult *Table 2* to interpret the test results. Water with more than 3 grains of hardness requires the installation of an ion-exchange softener or a reverse-osmosis unit.

3.2.0 Determining How to Troubleshoot Problems Caused by Discoloration

Iron and manganese particles dissolved in the water can cause discoloration. Sometimes the presence of iron- and manganese-eating bacteria has the same effect. Dissolved iron turns water red. Iron-contaminated water will stain clothes and porcelain plumbing fixtures, even at concentrations as low as 0.3 ppm, and will corrode steel pipe. Water with dissolved iron has a metal-

Table 2 Measure of Water Hardness Using Soap-Solution Test

Quantity of Soap Solution	Hardness
0–3 drops (0–3 grains per gallon) (approximately 50 ppm)	Soft
4–6 drops (4–6 grains per gallon) (50–100 ppm)	Moderately hard
7–12 drops (7–12 grains per gallon) (100–200 ppm)	Hard
13–18 drops (13–18 grains per gallon) (approximately 200–300 ppm)	Very hard
19 or more drops (19 or more grains per gallon) (over 300 ppm)	Extremely hard

lic taste. Freshly drawn water may appear clear at first, but after exposure to the air, iron particles will gradually settle out.

Test kits are available to test for iron contamination. You can purchase a test kit from a local plumbing supplier. The kit contains an acid solution, a color solution, and a color chart. Use the following steps to test the water for the presence of iron:

Step 1 Draw a water sample according to the instructions in the test kit. Add the acid solution to the sample according to the test instructions. The acid will dissolve any iron that has settled out of the water.

Step 2 Add the color solution to the water sample according to the test instructions. The water sample will turn a shade of pink.

Step 3 Match the color of the water sample against the standard color chart in the test kit. The chart will indicate the amount of iron in the water, based on the color. Charts are usually rated in ppm.

Step 4 Install a water conditioner if the test indicates the presence of more than 0.2 ppm of iron in the water (see *Table 3*).

Table 3 lists the various treatments for different levels of iron contamination. Note that there may be more than one treatment option for a given level of contamination. The selection will depend on product availability, installation and maintenance costs, and the plumber's familiarity with the systems.

Iron-eating bacteria are often associated with the presence of acid or other corrosive conditions. These bacteria create a red, slippery jelly on the inner surfaces of toilet tanks. To eliminate iron bacteria, heavily chlorinate the water source, pump, and piping system. This will kill the bacteria that are already in the system. Install a chlorinator and filter to prevent subsequent buildups.

Brownish-black water indicates the presence of both manganese and iron. Manganese and iron can stain fixtures and fabrics. They can also give a bitter taste to coffee and tea. Manganese bacteria also cause a slippery buildup on the inside of toilet tanks, but of a darker color. Iron test kits also detect manganese. Treat manganese contamination in the same way as iron contamination.

3.3.0 Determining How to Troubleshoot Problems Caused by Acidity

Decaying vegetable matter emits carbon dioxide gas. Carbon dioxide can also be found in the atmosphere. When water absorbs and reacts with carbon dioxide, a weak acid solution is created.

Table 3 Treatment Recommendations for Iron-Contaminated Water

Iron Concentration (ppm)	Treatment
0–0.02	No treatment necessary
1–2	Polyphosphate feeder
0.2–10	Zeolite softener
1–10	Oxidizing (manganese zeolite) filter, usually manganese-treated green sand
3 or more	Chlorinator and filter, usually a sand or carbon filter

This acid slowly eats away at metal pipes and the metal components in fixtures, tanks, and pumps. In rare instances, some water may contain sulfuric, nitric, or hydrochloric acids. These can cause damage even more quickly. If copper or brass pipes are being eaten by acid, the water may leave green stains on fixtures and faucets. Acids also reduce the effectiveness of iron treatments.

Acid test kits are available from your local plumbing supplier. They contain a chemical indicator and a color chart similar to those found in iron test kits. Draw a water sample and add the chemical indicator to it. Compare the color of the water with the kit's color chart. The colors are keyed to the level of acidity in the water, usually by pH level. Decreasing numbers indicate increasing acidity. If the pH is 7 or below, the water supply should be treated for acidity.

Acidity can be neutralized by using a soda-ash solution. Feed the solution into the water supply system using one of the following methods:

- Directly into a well
- Through a well-pump suction line
- Along with a chlorine solution
- Through a separate feeder unit

A neutralizing tank with limestone or marble chips will also remove acidity from the water. Note that this process will increase the water's hardness. Always install a water softener on the line when installing a neutralizing tank. The softener must be downstream of the tank to be effective.

3.4.0 Determining How to Troubleshoot Problems Caused by Foul Odors and Flavors

The following contaminants cause rotten-egg odor and flavor in water:

- Hydrogen sulfide gas
- Sulfate-reducing bacteria
- Sulfur bacteria

These contaminants eat metal in pumps, piping, and fixtures. If the water contains sulfur and iron, fine black particles may collect in fixtures. This condition is commonly called black water, and it can turn silverware black. Do not cook with water that has been contaminated by sulfur. The term *black water* is sometimes also used to refer to wastewater from toilets. Be sure not to confuse the two ways of using the term.

Consult with a water-conditioning equipment dealer on the level of sulfur contamination. The dealer will be able to test the water on the spot. If the water contains more than 1 ppm of sulfur, it should be treated. The water can be treated using chlorinators and filters. Manganese-treated green sand filters can also be used for contamination levels of 5 ppm or less.

The following factors can also cause off-flavors:

- Extremely high mineral content
- Presence of organic matter
- Excess chlorine
- Passage of water through areas with oily or salty waste

These contaminants cause water to taste bitter, brackish, oily, or salty. They can also give a chlorine odor or taste to the water. You can purchase tests for each of these contaminants from your local plumbing supplier. Install a water softener to eliminate metallic flavors. Salty flavors can be treated with reverse-osmosis units. Activated carbon filters are effective against most other tastes and odors, including chlorine.

3.5.0 Determining How to Troubleshoot Problems Caused by Turbidity

Salt, sediment, small organisms, and organic matter can give water a dirty or muddy appearance, called turbidity. Turbidity is generally measured in turbidity units. A turbidity unit is a measure of the amount of light that can pass through a water sample. Turbidity can range from less than 1 ppm up to 4,000 ppm. Turbidity tests require special laboratory equipment. Contact your local health department or a filter manufacturer to arrange for a test. Water should be treated if its turbidity is greater than 10 ppm. Use a sand or diatomaceous-earth filter. Other options include surface treatment of the water source with powdered gypsum or copper sulfate and installation of a settling tank filled with alum.

Additional Resources

Principles of Water Treatment. 2012. George Tchobanoglous, Kerry J. Howe, David W. Hand, and John C. Crittenden. New York: Wiley.

Water Treatment: Principles and Design, Second Edition. 2005. New York: Wiley.

3.0.0 Section Review

1. The amount of hardness in one grain is _____.
   a. 7 ppm
   b. 10.2 ppm
   c. 15 ppm
   d. 17.1 ppm

2. Iron test kits can also be used to detect _____.
   a. sodium hypochlorite
   b. turbidity
   c. manganese
   d. alum

3. If the pH of a water sample is 7 or below, the water supply system should be treated for _____.
   a. turbidity
   b. acidity
   c. salinity
   d. alkalinity

4. Treat the water supply system if testing reveals sulfur content greater than _____.
   a. 2 ppm
   b. 1.5 ppm
   c. 1 ppm
   d. 0.5 ppm

5. Water should be treated if its turbidity is greater than _____.
   a. 10 ppm
   b. 20 ppm
   c. 30 ppm
   d. 40 ppm

SUMMARY

Plumbers install water conditioners to provide safe, clean water. Conditioners remove harmful organisms and materials from the water. Water can be treated by disinfection, filtration, and softening. Disinfection kills bacteria and viruses. Filtration removes particles and chemicals. Softening removes the chemicals that can cause scale.

Disinfection kills those organisms that can cause sickness and death. Plumbers disinfect new water supply systems when they are completed. Common disinfection methods include chlorination, pasteurization, and ultraviolet light. Chlorine, found in bleach, kills harmful organisms in water. It also eliminates odors and foul taste.

Pasteurization heats water enough to kill any living creatures that are in the water. This method does not remove other harmful products from the water. Ultraviolet light also kills organisms using heat.

Filters remove particles and chemicals from water. Most filters work on the principle of adsorption. Activated carbon (AC) is an effective adsorbent and is a common form of water filter. Mechanical filters trap sand, silt, and clay by passing water through sandy material. Neutralizing filters trap acid wastes in marble, lime, or calcium carbonate. Oxidizing filters turn iron into rust.

Other common forms of filtration include precipitation, reverse osmosis, and distillation. Precipitation is an extra step that involves adding a chemical to the water. The chemical bonds with fine particles and chemicals in the water. The resulting large particle is called a precipitate. Precipitates are large enough to be caught by filters. Reverse osmosis forces water under pressure through a cellulose acetate membrane. The membrane traps everything except water and a few organic chemicals. Distillation treats water by boiling it. When water turns to steam, it leaves most contaminants behind.

Minerals that cause scale can be removed from water using an ion-exchange system.

This module reviewed commonly encountered problems in water supply systems, including hardness, discoloration, acidity, foul odors and flavors, and turbidity. Understanding how to identify and treat these problems is an essential part of troubleshooting.

Customers expect to have clean, safe water for drinking, washing, and cleaning. You can ensure that a customer's water is safe by testing it. You can then install water conditioners that will remove the problem. Learn to identify the organisms and materials that can pollute a water supply system. With experience, you will be able to provide systems that suit your customers' needs.

Review Questions

1. The process of scouring water to remove particles and chemicals is called _____.
   a. filtration
   b. disinfection
   c. aggregation
   d. dispersion

2. After a newly installed water supply system has been flushed and disinfected, water samples should be taken and laboratory-tested for _____.
   a. magnesium salts
   b. chlorine
   c. turbidity
   d. harmful bacteria

3. In situations where the water supply is from a private well, plumbers must install _____.
   a. water softening equipment
   b. a sterilization unit
   c. disinfection devices
   d. a sediment filter

4. One gallon of domestic laundry bleach has a sodium hypochlorite content of _____.
   a. 5.25 percent
   b. 11 percent
   c. 19 percent
   d. 32.75 percent

5. When calculating the contact time for chlorinating a water system to disinfect it, a plumber must determine the system's _____.
   a. C value
   b. F value
   c. K value
   d. R value

6. While chlorination will kill bacteria with an exposure of 30 minutes, the pasteurization process uses heated water to kill harmful organisms in as little as _____.
   a. 5 seconds
   b. 15 seconds
   c. 30 seconds
   d. 90 seconds

7. After pasteurization, the treated water is sent _____.
   a. directly into the water supply system
   b. to a reverse-osmosis system
   c. through an activated charcoal filter
   d. to a pressure tank

8. To allow the ultraviolet lamp to reach full intensity before water is admitted to the disinfecting chamber, units are equipped with a _____.
   a. photoelectric cell
   b. rheostat
   c. time-delay switch
   d. thermostatic valve

9. Scale occurs when the chemical structure of calcium, magnesium, or sodium atoms is changed by _____.
   a. polarization
   b. heat
   c. pressure
   d. sublimation

10. Resin pellets in an ion-exchange system are saturated with _____.
    a. zeolite
    b. calcium
    c. sodium
    d. carbon

11. An example of a highly effective adsorbent material is _____.
    a. charcoal
    b. activated carbon
    c. sodium hydroxide
    d. resin

12. Calcium carbonate is often used to trap acid in a(n) _____.
    a. neutralizing filter
    b. recirculating filter
    c. mechanical filter
    d. oxidizing filter

13. To give coagulants time to work, precipitation systems usually include a _____.

    a. buffer
    b. graduated filter
    c. recirculating pump
    d. retention tank

14. The principle of osmosis is that solutions tend to seek _____.

    a. lower alkalinity
    b. equilibrium
    c. greater acidity
    d. imbalance

15. The heating coils of a home distillation unit should be cleaned periodically with _____.

    a. denatured alcohol
    b. ammonia
    c. vinegar
    d. steel wool

16. Water hardness can be determined by using a(n) _____.

    a. soap solution
    b. pH tester
    c. electronic hardness meter
    d. alum solution

17. A water conditioner should be installed if the iron content tests higher than _____.

    a. 0.002 ppm
    b. 0.02 ppm
    c. 0.2 ppm
    d. 2.0 ppm

18. An unpleasant smell from water is often due to the presence of _____.

    a. hydrogen sulfide
    b. iron
    c. silt
    d. calcium

19. Water contaminated with iron and sulfur is often referred to as _____.

    a. red water
    b. sulfated water
    c. foul water
    d. black water

20. Most objectionable water tastes and odors, other than metallic flavors and saltiness, can be eliminated by using a(n) _____.

    a. ion-exchange system
    b. activated carbon filter
    c. diatomaceous-earth filter
    d. pasteurization system

Trade Terms Quiz

Fill in the blank with the correct term that you learned from your study of this module.

1. A(n) _____ is a measurement of the percentage of light that can pass through a sample of water.

2. A water conditioner that is located at an individual fixture is called a(n) _____.

3. _____ is the bonding that occurs between contaminants in the water and chemicals introduced to eliminate them.

4. A water conditioner that is located near the entry point of a water supply is called a(n) _____.

5. A(n) _____ or _____ is a particle created by coagulation that settles out of the water.

6. The type of disinfection in which a special lamp heats water as it flows through a chamber is called _____.

7. _____ is soil consisting of the microscopic skeletons of plants that are all locked together.

8. Heating water to kill harmful organisms in it is called _____.

9. A(n) _____ is an electronically operated plunger used to divert the flow of water in a water conditioner.

10. An impairment of potable water quality that creates a hazard to public health is called _____.

11. _____ is the liquid form of chlorine, commonly found in laundry bleach.

12. An atom with a positive or negative electrical charge is called a(n) _____.

13. A(n) _____ is a device used to remove harmful organisms and materials from water.

14. The use of chlorine to disinfect water is called _____.

15. _____ is the cleaning process of a fully automatic water softener.

16. Absorption that happens only on the surface of an object is called _____.

17. A(n) _____ is a device that distributes chlorine in a water supply system.

18. The process of removing impurities from water by boiling the water into steam is called _____.

19. _____ is the unequal back-and-forth flow of two different solutions as they seek equilibrium.

20. The process of removing contaminants from water by coagulation is called _____.

21. _____ is the destruction of harmful organisms in water.

22. The measure of the hardness of water is called _____.

23. _____ is the amount of time that a chlorine solution must remain in a water supply system in order to disinfect it.

24. The technique used in water softeners to remove hardness by neutralizing the electrical charge of the mineral atoms is called _____.

25. A(n) _____ is a chemical that is used for oxidation and is made up largely of oxygen.

26. The mathematical variable used to determine contact time based on the highest pH level and lowest temperature of the water in a water supply system is called the _____.

27. _____ is a chemical used for coagulation consisting of aluminum and sulfur.

28. The solid form of chlorine, available in powder and tablet form, is called _____.

29. _____ is an impairment of the potable water quality that does not create a hazard to public health but affects the water's taste and appearance.

Trade Terms

Adsorption
Alum
Backwashing
Calcium hypochlorite
Chlorination
Chlorinator
Coagulation
Contact time
Contamination
Diatomaceous earth

Disinfection
Distillation
Floc
Grains per gallon
Ion
Ion exchange
K value
Osmosis
Oxidizing agent
Pasteurization

Point-of-entry (POE) unit
Point-of-use (POU) unit
Pollution
Precipitate
Precipitation
Sodium hypochlorite
Solenoid valve
Turbidity unit
Ultraviolet light
Water conditioner

Trade Terms Introduced in This Module

Adsorption: A type of absorption that happens only on the surface of an object.

Alum: A chemical used for coagulation. It consists of aluminum and sulfur.

Backwashing: The cleaning process of a fully automatic water softener.

Calcium hypochlorite: The solid form of chlorine. It comes in powder and tablet form.

Chlorination: The use of chlorine to disinfect water.

Chlorinator: A device that distributes chlorine in a water supply system.

Coagulation: The bonding that occurs between contaminants in the water and chemicals introduced to eliminate them.

Contact time: The amount of time that a chlorine solution must remain in a water supply system in order to disinfect it, as determined by the type and amount of contamination in the system, the concentration of chlorine used, the pH (acidity or alkalinity) level of the water, and the temperature of the water.

Contamination: An impairment of potable water quality that creates a hazard to public health.

Diatomaceous earth: Soil consisting of the microscopic skeletons of plants that are all locked together.

Disinfection: The destruction of harmful organisms in water.

Distillation: The process of removing impurities from water by boiling the water into steam.

Floc: Another term for precipitate.

Grains per gallon: A measure of the hardness of water. One grain equals 17.1 ppm of hardness.

Ion: An atom with a positive or negative electrical charge.

Ion exchange: A technique used in water softeners to remove hardness by neutralizing the electrical charge of the mineral atoms.

K value: A mathematical variable used to determine contact time based on the highest pH level and lowest temperature of the water in a water supply system.

Osmosis: The unequal back-and-forth flow of two different solutions as they seek equilibrium.

Oxidizing agent: A chemical that is used for oxidation and is made up largely of oxygen.

Pasteurization: Heating water to kill harmful organisms in it.

Point-of-entry (POE) unit: A water conditioner that is located near the entry point of a water supply.

Point-of-use (POU) unit: A water conditioner that is located at an individual fixture.

Pollution: An impairment of the potable water quality that does not create a hazard to public health but affects the water's taste and appearance.

Precipitate: A particle created by coagulation that settles out of the water.

Precipitation: The process of removing contaminants from water by coagulation.

Sodium hypochlorite: The liquid form of chlorine, commonly found in laundry bleach.

Solenoid valve: An electronically operated plunger used to divert the flow of water in a water conditioner.

Turbidity unit: A measurement of the percentage of light that can pass through a sample of water.

Ultraviolet light: Type of disinfection in which a special lamp heats water as it flows through a chamber.

Water conditioner: Device used to remove harmful organisms and materials from water.

Additional Resources

This module presents thorough resources for task training. The following resource material is suggested for further study.

"Distillation for Home Water Treatment." Purdue University Cooperative Extension Service. **http://www.extension.purdue.edu**

International Plumbing Code®, Latest Edition. Falls Church, VA: International Code Council.

Practical Principles of Ion Exchange Water Treatment. 1985. Dean L. Owens. Littleton, CO: Tall Oaks Publishing.

Pipefitters Handbook. 1967. Forrest R. Lindsey. New York: Industrial Press Inc.

Principles of Water Treatment. 2012. George Tchobanoglous, Kerry J. Howe, David W. Hand, and John C. Crittenden. New York: Wiley.

"Treatment Systems for Household Water Supplies: Activated Carbon Filtration." University of Minnesota Extension. **http://www.extension.umn.edu**

"Treatment Systems for Household Water Supplies: Distillation." University of Minnesota Extension. **http://www.extension.umn.edu**

Water Treatment: Principles and Design, Second Edition. 2005. New York: Wiley.

Figure Credits

Answer	Section Reference	Objective
Section One		
1. b	1.0.0	1
2. d	1.2.0	1b
3. b	1.3.0	1c
Section Two		
1. b	2.1.0	2a
2. a	2.2.0	2b
3. b	2.3.0	2c
4. c	2.4.0	2d
5. a	2.5.0	2e
6. d	2.6.0	2f
Section Three		
1. d	3.1.0	3a
2. c	3.2.0	3b
3. b	3.3.0	3c
4. c	3.4.0	3d
5. a	3.5.0	3e

NCCER CURRICULA — USER UPDATE

NCCER makes every effort to keep its textbooks up-to-date and free of technical errors. We appreciate your help in this process. If you find an error, a typographical mistake, or an inaccuracy in NCCER's curricula, please fill out this form (or a photocopy), or complete the online form at **www.nccer.org/olf**. Be sure to include the exact module ID number, page number, a detailed description, and your recommended correction. Your input will be brought to the attention of the Authoring Team. Thank you for your assistance.

Instructors – If you have an idea for improving this textbook, or have found that additional materials were necessary to teach this module effectively, please let us know so that we may present your suggestions to the Authoring Team.

NCCER Product Development and Revision

13614 Progress Blvd., Alachua, FL 32615

Email: curriculum@nccer.org
Online: www.nccer.org/olf

❏ Trainee Guide   ❏ Lesson Plans   ❏ Exam   ❏ PowerPoints   Other _____

Craft / Level: _____   Copyright Date: _____

Module ID Number / Title: _____

Section Number(s): _____

Description: _____

_____

_____

_____

Recommended Correction: _____

_____

_____

_____

Your Name: _____

Address: _____

_____

Email: _____   Phone: _____

02305-14

Types of Venting

OVERVIEW

Without vents, plumbing systems would not work. Vents allow air to enter the system when wastewater flows out of drainpipes. They prevent trap seals from draining by maintaining equalized air pressure throughout the drainage system. Plumbers must custom-design every vent system to provide the most efficient venting of all the fixtures in a building. Plumbers can choose from many different types of vents to create the best drain, waste, and vent (DWV) system possible.

Module Four

Trainees with successful module completions may be eligible for credentialing through the NCCER Registry. To learn more, go to **www.nccer.org** or contact us at **1.888.622.3720**. Our website has information on the latest product releases and training, as well as online versions of our *Cornerstone* magazine and Pearson's product catalog.

Your feedback is welcome. You may email your comments to **curriculum@nccer.org,** send general comments and inquiries to **info@nccer.org**, or fill in the User Update form at the back of this module.

This information is general in nature and intended for training purposes only. Actual performance of activities described in this manual requires compliance with all applicable operating, service, maintenance, and safety procedures under the direction of qualified personnel. References in this manual to patented or proprietary devices do not constitute a recommendation of their use.

Objectives

When you have completed this module, you will be able to do the following:

1. Describe the principles and components of vent systems and their code requirements.
   a. Describe the principles of venting.
   b. Describe the components of a vent system.
   c. Describe how to grade vents properly.
2. Describe the different types of vent systems that plumbers install.
   a. Describe the characteristics and requirements of individual and common vents.
   b. Describe the characteristics and requirements of battery vents.
   c. Describe the characteristics and requirements of wet vents.
   d. Describe the characteristics and requirements of air admittance and island vents.
   e. Describe the characteristics and requirements of relief and Sovent® vents.

Performance Task

Under the supervision of your instructor, you should be able to do the following:

1. Install different types of vents.

Trade Terms

Air admittance vent
Back pressure
Branch interval
Branch vent
Circuit vent
Combination waste-and-vent
   system
Common vent

Continuous vent
Horizontal vent
Hydraulic gradient
Indirect or momentum siphonage
Individual vent
Loop vent
Relief vent
Re-vent

Self-siphonage
Sovent® system
Stack vent
Vent
Vent stack
Vent terminal
Wet vent
Yoke vent

Industry-Recognized Credentials

If you're training through an NCCER-accredited sponsor, you may be eligible for credentials from NCCER's Registry. The ID number for this module is 02305-14. Note that this module may have been used in other NCCER curricula and may apply to other level completions. Contact the NCCER Registry at 888.622.3720 or go to **www.nccer.org** for more information.

Code Note

Codes vary among jurisdictions. Because of the variations in code, consult the applicable code whenever regulations are in question. Referring to an incorrect set of codes can cause as much trouble as failing to reference codes altogether. Obtain, review, and familiarize yourself with your local adopted code.

Contents

Topics to be presented in this module include:

Figures

1.0.0 PRINCIPLES AND COMPONENTS OF VENT SYSTEMS

Objective

Describe the principles and components of vent systems and their code requirements.

a. Describe the principles of venting.
b. Describe the components of a vent system.
c. Describe how to grade vents properly.

Trade Terms

Back pressure: Excess air pressure in vent piping that can blow trap seals out through fixtures. Back pressure can be caused by the weight of the water column or by downdrafts through the stack vent.

Horizontal vent: A vent that runs at an angle no greater than 44 degrees to the horizontal.

Hydraulic gradient: The maximum degree of allowable fall between a trap weir and the opening of a vent pipe. The total fall should not exceed one pipe diameter from weir to opening.

Indirect or momentum siphonage: The drawing of a water seal out of the trap and into the waste pipe by lower-than-normal pressure. It is the result of discharge into the drain by another fixture in the system.

Self-siphonage: A condition whereby lower-than-normal pressure in a drainpipe draws the water seal out of a fixture trap and into the drain.

Stack vent: An extension of the main soil-and-waste stack above the highest horizontal drain. It allows air to enter and exit the plumbing system through the building roof and is a terminal for other vent pipes.

Vent: A pipe in a DWV system that allows air to circulate in the drainage system, thereby maintaining equalized pressure throughout.

Vent stack: A stack that serves as a building's central vent, to which other vents may be connected. The vent stack may connect to the stack vent or have its own roof opening. It is also called the main vent.

Vent terminal: The point at which a vent terminates. For a fixture vent, it is the point where it connects to the vent stack or stack vent. For a stack vent, it is the point where the vent pipe ends above the roof.

Vents allow air to enter and exit a drain, waste, and vent (DWV) stack and protect the trap seal from siphoning. Vents work in conjunction with drains to enable wastes to flow from fixtures into the waste stack, where they are carried into the sewer. Without vents, plumbing systems would not work. Odors and contaminants would build up, and people would become sick. With proper venting, wastes are carried away efficiently and effectively.

In this module you will learn how vents are configured. You will also review the different types of vents that can be installed in a DWV system. Many types of vents can be grouped together into broad categories. This will make it easier for you to remember what they are called and how they are used.

A vent system must provide adequate venting for all the fixtures in a building. Each building has its own special requirements, which are dictated by the building's location, use, and level of occupancy. Vent systems must also be specially designed. However, most vent systems, regardless of where and how they are installed, have the same basic components. Vents can be installed using procedures that are already familiar to you.

1.1.0 Understanding the Principles of Venting

Vents prevent trap seals from draining when water flows through the DWV system. They do this by maintaining equalized air pressure throughout the drainage system. Air is used to equalize pressure that is exerted when wastewater drains from a fixture into the system.

You can visualize the principle of air pressure by observing the behavior of water in a straw. If you fill a straw with water and put your thumb over the top, the water remains in the straw. As soon as you remove your thumb, the water drains out. This is because water cannot flow out of the straw unless air can flow into the straw at the same rate, filling the space left by the draining water. In other words, by replacing the water in the straw with an equivalent amount of air, an equal pressure is maintained inside the straw.

To achieve the same effect in a DWV installation, plumbers install vents. Vents allow air to enter the system when wastewater flows out of the drainpipe. They ventilate the drainage system to the outside air and also keep the pipes from clogging and siphoning the trap dry. Air pressure, therefore, is important for creating and maintaining proper flow in drainage pipes. Also, normal air pressure throughout the DWV system helps maintain the water seal in fixture traps. System pressure should be equal to air pressure.

Too much or too little air pressure in the vent pipes can cause the DWV system to malfunction. Low pressure in the waste piping (also called a partial vacuum or negative pressure) can suck the water seal out of a trap and into the drain. This is called self-siphonage (see *Figure 1*). When the discharge from another drain creates lower pressure in the system, it can draw a trap seal into the waste pipe (see *Figure 2*). This is called indirect or momentum siphonage. Properly placed vents prevent both types of siphonage. To further protect against indirect siphonage, ensure that the stack is properly sized.

Likewise, excess pressure in vent piping can have the reverse effect, blowing the trap seal out through the fixture. Air in the stack that is compressed by the weight of water above can force the water out of a trap as the slug of air passes a fixture. This phenomenon, called back pressure, has a tendency to happen in taller buildings (see *Figure 3*). Use a vent near the fixture traps or at the point where the piping changes direction, to prevent back pressure. In some cases, downdrafts of air entering a vent on a roof can create a condition similar to back pressure. Install roof vents away from roof ridges and valleys that stir up downdrafts.

1.2.0 Understanding the Components of a Vent System

The relationship among the vents in a DWV system is illustrated in *Figure 4*. The main soil stack runs between the building drain and the highest horizontal drain in the system. Above the highest horizontal drain is the stack vent, which is an extension of the main soil-and-waste stack. The stack vent allows air to enter and exit the plumbing system through the roof. The stack vent also serves as the vent terminal for other vent pipes.

The central vent in a building is called the vent stack. The vent stack permits air circulation between the drainage system and the outside air. Other vents may be connected to it. The vent stack, which is also called the main vent, runs vertically. It is usually located within a few feet of the main soil-and-waste stack. The vent stack begins at the base of the main stack and continues

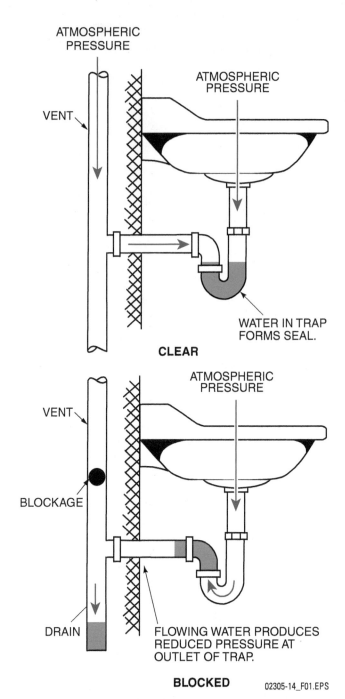

02305-14_F01.EPS

Figure 1 Self-siphonage resulting from blocked venting.

until it connects with the stack vent. It may also extend through the roof on an independent path. Branches connect one or more individual fixture vents to the vent stack. Codes require that vents go out through the roof unobstructed.

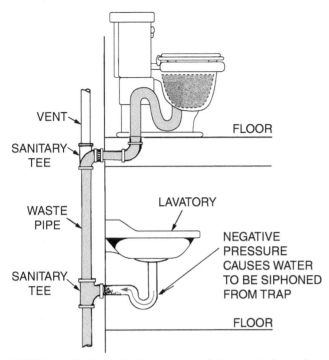

VENT

SANITARY TEE

WASTE PIPE

SANITARY TEE

FLOOR

LAVATORY

NEGATIVE PRESSURE CAUSES WATER TO BE SIPHONED FROM TRAP

FLOOR

NOTE: Installation is for the purpose of demonstration only.

02305-14_F02.EPS

Figure 2 Indirect or momentum siphonage.

Vents that run horizontally to a drain (that is, at an angle no greater than 44 degrees to the horizontal) are called horizontal vents. Horizontal vents should not be used in floor-mounted fixtures. They can cause waste to back up into the vent and block the fixture drain (see *Figure 5*). Codes prohibit vents from being turned from vertical to horizontal until at least 6 inches above the flood-level rim of the fixture.

The vent terminal is the point at which a vent terminates. Fixture vents terminate at the point where they connect to the vent stack or directly to the stack vent. Stack vents and some vent stacks terminate above the roof. Local codes govern the location, size, and type of roof vents (see *Figure 6*). Follow the code closely when selecting and install-

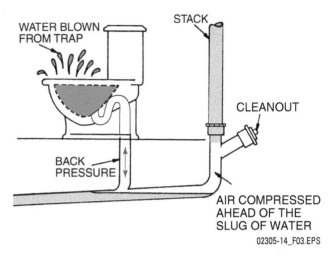

WATER BLOWN FROM TRAP

STACK

CLEANOUT

BACK PRESSURE

AIR COMPRESSED AHEAD OF THE SLUG OF WATER

02305-14_F03.EPS

Figure 3 Back pressure.

ing a roof vent terminal. When installing a roof vent, consider the following issues:

- Code requirements for freeze protection
- The use of permitted prefabricated vent flashing
- Pitched versus flat roof
- Code requirements for flashing collars

When vent terminals are installed, they should be protected from being plugged by snow accumulation or other means. Some codes require that the dimensions of the vent terminal be increased before exiting the roof. Your local code will provide specific information on vent terminals that are permitted for use in your area. Your code will also include dimensions for property lines; air intakes for heating, ventilating, and air conditioning (HVAC) systems; and air-compressor intakes.

1.3.0 Grading Vents

Grade is the slope or fall of a line of pipe in reference to a horizontal plane. In a DWV system, grade allows solid and liquid wastes to flow out under the force of gravity.

Horizontal vent pipes also need to be installed with proper grade. This will allow condensation

Purpose of Vents

Despite their different uses, plumbing vents are designed for the same purpose—to allow air to circulate in a DWV system. Without air circulation, a DWV system could not work properly because a pressure differential would be created within the system when a fixture is operated, causing trap seals to be sucked out.

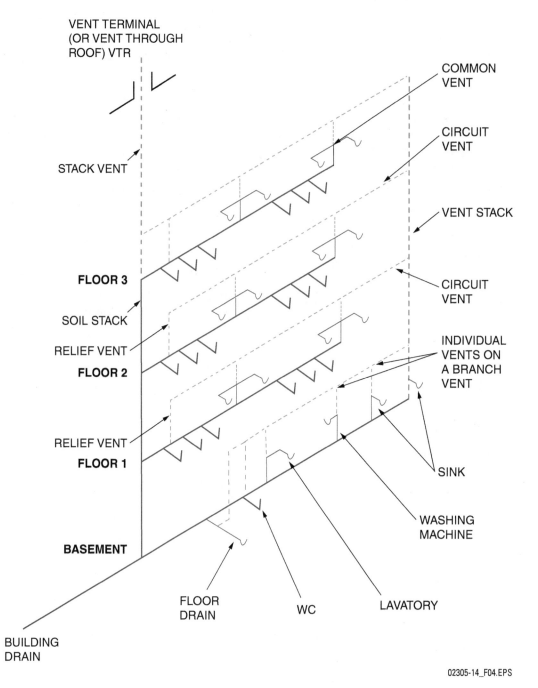

VENT TERMINAL
(OR VENT THROUGH
ROOF) VTR

STACK VENT

COMMON
VENT

CIRCUIT
VENT

VENT STACK

CIRCUIT
VENT

INDIVIDUAL
VENTS ON
A BRANCH
VENT

FLOOR 3

SOIL STACK

RELIEF VENT

FLOOR 2

RELIEF VENT

FLOOR 1

SINK

WASHING
MACHINE

BASEMENT

FLOOR
DRAIN

WC

LAVATORY

BUILDING
DRAIN

02305-14_F04.EPS

Figure 4 Vents in a DWV system.

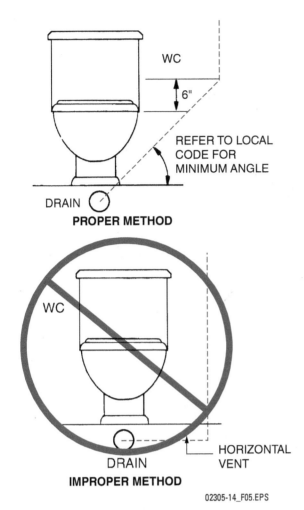

WC

6"

REFER TO LOCAL
CODE FOR
MINIMUM ANGLE

DRAIN

PROPER METHOD

WC

DRAIN

HORIZONTAL
VENT

IMPROPER METHOD

02305-14_F05.EPS

Figure 5 Improper and proper vent installation for a floor-mounted fixture.

to drain without blocking the vent pipe. Your local plumbing code will specify the degree of slope required for a horizontal vent. Most codes require that the vent be taken off above the soil-pipe center line. Then it must rise at an angle, typically not less than 45 degrees from horizontal, to a point 6 inches above the fixture's flood-level rim (see *Figure 7*). That way, if the drainpipe becomes blocked, solids cannot enter and potentially clog the vent instead.

The maximum grade between the trap weir and the vent pipe opening should not exceed one pipe diameter (see *Figure 8*). This grade is called the hydraulic gradient. The gradient determines the distance from the weir to the vent. If the vent is placed too far from the fixture trap, the vent opening will end up below the weir. This could cause self-siphonage. Codes include tables that specify the maximum distance the weir can be positioned from the vent based on pipe fall and diameter.

WARNING!

Sewer gas contains methane. Methane can build up to explosive levels of concentration if a DWV system is vented improperly. Review your local code and all building plans to ensure that the system design has adequate venting.

The Discovery of Air Pressure

In seventeenth-century Italy, the Grand Duke of Tuscany ordered his engineers to build a giant pump. The pump was designed to raise water more than 40 feet above the ground. It worked by drawing water into a tall column by creating a partial vacuum. However, the engineers found that the pump could not raise water higher than 32 feet. The Grand Duke asked Evangelista Torricelli to solve the mystery. Torricelli (1608–1647) was a famous mathematician and a physicist.

Torricelli discovered that the pump worked fine; something else was preventing the water from reaching a higher level. To find the answer, he conducted a simple experiment. He filled a 3-foot-long test tube with liquid mercury. Then he quickly turned the tube upside down into a bowl full of more mercury, making sure the mouth of the tube was below the surface of the mercury in the bowl.

Some of the mercury in the tube flowed into the bowl, but most stayed inside the upside-down tube. Between the column of mercury in the tube and the bottom—now the top—of the tube, there was a gap, a vacuum like the one created in the Duke's pump.

Torricelli reasoned that the air in the atmosphere was pushing down on the mercury in the bowl, forcing some of the mercury to stay in the tube. Torricelli's pushing force is what we call air pressure. The same thing was happening to the Duke's pump. The water could only rise to 32 feet because the weight of the surrounding air could not push it any higher.

We now know that air pressure at sea level exerts a pressure of approximately 14.7 pounds per square inch (psi). In other words, at any given moment you are being gently squeezed on all sides by the weight of the air around you. Air pressure affects things that most people may never think about: how long it takes water to boil, how birds fly, or how a pump works.

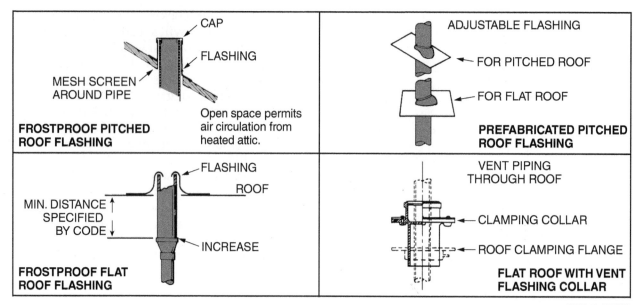

Figure 6 Types of vent terminals.

02305-14_F06.EPS

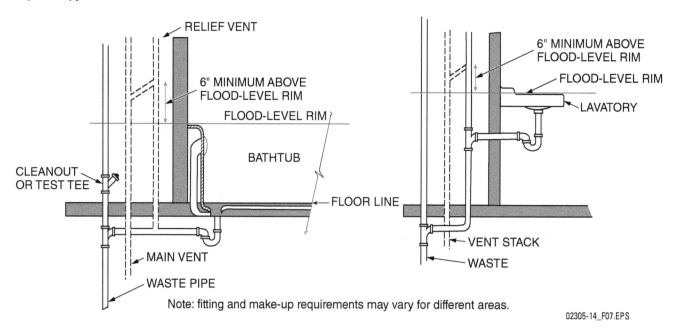

Note: fitting and make-up requirements may vary for different areas.

02305-14_F07.EPS

Figure 7 Connection to the vent stack.

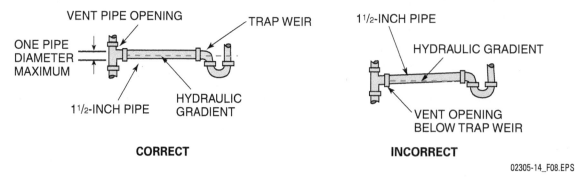

02305-14_F08.EPS

Figure 8 Determining the hydraulic gradient.

Crown Vent

A vent opening that is located within two pipe diameters of a trap is called a crown vent (see illustration). Crown vents are prohibited by code. Relocate such vents farther away from the trap. Never rush to save time; you might end up wasting even more time as a result.

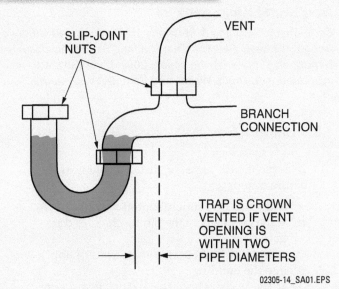

SLIP-JOINT NUTS

VENT

BRANCH CONNECTION

TRAP IS CROWN VENTED IF VENT OPENING IS WITHIN TWO PIPE DIAMETERS

02305-14_SA01.EPS

Plumbing Resources on the Web

You can find a lot of useful information on the web that you can access easily and quickly using computers, tablets, and smartphones. Web sources are constantly updated to reflect changes in the industry and the profession. Here is a small sampling of websites that contain valuable information for plumbers:

Code Organizations

International Association of Plumbing & Mechanical Officials (IAPMO)—**www.iapmo.org**
International Code Council (ICC)—**www.iccsafe.org**

Standards and Testing

Home Innovations Research Labs—**www.homeinnovation.com**
American Society of Sanitary Engineering—**www.asse-plumbing.org**

Mobile Apps

Construction Master Pro from Calculated Industries—
www.calculated.com/prd296/Construction-Master-Pro-for-iPhone.html (for iPhone),
www.calculated.com/prd314/Construction-Master-Pro-for-Android.html (for Android)
BIMx from GRAPHISOFT—**www.graphisoft.com/bimx/**
Home Builder Pro Calcs from Double Dog Studios—**doubledogstudios.com/apps/homebuilderprocalcs/**
iBlueprint from Local Slacks—**www.localslacks.com/LS/iBlueprint.html**

Books and Online Magazines

Amazon.com—**www.amazon.com**
Pipefitter.com—**www.pipefitter.com**
Plumbing & Mechanical Magazine—**www.pmmag.com**Reference and Directories
PlumbingNet—**www.plumbingnet.com**
PlumbingWeb—**www.plumbingweb.com**
ThePlumber.com—**www.theplumber.com**

Also, be sure to bookmark the website of your local authority having jurisdiction (AHJ). It will contain useful information.

Remember, just because you found it on the web doesn't mean that it is accurate. Find out if an author or an organization is an established authority on the subject. With information, just like with anything in plumbing, it is better to be safe than sorry.

Additional Resources

Code Check Plumbing & Mechanical: An Illustrated Guide to the Plumbing and Mechanical Codes. 2011.
Douglas Hansen, Redwood Kardon, and Paddy Morrisey. Newtown, CT: Taunton Press
Handbook of Materials Selection. 2002. Myer Kutz, ed. New York: J. Wiley.
Introduction to Drain, Waste, and Vent (DWV) Systems, Fourth Edition. 2012. Upper Saddle River, NJ: Pearson.

1.0.0 Section Review

1. To prevent back pressure, install a vent near the fixture trap or _____.

   a. at the fixture's supply shutoff valve
   b. at the point where the piping changes direction
   c. at the point of entry for the water supply into the building
   d. at the point where the fixture drain connects with the main stack

2. The vent stack begins at the base of the main stack and continues until it connects with _____.

   a. the vent for the lowest fixture in the building
   b. another vent
   c. the stack vent
   d. the crown vent

3. Most codes require that a vent be taken off _____.

   a. above the soil-pipe center line
   b. at a grade that is not less than one pipe diameter below the trap weir
   c. below the soil-stack center line
   d. greater than 45 degrees from the horizontal

Section Two

2.0.0 Types of Vent Systems

Objective

Describe the different types of vent systems that plumbers install.

    a. Describe the characteristics and requirements of individual and common vents.

    b. Describe the characteristics and requirements of battery vents.

    c. Describe the characteristics and requirements of wet vents.

    d. Describe the characteristics and requirements of air admittance and island vents.

    e. Describe the characteristics and requirements of relief and Sovent® vents.

Performance Task

Install different types of vents.

Trade Terms

Air admittance vent: A valve-type vent that ventilates a stack using air inside a building. The vent opens when exposed to reduced pressure in the vent system, and it closes when the pressure is equalized.

Branch interval: The space between two branches connecting with a main stack. The space between branch intervals is usually story height but cannot be less than 8 feet.

Branch vent: A vent that connects fixtures to the vent stack.

Circuit vent: A vent that connects a battery of fixtures to the vent stack.

Combination waste-and-vent system: A line that serves as a vent and also carries wastewater. Sovent® systems are an example of a combination waste-and-vent system.

Common vent: A vent that is shared by the traps of two similar fixtures installed back-to-back at the same height. It is also called a unit vent.

Continuous vent: A vertical continuation of a drain that ventilates a fixture.

Individual vent: A vent that connects a single fixture directly to the vent stack.

Loop vent: A vent that connects a battery of fixtures with the stack vent at a point above the waste stack.

Relief vent: A vent that increases circulation between the drain and vent stacks, thereby preventing back pressure by allowing excess air pressure to escape.

Re-vent: To install an individual vent in a fixture group.

Sovent® system: A combination waste-and-vent system used in high-rise buildings that eliminates the need for a separate vent stack. The system uses aerators on each floor to mix waste with air in the stack and a de-aerator at the base of the stack to separate the mixture.

Wet vent: A vent pipe that also carries liquid waste from fixtures with low flow rates. It is most frequently used in residential bathrooms.

Yoke vent: A relief vent that connects the soil stack with the vent stack.

Every vent system must be custom-designed according to your local applicable code to provide the most efficient venting, taking into account the building's function, location, and physical layout.

There are many different types of vents from which to choose. This gives plumbers the flexibility to create the best DWV system possible. Each vent is constructed to serve a particular purpose. Always refer to your local code to identify the vents that are approved for use in each individual application.

Vents are used in a variety of combinations. Many vents have more than one name, depending on the code being used. To make it easier to remember all the different types and names of vents, it will help to begin by grouping them together into a few broad categories.

2.1.0 Understanding Individual and Common Vents

A vent that runs from a single fixture directly to the vent stack is called an individual vent. Plumbers install, or re-vent, individual vents to provide correct air pressure at each trap. Because of this, they are generally considered to be the most effective means of venting a fixture. A continuous vent is a type of individual vent (see *Figure 9*). It is a vertical extension of the drain to which it is connected. Install continuous vents far enough away from fixture traps to avoid crown venting. Connect continuous vents to the stack at least 6 inches above the fixture's flood-level rim or overflow line. A branch vent is any vent that connects a single fixture or a group of fixtures to a vent stack at least 6 inches above the flood-level rim of the fixture. Refer to

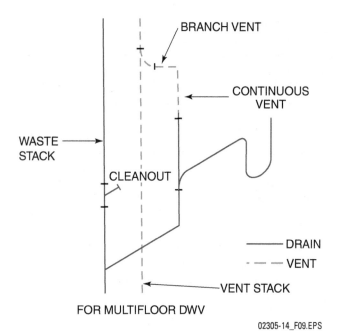

FOR MULTIFLOOR DWV

02305-14_F09.EPS

Figure 9 Branch vent incorporating a continuous vent.

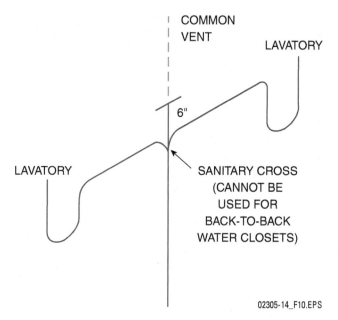

02305-14_F10.EPS

Figure 10 Common or unit vent.

your local applicable code before installing branch vents, as the requirements may vary.

Use a common vent where two similar fixtures are installed with all three of the following characteristics:

- The fixtures are back-to-back.
- The fixtures are on opposite sides of a wall.
- The fixtures are at the same height.

Common vents are also called unit vents. In a common vent, both fixture traps are connected to the drain with a short-pattern sanitary cross (see *Figure 10*). The two fixtures share a single vent. The vent extends above the drain from the top connection of the sanitary cross. Common vents are used in apartments or hotels where the design calls for back-to-back fixtures and shared vents.

2.2.0 Understanding Battery Vents

Battery vents, usually used in commercial buildings, connect a series of fixtures to the vent stack or stack vent. Many commercial buildings have large numbers of fixtures. It may be impractical or even impossible to provide each fixture with an individual vent. Horizontal vents connect a series, or battery, of fixtures to the vent stack (see *Figure 11*). Circuit vents and loop vents connect horizontal vents to the vent stack or stack vent. A circuit vent is a vent that connects a battery of fixtures to the vent stack. In this arrangement, the circuit vent is connected to the drain before the first fixture and the last fixture and to the vent stack. The loop vent is a special type of circuit vent. Loop vents are installed from the fixture group to the stack vent. Some

codes require that fixture batteries on the top floors of a building be vented with loop vents.

Select the proper type of battery vent for the application, and maintain the same pipe sizes throughout the vent. Local codes vary in their requirements for battery vents. Refer to your local code before venting a fixture battery.

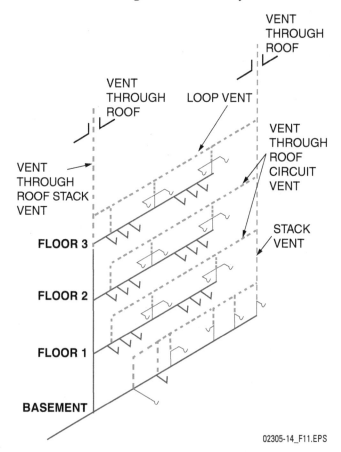

02305-14_F11.EPS

Figure 11 Components of a vent system.

2.3.0 Understanding Wet Vents

Vent piping that also carries liquid waste is a wet vent. It eliminates the need to install separate waste and vent stacks. Because of this, wet vents are a popular option for venting residential bathroom fixtures. Wet vents are installed either vertically or horizontally (see *Figure 12*). Only use wet vents with bathroom fixture groups.

When installing wet vents on toilets or water closets, follow the local code carefully. Make sure that the wet-vented fixture traps are capable of resealing. Maintain the proper hydraulic gradient and correct distances to ensure resealing.

Wet vents are considered acceptable by most model codes. Model codes cite the maximum vertical and horizontal run allowed for wet-vented pipe. The *Uniform Plumbing Code*™ (UPC), for example, allows the use of wet vents for single and back-to-back bathroom fixtures. Check your local code to determine the proper sizing and placement of a wet vent.

> ⚠️ **WARNING!**
> Be sure to follow your local code very carefully when installing wet vents. Ensure that the slope is correct. A mistake could cause serious drainage problems that could endanger public health.

2.4.0 Understanding Air Admittance and Island Vents

Air admittance vents ventilate the DWV stack using the building's own air (see *Figure 13*). Air admittance vents are also called air admittance valves. They are an alternative to vent stacks that penetrate building roofs. Air admittance vents operate like valves by allowing a controlled one-way flow of air into the drainage system. Reduced pressure causes the vent to open and allows air to enter the drain-

pipes. When the pressure is equalized, the vent closes by gravity. If back pressure occurs, the vent seals tightly to prevent sewer gases from escaping. Refer to your local applicable code before installing air admittance vents. The code may not allow them to be installed in your jurisdiction, or may have requirements that differ from other codes.

Two types of air admittance vent are available. One type, which is larger, is installed on stacks and is used to vent entire systems. A smaller type vents individual fixtures. Local codes permit air admittance valves in residential buildings. These two types of air admittance vents are not interchangeable. Always ensure that the air admittance valve you are installing is suitable for the particular application. Check your local code for the proper sizing and placement of air admittance vents. Install air admittance vents as high as possible. Use air admittance vents to vent fixtures in the center of a room, such as kitchen islands and wet bars.

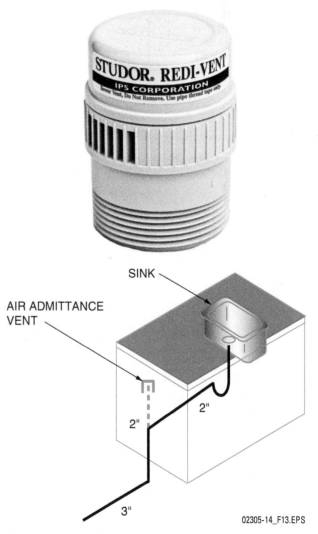

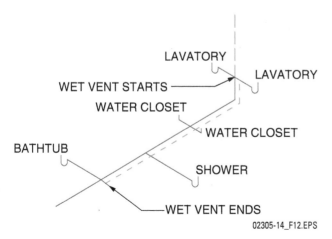

Figure 12 Wet vents.

Figure 13 Air admittance vent.

Other options for venting fixtures in an island are to use a loop vent or a combination waste-and-vent system under the fixture (see *Figure 14*). This is a pipe that acts as a vent and also drains wastewater. If a combination waste and vent is used, ensure that the drainpipe is one size larger than the trap. The options for venting island fixtures depend on local codes. Codes cite the maximum permitted distances between an island fixture and a main vent.

Island vents provide venting to sinks and other fixtures in a stand-alone installation such as a kitchen island. The *International Plumbing Code®* permits island venting to be used on sinks that have dishwasher drain connections and food-waste disposers. The island vent must rise vertically to a point above the fixture's drainage outlet before being offset downward (see *Figure 15*). If using a branch vent to vent multiple island fixtures, the branch must be a minimum of 6 inches above the highest fixture being vented before connecting to the vent stack. The lowest point of the fixture vent is connected to the drainage system. Therefore, the vent pipe must be the same diameter as the drainage pipe. Install cleanouts in the island vent below the flood-level rim of the fixture or fixtures being vented.

2.5.0 Understanding Relief and Sovent® Vents

A relief vent (*Figure 16*) allows excess air pressure in the drainage system to escape. This prevents back pressure from blowing out the seals in the fixture traps. Relief vents also provide air to prevent trap loss due to pressure differential.

Use a relief vent to increase circulation between the drain and vent systems. It can also serve as an auxiliary vent. A relief vent that connects the soil stack to the vent stack is called a yoke vent. Relief vents can also be connected between the soil stack and the fixture vents. In either case, connect the relief vent to the soil stack in a branch interval, which is the space between two branches entering a main stack. The space between branch intervals is usually story height but cannot be less than 8 feet.

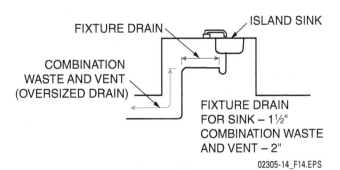

02305-14_F14.EPS

Figure 14 Combination waste-and-vent system.

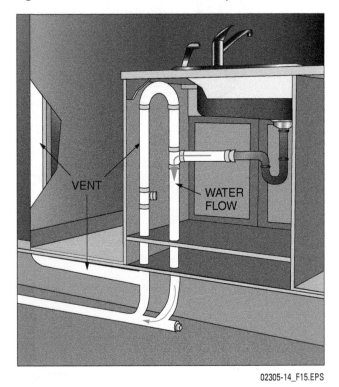

02305-14_F15.EPS

Figure 15 Island vent installation.

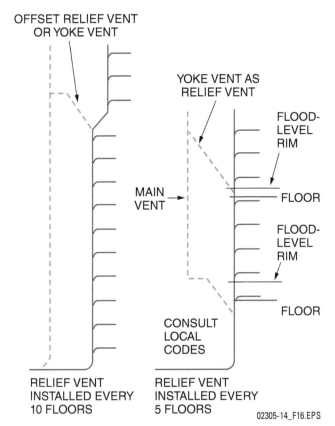

02305-14_F16.EPS

Figure 16 Relief vents.

Relief vents are very important in tall buildings. The weight of the water column in the stack can compress air trapped below it. Install relief vents on the floors of a high-rise building as called for in the design. A common practice is to install offset relief vents and yoke vents every 10 floors. Review your local code for specific details regarding the proper installation of relief vents.

In multistory buildings, a relief vent can often be combined with an offset in the waste stack to slow down the waste (see *Figure 17*). The installation will depend on the design of the stack. Local codes will specify the appropriate use of an offset relief vent in your area.

Also used in tall buildings, the innovative Sovent® system combines vent and waste stacks into a single stack (see *Figure 18*). Where a branch joins the single stack, an aerator mixes waste from the branch with the air in the stack. This slows the velocity of the resulting mixture and keeps the stack from being plugged by wastewater. At the base of the stack, a de-aerator separates the air and the liquid. This relieves the pressure in the stack and allows the waste to drain out of the system. The Sovent® system is intended mostly for use in high-rise buildings.

Note that Sovent® systems are engineered vent systems. They are not sized and installed using the standard techniques that apply to other types of vent systems. Some codes require approval by the project engineer before an engineered vent can be installed. Refer to your local code for the proper procedure in your area.

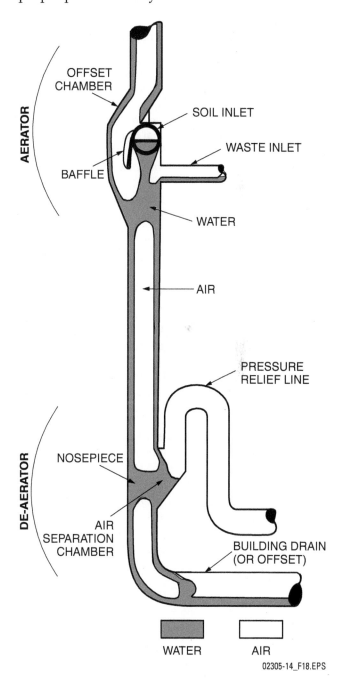

02305-14_F18.EPS

Figure 18 Cut-away of Sovent® vent and waste system.

VENT
STACK

WASTE
STACK

02305-14_F17.EPS

Figure 17 A relief vent at an offset in the waste stack.

Sovent® Vent Systems

Swiss professor Fritz Sommer invented a combination waste and vent system in 1959. Extensive tests proved the soundness of the basic design. Within a few years, Sovent® systems appeared in new buildings in Europe and the United States. Cast-iron, hubless fittings appeared in 1977. These cost much less than previous copper pipe and fittings. Advocates of Sovent® systems praise their simplicity, efficiency, and quiet operation.

Pneumatic-Ejector Relief Vents

Do not connect a relief vent serving a pneumatic ejector to a fixture branch vent. Instead, carry the vent separately to a main vent or stack vent. Pneumatic-ejector relief vents can also be vented to the open air.

Additional Resources

Estimator's Man-Hour Manual on Heating, Air Conditioning, Ventilating, and Plumbing. 1978. John S. Page. Woburn, MA: Gulf Professional Publishing Company.
International Plumbing Code®, Latest Edition. Falls Church, VA: International Code Council.
Plumbers and Pipefitters Handbook. 1996. William J. Hornung. Englewood Cliffs, NJ: Prentice Hall College Division.
Uniform Plumbing Code™, Latest Edition. Ontario, CA. International Association of Plumbing and Mechanical Officials.

2.0.0 Section Review

1. Continuous vents should be connected to the stack _____.
   a. at least 6 inches above the fixture's flood-level rim or overflow line
   b. at a 45-degree angle
   c. when fixtures are back-to-back
   d. before the first fixture and the last fixture in the battery

2. Some codes require that fixture batteries on the top floors of a building be vented with _____.
   a. circuit vents
   b. loop vents
   c. branch vents
   d. common vents

3. Wet vents should only be used with _____.
   a. fixtures installed in an island
   b. fixtures with manual flush valves
   c. commercial fixture groups
   d. bathroom fixture groups

4. Install air admittance vents _____.
   a. as high as possible
   b. as low as possible
   c. as close to the fixture as possible
   d. as far from the fixture as possible

5. The space between branch intervals cannot be less than _____.
   a. 4 feet
   b. 6 feet
   c. 8 feet
   d. 10 feet

Summary

Vents allow air to enter and exit a DWV stack to prevent pressure or a vacuum from disturbing fixture traps. Vents may ventilate single fixtures or a battery of them. They can provide relief venting to the main stack or connect with other vents. Regardless of how they are used, their purpose is to enable DWV systems to work by keeping an equal pressure inside the system. Vents come in many forms, and many can be used in combination with each other.

Like other plumbing installations, vents must be installed with proper grade and the correct pipe. Many codes differ as to how vents should be installed and where different types of vents should be used. Always consult your local code and the construction drawings prior to installing vents.

1. As wastewater flows out of a DWV system, equal pressure in the system is maintained by allowing the entry of _____.

   a. air
   b. additional wastes
   c. contaminants
   d. odors

2. A building's venting requirements depend upon its location, use, and _____.

   a. number of floors
   b. types of fixtures
   c. level of occupancy
   d. water pressure

3. Roof vents should be installed away from roof ridges and valleys that stir up _____.

   a. updrafts
   b. debris
   c. heated air
   d. downdrafts

4. The highest horizontal drain in the system connects to the venting system at the point where the stack vent joins the _____.

   a. vent stack
   b. building drain
   c. main soil-and-waste stack
   d. relief vent

5. The grade of a pipe allows the flow of wastes as a result of _____.

   a. unequal air pressure
   b. gravity
   c. sewer gas generation
   d. hydraulic pressure

6. Horizontal vents should *not* be used with _____.

   a. wall-hung fixtures
   b. back-to-back fixtures
   c. floor-mounted fixtures
   d. fixtures having a high rate of flow

7. To prevent the blocking of horizontal vent pipes by condensation, the pipes are _____.

   a. sloped
   b. heated
   c. under positive pressure
   d. lined with smooth plastic

8. If a vent is too far from a fixture, so that the vent opening is lower than the trap weir, it could cause _____.

   a. back pressure
   b. indirect siphonage
   c. a blown trap seal
   d. self-siphonage

9. An individual vent that is a vertical extension of a drain is known as a(n) _____.

   a. branch vent
   b. continuous vent
   c. common vent
   d. extension vent

10. A loop vent connects the stack vent to _____.

    a. two or more wet vents
    b. an individual fixture
    c. a group of fixtures
    d. fixtures located below grade

11. Wet vents may be installed _____.

    a. either vertically or horizontally
    b. vertically only
    c. horizontally only
    d. only at 45 degrees

12. For ventilating a fixture in a kitchen island, a combination waste-and-vent system is an alternative to using a(n) _____.

    a. horizontal vent
    b. air admittance vent
    c. continuous vent
    d. branch vent

13. Before turning downward, the vent on an island sink must rise vertically to a point above _____.

    a. the fixture's drainage outlet
    b. the sink strainer
    c. the lowest point of the trap
    d. the fixture's flood-level rim

14. Preventing back pressure from blowing out the seals in the fixture traps is a primary reason for installing _____.

    a. continuous vents
    b. loop vents
    c. air admittance vents
    d. relief vents

15. Sovent® engineered vent systems are installed primarily in _____.

    a. high-end residential buildings
    b. high-rise buildings
    c. industrial plants
    d. very large single-story buildings

Trade Terms Quiz

Fill in the blank with the correct term that you learned from your study of this module.

1. A(n) _____ is a pipe in a DWV system that allows air to circulate in the drainage system, thereby maintaining equalized pressure throughout.

2. A vent that connects a single fixture directly to the vent stack is called a(n) _____.

3. The _____ is a combination waste-and-vent system that uses aerators on each floor to mix waste with air in the stack and a de-aerator at the base of the stack to separate the mixture.

4. A vent that runs at an angle no greater than 44 degrees to the horizontal is called a(n) _____.

5. A(n) _____ is a vent that increases circulation between the drain and vent stacks, thereby preventing back pressure by allowing excess air pressure to escape.

6. A vertical continuation of a drain that ventilates a fixture is called a(n) _____.

7. A(n) _____ is a vent that is shared by the traps of two similar fixtures installed back-to-back at the same height.

8. Excess air pressure in vent piping that can blow trap seals out through fixtures is called _____.

9. A(n) _____ is an extension of the main soil-and-waste stack above the highest horizontal drain that allows air to enter and exit the plumbing system through the building roof and is a terminal for other vent pipes.

10. To _____ is to install an individual vent in a fixture group.

11. The space between two branches connecting with a main stack is called the _____.

12. A(n) _____ is a vent that connects fixtures to the vent stack.

13. A vent pipe that also carries liquid waste from fixtures with low flow rates is called a(n) _____.

14. A(n) _____ is a valve-type vent that ventilates a stack using air inside a building.

15. A vent that connects a battery of fixtures with the stack vent at a point above the waste stack is called a(n) _____.

16. The _____ is the point at which a vent terminates.

17. A line that serves as a vent and also carries wastewater is called a(n) _____.

18. A(n) _____ is a vent that connects a battery of fixtures to the vent stack.

19. The maximum degree of allowable fall between a trap weir and the opening of a vent pipe is called the _____.

20. A(n) _____ is a relief vent that connects the soil stack with the vent stack.

21. A stack that serves as a building's central vent, to which other vents may be connected, is called a(n) _____.

22. _____ is the drawing of a water seal out of the trap and into the waste pipe by lower-than-normal pressure.

23. The condition whereby lower-than-normal pressure in a drainpipe draws the water seal out of a fixture trap and into the drain is called _____.

Trade Terms

Air admittance vent
Back pressure
Branch interval
Branch vent
Circuit vent

Combination waste-
    and-vent system
Common vent
Continuous vent
Horizontal vent

Hydraulic gradient
Indirect or
    momentum
    siphonage
Individual vent

Loop vent
Relief vent
Re-vent
Self-siphonage
Sovent® system

Stack vent
Vent
Vent stack
Vent terminal
Wet vent
Yoke vent

Air admittance vent: A valve-type vent that ventilates a stack using air inside a building. The vent opens when exposed to reduced pressure in the vent system, and it closes when the pressure is equalized.

Back pressure: Excess air pressure in vent piping that can blow trap seals out through fixtures. Back pressure can be caused by the weight of the water column or by downdrafts through the stack vent.

Branch interval: The space between two branches connecting with a main stack. The space between branch intervals is usually story height but cannot be less than 8 feet.

Branch vent: A vent that connects fixtures to the vent stack.

Circuit vent: A vent that connects a battery of fixtures to the vent stack.

Combination waste-and-vent system: A line that serves as a vent and also carries wastewater. Sovent® systems are an example of a combination waste-and-vent system.

Common vent: A vent that is shared by the traps of two similar fixtures installed back-to-back at the same height. It is also called a unit vent.

Continuous vent: A vertical continuation of a drain that ventilates a fixture.

Horizontal vent: A vent that runs at an angle no greater than 44 degrees to the horizontal.

Hydraulic gradient: The maximum degree of allowable fall between a trap weir and the opening of a vent pipe. The total fall should not exceed one pipe diameter from weir to opening.

Indirect or momentum siphonage: The drawing of a water seal out of the trap and into the waste pipe by lower-than-normal pressure. It is the result of discharge into the drain by another fixture in the system.

Individual vent: A vent that connects a single fixture directly to the vent stack.

Loop vent: A vent that connects a battery of fixtures with the stack vent at a point above the waste stack.

Relief vent: A vent that increases circulation between the drain and vent stacks, thereby preventing back pressure by allowing excess air pressure to escape.

Re-vent: To install an individual vent in a fixture group.

Self-siphonage: A condition whereby lower-than-normal pressure in a drainpipe draws the water seal out of a fixture trap and into the drain.

Sovent® system: A combination waste-and-vent system used in high-rise buildings that eliminates the need for a separate vent stack. The system uses aerators on each floor to mix waste with air in the stack and a de-aerator at the base of the stack to separate the mixture.

Stack vent: An extension of the main soil-and-waste stack above the highest horizontal drain. It allows air to enter and exit the plumbing system through the building roof and is a terminal for other vent pipes.

Vent: A pipe in a DWV system that allows air to circulate in the drainage system, thereby maintaining equalized pressure throughout.

Vent stack: A stack that serves as a building's central vent, to which other vents may be connected. The vent stack may connect to the stack vent or have its own roof opening. It is also called the main vent.

Vent terminal: The point at which a vent terminates. For a fixture vent, it is the point where it connects to the vent stack or stack vent. For a stack vent, it is the point where the vent pipe ends above the roof.

Wet vent: A vent pipe that also carries liquid waste from fixtures with low flow rates. It is most frequently used in residential bathrooms.

Yoke vent: A relief vent that connects the soil stack with the vent stack.

Additional Resources

This module presents thorough resources for task training. The following resource material is suggested for further study.

Code Check Plumbing & Mechanical: An Illustrated Guide to the Plumbing and Mechanical Codes. 2011. Douglas Hansen, Redwood Kardon, and Paddy Morrisey. Newtown, CT: Taunton Press

Estimator's Man-Hour Manual on Heating, Air Conditioning, Ventilating, and Plumbing. 1978. John S. Page. Woburn, MA: Gulf Professional Publishing Company.

Handbook of Materials Selection. 2002. Myer Kutz, ed. New York: J. Wiley.

International Plumbing Code®, Latest Edition. Falls Church, VA: International Code Council.

Introduction to Drain, Waste, and Vent (DWV) Systems, Fourth Edition. 2012. Upper Saddle River, NJ: Pearson.

Plumbers and Pipefitters Handbook. 1996. William J. Hornung. Englewood Cliffs, NJ: Prentice Hall College Division.

Uniform Plumbing Code™, Latest Edition. Ontario, CA. International Association of Plumbing and Mechanical Officials.

Figure Credits

Answer	Section Reference	Objective
Section One		
1. b	1.1.0	1a
2. c	1.2.0	1b
3. a	1.3.0	1c
Section Two		
1. a	2.1.0	2a
2. b	2.2.0	2b
3. d	2.3.0	2c
4. a	2.4.0	2d
5. c	2.5.0	2e

NCCER CURRICULA — USER UPDATE

NCCER makes every effort to keep its textbooks up-to-date and free of technical errors. We appreciate your help in this process. If you find an error, a typographical mistake, or an inaccuracy in NCCER's curricula, please fill out this form (or a photocopy), or complete the online form at **www.nccer.org/olf**. Be sure to include the exact module ID number, page number, a detailed description, and your recommended correction. Your input will be brought to the attention of the Authoring Team. Thank you for your assistance.

Instructors – If you have an idea for improving this textbook, or have found that additional materials were necessary to teach this module effectively, please let us know so that we may present your suggestions to the Authoring Team.

NCCER Product Development and Revision
13614 Progress Blvd., Alachua, FL 32615

Email: curriculum@nccer.org
Online: www.nccer.org/olf

❏ Trainee Guide ❏ Lesson Plans ❏ Exam ❏ PowerPoints Other _____

Craft / Level: _____ Copyright Date: _____

Module ID Number / Title: _____

Section Number(s): _____

Description: _____

_____

_____

_____

Recommended Correction: _____

_____

_____

_____

_____

Your Name: _____

Address: _____

_____

Email: _____ Phone: _____

02306-14

Sizing DWV and Storm Systems

OVERVIEW

Safe and efficient drainage of wastes is one of the most important functions of plumbing systems. Drain, waste, and vent (DWV) systems and storm sewers remove these wastes. Codes require DWV and storm drainage systems in all commercial and residential buildings; specific requirements vary from region to region. Plumbers are responsible for sizing and installing both types of systems, and must custom-design each one.

Trainees with successful module completions may be eligible for credentialing the NCCER Registry. To learn more, go to **www.nccer.org** or contact us at **1.888.622.3720**. Our website has information on the latest product releases and training, as well as online versions of our *Cornerstone* magazine and Pearson's product catalog.

Your feedback is welcome. You may email your comments to **curriculum@nccer.org,** send general comments and inquiries to **info@nccer.org**, or fill in the User Update form at the back of this module.

This information is general in nature and intended for training purposes only. Actual performance of activities described in this manual requires compliance with all applicable operating, service, maintenance, and safety procedures under the direction of qualified personnel. References in this manual to patented or proprietary devices do not constitute a recommendation of their use.

SIZING DWV AND STORM SYSTEMS

Objectives

When you have completed this module, you will be able to do the following:

1. Describe how to size drain, waste, and vent systems.
   a. Describe how to size drains.
   b. Describe how to size vents.
2. Describe how to size storm drainage systems.
   a. Describe how to calculate rainfall conversions.
   b. Describe how to size roof storage and drainage systems.
   c. Describe how to size above-grade and below-grade drainage systems.

Performance Tasks

Under the supervision of your instructor, you should be able to do the following:

1. Calculate drainage fixture units for a plumbing system.
2. Size branch lines for plumbing fixtures.
3. Size waste stacks.
4. Size building drains and sewers.
5. Size vents according to local code.
6. Determine annual rainfall and 10-, 25-, 50-, and 100-year averages per your local code.
7. Calculate the surface area of a roof for storm-system sizing.
8. Size conventional roof drainage systems for stormwater removal.

Trade Terms

Conductor
Controlled-flow roof drainage system
Conventional roof drainage system

Drainage fixture unit (DFU)
Leader
Percolate
Permeability
Ponding

Scupper
Secondary drain
Siphonic roof drain system

Industry-Recognized Credentials

If you're training through an NCCER-accredited sponsor, you may be eligible for credentials from the NCCER Registry. The ID number for this module is 02306-14. Note that this module may have been used in other NCCER curricula and may apply to other level completions. Contact NCCER's Registry at 888.622.3720 or go to **www.nccer.org** for more information.

Code Note

Codes vary among jurisdictions. Because of the variations in code, consult the applicable code whenever regulations are in question. Referring to an incorrect set of codes can cause as much trouble as failing to reference codes altogether. Obtain, review, and familiarize yourself with your local adopted code.

Contents

Topics to be presented in this module include:

Figures and Tables ────────────────

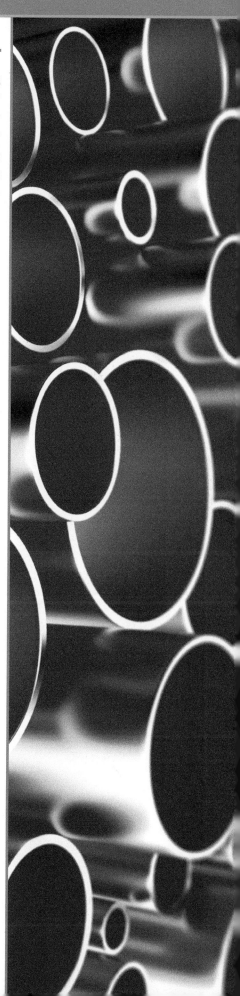

1.0.0 SIZING DRAIN AND VENT SYSTEMS

Objective

Describe how to size drain, waste, and vent systems.
   a. Describe how to size drains.
   b. Describe how to size vents.

Performance Tasks

Calculate drainage fixture units for a plumbing system.

Size branch lines for plumbing fixtures.

Size waste stacks.

Size building drains and sewers.

Size vents according to local code.

Trade Term

Drainage fixture unit (DFU): A measure of the waste discharge of a fixture in gallons per minute. DFUs are calculated from the fixture's rate of discharge, the duration of a single discharge, and the average time between discharges.

Safe and efficient drainage is one of the most important functions of plumbing systems. Public health depends on the removal of sewage and storm wastes. Drain, waste, and vent (DWV) systems and storm sewers remove these wastes. Plumbers are responsible for sizing and installing both types of systems. You have already learned how to install different types of pipe. In this module, you will learn to apply your math and code-reading skills to select the proper size of pipe for sewage and stormwater drainage.

Codes require DWV and storm drainage systems in all commercial and residential buildings. As with other types of plumbing installations, each system is unique. Plumbers custom-design systems for each installation. Code requirements for DWV and storm drainage vary from region to region. Take the time to study your local code.

Note that the tables shown in this module are specific examples from real model codes. They are provided for reference only. Do not use the tables in this module to size actual DWV and storm drainage systems in the field. Always refer to the appropriate information in your local applicable code.

You learned how to install and test DWV systems in the Plumbing Level Two module *Installing and Testing DWV Piping*. In this section, you will learn how to size the pipe used in those systems. Sizing drains, stacks, sewer lines, and vents is generally a paper-and-pencil task. You have to sit down and calculate the proper pipe sizes before you can begin to build the system. You do not have to be an engineer to be able to size a DWV system so that it works. Local plumbing codes provide many tables and guidelines that allow you to size the system components. You will learn about those guidelines and practice using them as part of this module. Terms and definitions vary from jurisdiction to jurisdiction. Be sure to consult your local code for the proper terms in your area.

You have already learned that vents work in conjunction with drains to enable wastes to flow from fixtures into the waste pipe, where they are carried into the sewer. Without vents, plumbing systems would not work. Every structure with a plumbing system requires at least one main vent. Regardless of the type of installation, consider the following factors when sizing vents:

- The number and types of fixtures attached to the system
- The number of drainage fixture units (DFUs) being transferred to the drainage system
- The length of the vent pipe

Pipe size in a DWV system is based on the number of drainage fixture units the system carries. The larger the number of DFUs, the greater the pipe diameter required. DFUs measure the waste discharge of fixtures (see *Table 1*). DFUs are determined based on the following variables:

- The fixture's rate of discharge
- The duration of a single discharge
- The average time between discharges

When a DFU value has not been determined, use the standard value of 7.5 gallons of discharge per minute. Always consult your local applicable codes to find the values in use in your area.

The more DFUs a fixture discharges, the larger the fixture drainpipe needs to be. The fixtures listed in *Table 1* pass a variable quantity of water, depending on how frequently they are used. For fixtures that drain waste continuously, such as air conditioners, codes will assign a minimum value for gallons per minute (gpm) of discharge.

1.1.0 Sizing Drains

After determining the DFUs for all the fixtures in the structure, you can begin to size the drainage

Table 1 DFUs for Fixtures and Groups

Fixture Type	Drainage Fixture Unit Value as Load Factors	Minimum Size of Trap (inches)
Automatic clothes washers, commercial	3	2
Automatic clothes washers, residential	2	2
Bathroom group as defined in Section 202 (1.6 gpf water closet)[d]	5	—
Bathroom group as defined in Section 202 (water closet flushing greater than 1.6 gpf)[d]	6	—
Bathtub[a] (with or without overhead shower or whirlpool attachments)	2	1½
Bidet	1	1¼
Combination sink and tray	2	1½
Dental lavatory	1	1¼
Dental unit or cuspidor	1	1¼
Dishwashing machine, domestic	2	1½
Drinking fountain	½	1¼
Emergency floor drain	0	2
Floor drains	2	2
Floor sinks		2
Kitchen sink, domestic	2	1½
Kitchen sink, domestic with food waste grinder and/or dishwasher	2	1½
Laundry tray (1 or 2 compartments)	2	1½
Lavatory	1	1¼

Shower (based on the total flow rate through showerheads and body sprays) Flow rate:
   5.7 gpm or less
   Greater than 5.7 gpm to 12.3 gpm
   Greater than 12.3 gpm to 25.8 gpm
   Greater than 25.8 gpm to 55.6 gpm
	
	2
	3
	5
	6
		
		1½
		2
		3
		4

Service sink	2	1½
Sink	2	1½
Urinal	4	Note b
Urinal, 1 gallon per flush or less	2[c]	Note b
Urinal, nonwater supplied	½	Note b
Wash sink (circular or multiple) each set of faucets	2	1½
Water closet, flushometer tank, public or private	4[c]	Note b
Water closet, private (1.6 gpf)	3[c]	Note b
Water closet, private (flushing greater than 1.6 gpf)	4[c]	Note b
Water closet, public (1.6 gpf)	4[c]	Note b
Water closet, public (flushing greater than 1.6 gpf)	6[c]	Note b

a. A showerhead over a bathtub or whirlpool bathtub attachment does not increase the drainage fixture unit value.

b. Trap size shall be consistent with the fixture outlet size.

c. For the purpose of computing loads on building drains and sewers, water closets and urinals shall not be rated at a lower drainage fixture unit unless the lower values are confirmed by testing.

d. For fixtures added to a bathroom group, add the DFU value of those additional fixtures to the bathroom group fixture count.

system. Your local code will provide tables that show the correct pipe size based on the number of DFUs in the system (see *Table 2*). Use this information to determine pipe sizes for both vertical stacks and horizontal fixture branches. Typically, the sizing process begins at the farthest branch lines in the system and continues to the next-farthest branch lines and so on, ending with the fixtures that are closest to the main stack. Plumbers refer to this technique as following the flow.

Table 2 Sample Table for Calculating the Size of Horizontal Branches and Vertical Stacks

Diameter of Pipe (inches)	Total for Horizontal Branch	Maximum Number of Drainage Fixture Units (DFU)		
		Stacks		
		Total Discharge into One Branch Interval	Total for Stack of Three Branch Intervals or Less	Total for Stack Greater than Three Branch Intervals
1½	3	2	4	8
2	6	6	10	24
2½	12	9	20	42
3	20	20	48	72
4	160	90	240	500
5	360	200	540	1,100
6	620	350	960	1,900
8	1,400	600	2,200	3,600
10	2,500	1,000	3,800	5,600
12	3,900	1,500	6,000	8,400
15	7,000	Note a	Note a	Note a

a. Sizing load based on design criteria.

For example, refer to *Table 1*. Note that it assigns 2 DFUs to a bathtub with or without attachments. Assume that the bathtub is the only fixture on the horizontal branch. Use *Table 2* to calculate the correct pipe size. In this case, the pipe should be 1½ inches (because the DFU rating of 2 is not listed, size up to the next listed DFU rating). *Table 1* also requires that the minimum trap size for a bathtub be 1½ inches. Note that you can also use fixture-group sizing methods for bathroom groups. These values are based on a probability of usage. This is because the fixtures are not likely to be used all at the same time.

The number of water closets on a line and the number of floors in the structure can also affect drainage pipe sizing. If a single bathroom group has more than one water closet, you must increase the size of the drain stack. That way, the stack can handle the extra surge if both water closets are flushed at the same time. Follow your local code closely when installing stacks in buildings that require three or more branch intervals. Remember also to consider future fixture installations when sizing pipes. Install branch stubs for these. Cap the stubs with approved caps or plugs.

1.1.1 Exceptions and Limitations

The guidelines given in the previous sections have many exceptions. In every case, consult your local code. Discuss tips and techniques with experienced plumbers. Learn about the limitations and exceptions as well as the general rules. Because the health of the public is at stake, take the time to size each system correctly.

Many codes require drainage pipes that run underground or below a basement or cellar to be at least 2 inches in diameter. These codes require this regardless of the DFU values that have been calculated for the system. Horizontal drains should be sloped to allow the wastewater to flow at least 2 feet per second. If a drain line is less than 4 inches in diameter, the slope must be ¼ inch per foot of pipe. If a drain line is larger than 4 inches in diameter, the slope must be ⅛ inch per foot. Water closets require a minimum drain size of 3 inches. Some codes require 4-inch pipe if three or more water closets are installed on a single drain line.

Use at least a 3-inch-diameter pipe when installing soil stacks. Codes prohibit using branch piping that has a larger diameter than the soil stack to which it connects. Many codes also limit

The Dowd Family, Pipe Pioneers

At the dawn of the twentieth century, W. Frank Dowd opened a foundry in Charlotte, North Carolina, to produce cast-iron soil pipe. Originally, Charlotte Pipe employed 25 men. Dowd led the company until 1926, when his son Frank Dowd II took over and expanded the business. In the 1950s, Frank II's sons Frank Jr. and Roddey took over. They explored new production technologies and experimented with new pipe materials, including plastics. Today, Charlotte Pipe is the country's largest producer of DWV pipe and fittings. It also holds the record for the longest period of any pipe manufacturer under single ownership—the Dowd family, innovators in the pipe manufacturing field.

how many branches can be connected to a 3-inch stack. Usually, no more than two branches per floor or six branches per stack are permitted.

Often, the location of a building's structural components makes it impossible to install drainage stacks that run straight from top to bottom. Install offsets to route the stacks around obstacles. In most codes, if the offset is 45 degrees or less, the pipe can be the same size as the vertical stack. If the offset is greater than 45 degrees, size the pipe according to the rules for sizing horizontal pipe. If the offset is installed above the highest fixture branch, size the offset according to the rules for sizing vents. Before connecting horizontal branch lines above and below offsets, consult your local code.

1.1.2 Sizing Building Drains and Sewers

Like horizontal branches and vertical stacks, building drains and sewers are sized according to the DFUs assigned to them (see *Table 3*). Additionally, consider the pipe slope when sizing building drains and sewers. The slope may range from $\frac{1}{16}$ to $\frac{1}{2}$ inch per foot. The combination of pipe diameter and slope determines the carrying capacity of every building drain and sewer.

Local codes limit the number of branches and water closets on a building drain or sewer. Most codes require that building sewer lines be at least 4 inches in diameter. Some codes allow the slope of the building drain or sewer to be increased so that it can handle a greater number of DFUs. For example, referring to *Table 3*, an 8-inch pipe can handle 1,400 DFUs at a slope of $\frac{1}{16}$ inch per foot. If the slope is increased to $\frac{1}{2}$ inch per foot, however, the same-size pipe can handle 2,300 DFUs. Refer to your local code for the standards that apply in your area.

1.2.0 Sizing Vents

In the *Types of Venting* module, you learned about the different types of vents and how they work. In this section, you will learn the general techniques for sizing vents. Always refer to your local code for specific requirements in your area. Always ensure that all joints in vent piping are sealed cor-

Table 3 Sample Table for Calculating the Size of Building Drains and Sewers

Diameter of Pipe (inches)	Maximum Number of Drainage Fixture Units Connected to any Portion of the Building Drain or the Building Sewer, Including Branches of the Building Drain[a]			
	Slope per Foot			
	¹⁄₁₆ inch	⅛ inch	¼ inch	½ inch
1¼	—	—	1	1
1½	—	—	3	3
2	—	—	21	26
2½	—	—	24	31
3	—	36	42	50
4	—	180	216	250
5	—	390	480	575
6	—	700	840	1,000
8	1,400	1,600	1,920	2,300
10	2,500	2,900	3,500	4,200
12	3,900	4,600	5,600	6,700
15	7,000	8,300	10,000	12,000

a. The minimum size of any building drain serving a water closet shall be 3 inches.

rectly. Leaks allow odors to escape, and escaping sewer gases can cause explosions, contamination, and illness.

In cold climates and regions with heavy snow, vent terminals can become plugged by frost or covered by a snow blanket. This is caused by water in the vent air condensing and building up on the inside of the vent pipe as it comes into contact with the cold outside air. To prevent frost buildup in the vent pipe, enlarge the section of pipe that extends above the roof (see *Figure 1*). A pipe at least 3 inches in diameter is less likely to completely close up with frost than a narrower pipe. Unless otherwise specified by the local applicable code, the enlargement should extend at least 1 foot inside the roof. Ensure that the location of the vent and the vent pipe materials are suitable for the climate conditions. Otherwise, the same problem may happen in the future.

In most cases, the diameter of a vent should be one-half the diameter of the drain being vented.

Correct Pipe Sizing and Installation

You have learned that undersizing a pipe can be a costly mistake. However, oversizing can be just as costly. If an installer uses a larger pipe size than necessary, water may not flow fast enough in the horizontal section to scour the pipe and wash away wastes. When installed at an incorrect slope, an oversized pipe can clog even faster than an undersized one. Take the time to review your local code requirements for pipe sizing. Calculate pipe sizes using the most recent standards. The results will help your customers.

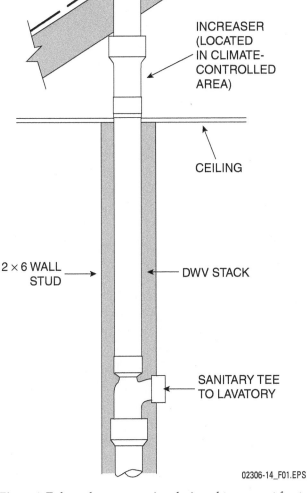

Figure 1 Enlarged vent extension designed to prevent frost closure.

ROOF

INCREASER (LOCATED IN CLIMATE-CONTROLLED AREA)

CEILING

2 × 6 WALL STUD

DWV STACK

SANITARY TEE TO LAVATORY

02306-14_F01.EPS

However, the vent diameter should never be less than 1¼ inches. If the vent is longer than 40 feet, use one pipe size larger than the calculated size for the entire length. Measure the developed length from the farthest drainage connection to the stack or terminal connection. Review your local code to find out how to calculate vent sizes in your area.

Calculate the diameter for a vent stack or stack vent by measuring the developed length of the vent and the total number of DFUs connected to it. You will also need to know the diameter of the soil or waste stack. Your local code should have a table that allows you to calculate this (see Table 4).

1.2.1 Main Vents

The main vent is the vent stack that carries the largest number of DFUs in the system. Usually, the location of the water closet determines the location of the main vent. Most installations require a main vent that is at least 3 inches in diameter. Ensure that the main vent meets local code requirements. If the DFUs exceed the code specifications, install a larger vent.

If a vent for a washing machine or laundry tray is installed in an unattached garage, use 1½-inch pipe for the main vent. Buildings that have many separate units, such as shopping malls and condominiums, may also require smaller vents. Install a main vent for each individual unit. Consult your local code for sizing requirements.

Use the same-diameter pipe for the entire length of the main vent. Measure the developed length from the point where the vent intersects the stack or drain to the point where it terminates through the roof. Note that in some designs, the main vent may intersect the main stack before it exits the structure. In such cases, be sure to include the length of all sections that are required for the vent to reach the open air (see Figure 2).

1.2.2 Individual Vents

Individual vents connect single fixtures to the main vent (see Figure 3). They are very dependable and can be used on remote fixtures where reventing is not practical. The size of an individual vent depends on the size and DFU rating of the fixture to be attached. Individual vents must be at least half the size of the fixture drain and are never less than 1¼ inches in diameter. Consult your local code for other requirements.

1.2.3 Common Vents

Common vents are used to vent two fixtures that are set back to back or side by side on the same floor (see Figure 4). The fixture vents can be connected to the common vent at the same level or at different levels. If the vents are connected at the same level (see Figure 4A), locate the vents at the intersection of the fixture drains or downstream of them. If the vents are connected at different levels, the upper fixture drain must be one pipe diameter larger than the vertical drain. The lower fixture drain, however, should not be larger than the vertical drain (see Figure 4B). As a rule of thumb, do not allow more than 1 DFU on a 1½-inch pipe or more than 4 DFUs on a 2-inch pipe when sizing common vents.

Table 4 Size and Developed Length of Stack Vents and Vent Stacks

Diameter of Soil or Waste Stack (inches)	Total Fixture Units Being Vented (DFU)	Maximum Developed Length of Vent (feet)										
		Diameter of Vent (inches)										
		1¼	1½	2	2¼	3	4	5	6	8	10	12
1¼	2	30	150	—	—	—	—	—	—	—	—	—
1½	8	50	100	—	—	—	—	—	—	—	—	—
1½	10	30	100	—	—	—	—	—	—	—	—	—
2	12	30	75	200	—	—	—	—	—	—	—	—
2	20	26	50	150	300	—	—	—	—	—	—	—
2½	42	—	30	100	—	—	—	—	—	—	—	—
3	10	—	42	150	360	1,040	—	—	—	—	—	—
3	21	—	32	110	270	810	—	—	—	—	—	—
3	53	—	27	94	230	680	—	—	—	—	—	—
3	102	—	25	86	210	620	—	—	—	—	—	—
4	43	—	—	35	85	250	980	—	—	—	—	—
4	140	—	—	27	65	200	750	—	—	—	—	—
4	320	—	—	—	23	55	170	640	—	—	—	—
4	540	—	—	—	21	50	150	580	—	—	—	—
5	190	—	—	—	—	28	82	320	990	—	—	—
5	490	—	—	—	—	21	63	250	760	—	—	—
5	940	—	—	—	—	18	53	210	670	—	—	—
5	1,400	—	—	—	—	16	49	190	590	—	—	—
6	500	—	—	—	—	33	130	400	1,000	—	—	—
6	1,100	—	—	—	—	26	100	310	780	—	—	—
6	2,000	—	—	—	—	22	84	260	660	—	—	—
6	2,900	—	—	—	—	20	77	240	600	940	—	—
8	1,800	—	—	—	—	—	31	95	240	940	—	—
8	3,400	—	—	—	—	—	24	73	190	729	—	—
8	5,600	—	—	—	—	—	20	62	160	610	—	—
8	7,600	—	—	—	—	—	18	56	140	560	960	—
10	4,000	—	—	—	—	—	—	31	78	310	960	—
10	7,200	—	—	—	—	—	—	24	60	240	740	—
10	11,000	—	—	—	—	—	—	20	51	200	630	—
10	15,000	—	—	—	—	—	—	18	46	180	571	—
12	7,300	—	—	—	—	—	—	31	120	380	940	940
12	13,000	—	—	—	—	—	—	24	94	300	720	720
12	20,000	—	—	—	—	—	—	20	79	250	610	610
12	26,000	—	—	—	—	—	—	—	18	72	230	500
15	15,000	—	—	—	—	—	—	—	—	40	130	310
15	25,000	—	—	—	—	—	—	—	—	31	96	240
15	38,000	—	—	—	—	—	—	—	—	26	81	200
15	50,000	—	—	—	—	—	—	—	—	24	74	180

Note: The developed length shall be measured from the vent connection to the open air.

1.2.4 Wet Vents

To determine the correct pipe size for a wet vent, add the DFUs for all the fixtures in the fixture group. Refer to *Table 1* or the appropriate table in your local code to determine the DFU load. Then refer to *Table 5* or your local code to select the proper vent size. Do not install wet vents on wall-hung water closets.

For example, if a lavatory drain is being used to vent a shower and water closet (see *Figure 5*), the lavatory drain must be larger than the normal required size. Calculate the total DFUs for all three fixtures. According to *Table 1*, a total of 6 DFUs should be assigned to the group. According to *Table 5*, a 2½-inch pipe will be required to vent the group.

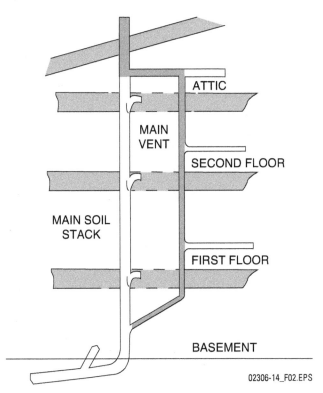

Figure 2 Measuring the developed length of a vent.

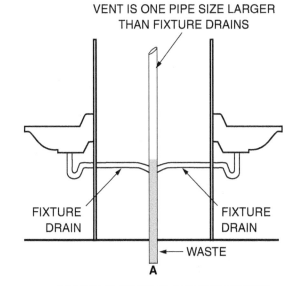

Figure 4 Typical common vent installation.

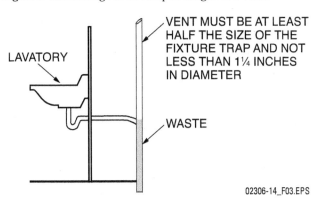

Figure 3 Individual vent.

1.2.5 Relief Vents

Relief vents provide adequate airflow in cases where the number of fixtures installed becomes too large for conventional venting methods. Install relief vents, also called yoke vents, in all soil and waste stacks that have more than 10 branch intervals. Install relief vents at every tenth branch, counting from the top floor down. The relief vent must be the same size as the vent stack to which it connects.

1.2.6 Battery Vents

Use a circuit vent to vent a battery of fixtures (see *Figure 6*). Consult your local code for sizing information. Circuit vents are sized differently than other types of vents. A circuit vent should be at

least one-half the size of the horizontal waste line that it is venting, but it must never be less than 1½ inches in diameter. Do not install more than eight floor-outlet water closets, floor-outlet urinals, showers, bathtubs, or floor drains to a single circuit vent. However, if the water closets are blowout types, more than eight can be installed. If lavatories are installed above a circuit vent, install a continuous vent on each vertical branch.

1.2.7 Sump Vents

The venting of drainpipes that are below the sewer level is similar to the venting of gravity-fed systems. Consult your local code for the correct sizes and lengths (see *Table 6*). For pneumatic ejectors, the vent should allow the ejector to reach normal air pressure. Connect vents for pneumatic ejectors to an independent vent stack

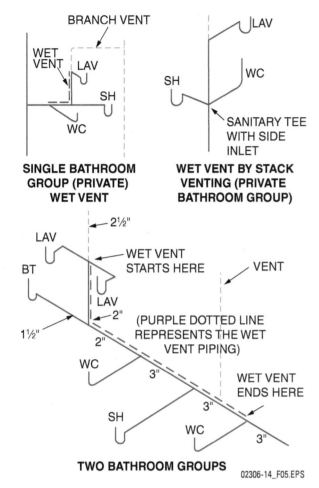

Figure 5 Examples of wet-vent installation in a bathroom fixture group.

Table 5 Sample Table for Sizing Wet Vents

Pipe Diameter (inches)	Maximum Number of Drainage Fixture Units (DFU)	
	Connecting to a horizontal branch or stack	Connecting to a building drain or building subdrain
2	3	4
2½	6	26
3	12	31
4	20	50
5	160	250
6	360	575

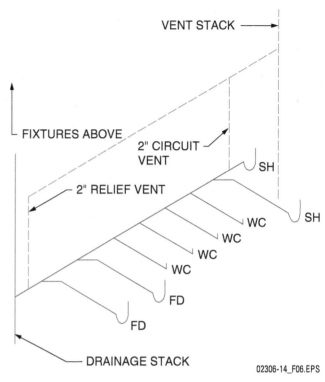

Figure 6 Typical installation of circuit vents.

that terminates through the roof. Do not size the vent smaller than 1¼ inches in diameter. Connect vents for pneumatic sewage ejectors to an independent vent stack. This vent stack must be at least 1¼ inches in diameter.

1.2.8 Combination Drain-and-Vent Systems

Use combination drain-and-vent systems (also called combination waste-and-vent systems) for floor drains, standpipes, sinks, and lavatories (see *Figure 7*). Do not vent urinals or water closets using a combination drain and vent. Kitchen-sink and washing-machine traps may not be installed on a 2-inch combination drain and vent. Consult your local code for sizing tables for combination drain-and-vent installations. On larger DWV systems, the combination drain-and-vent systems are required to be engineered systems, rather than sized by the plumber installing them.

Table 6 Sample Sizing Table for Sump Vents

Discharge Capacity of Pump (gpm)	Maximum Developed Length of Vent (feet)ᵃ					
	Diameter of Vent (inches)					
	1¼	1½	2	2½	3	4
10	No limitᵇ	No limit	No limit	No limit	No limit	No limit
20	270	No limit	No limit	No limit	No limit	No limit
40	72	160	No limit	No limit	No limit	No limit
60	31	75	270	No limit	No limit	No limit
80	16	41	150	380	No limit	No limit
100	10ᶜ	25	97	250	No limit	No limit
150	Not permitted	10ᶜ	44	110	370	No limit
200	Not permitted	Not permitted	20	60	210	No limit
250	Not permitted	Not permitted	10	36	132	No limit
300	Not permitted	Not permitted	10ᶜ	22	88	380
400	Not permitted	Not permitted	Not permitted	10ᶜ	44	210
500	Not permitted	Not permitted	Not permitted	Not permitted	24	130

a. Developed length plus an appropriate allowance for entrance losses and friction due to fittings, changes in direction and diameter. Suggested allowances shall be obtained from NBS Monograph 31 or other approved sources. An allowance of 50 percent of the developed length shall be assumed if a more precise value is not available.
b. Actual values greater than 500 feet.
c. Less than 10 feet.

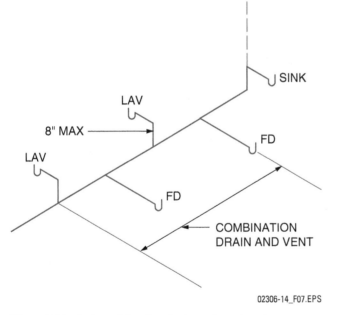

SINK

LAV

8" MAX

FD

LAV

FD

COMBINATION DRAIN AND VENT

02306-14_F07.EPS

Figure 7 Typical combination drain-and-vent system.

The National Pollutant Discharge Elimination System (NPDES) Program

GOING GREEN

The US Environmental Protection Agency (EPA) controls water pollution through its NPDES program. NPDES regulates point sources, such as pipes and ditches, that discharge pollutants into surface water. Industrial, municipal, and other facilities must obtain permits from the EPA if they discharge directly into surface water. In most cases, the NPDES permit program is administered at the state level.

The NPDES permit program has helped to improve the country's water quality significantly. Twenty-five years ago, only about one-third of US surface waters were safe for fishing and swimming. Today, the percentage has risen to two-thirds. Loss of wetlands has been reduced by more than 80 percent. Soil erosion from agricultural runoff has been nearly cut in half.

The Mechanics of Fluid Flow

In horizontal branch lines, water flows along the lower portion of the pipe. Air circulates in the upper portion. Horizontal pipes are sloped so that water can flow through the pipe at a rate of at least 2 feet per second. This rate of flow sets up a scouring action in the pipe and prevents sediment from building up in the pipe.

The mechanics of vertical flow are different from those of horizontal flow. When water enters the vertical stack and flows down to the building drain, it runs along the wall of the pipe. The velocity of water in a vertical stack is usually about 10 to 15 feet per second.

As the flow in the stack increases, the space for air decreases. When the stack is about one-third full of water, it reaches its maximum flow capacity. Above this level, slugs of water begin to drop away from the pipe wall and fall through the middle of the stack. The slugs eventually land back on the pipe wall further down. This is called diaphragming. Diaphragming causes many of the noises heard in soil stacks.

Additional Resources

Handbook of Materials Selection. 2002. Myer Kutz, ed. New York: J. Wiley.
Introduction to Drain, Waste, and Vent (DWV) Systems, Fourth Edition. 2012. Upper Saddle River, NJ: Pearson.
Water, Sanitary, and Waste Services for Buildings. 2002. Alan F. E. Wise and J. A. Swaffield. Boston, MA: Butterworth-Heinemann.

1.0.0 Section Review

1. Typically, the sizing process begins at the branch lines that are _____.

    a. lowest in the system
    b. highest in the system
    c. closest to the main stack
    d. farthest from the main stack

2. When installing a vent for a washing machine in an unattached garage, the diameter of the pipe used for the main vent should be _____.

    a. ½ inch
    b. 1 inch
    c. 1½ inches
    d. 2 inches

SECTION TWO

2.0.0 SIZING STORM DRAINAGE SYSTEMS

Objective

Describe how to size storm drainage systems.
  a. Describe how to calculate rainfall conversions.
  b. Describe how to size roof storage and drainage systems.
  c. Describe how to size above-grade and below-grade drainage systems.

Performance Tasks

Determine annual rainfall and 10-, 25-, 50-, and 100-year averages per your local code.

Calculate the surface area of a roof for storm-system sizing.

Size conventional roof drainage systems for stormwater removal.

Trade Terms

Conductor: An interior vertical spout or pipe that drains stormwater into a main drain.

Controlled-flow roof drainage system: A roof drainage system that retains stormwater for an average of 12 hours before draining it. Controlled-flow systems help prevent overloads in the storm-sewer system during heavy storms.

Conventional roof drainage system: A roof drainage system that drains stormwater as soon as it falls.

Leader: An exterior vertical spout or pipe that drains stormwater into a main drain.

Percolate: To seep into the soil and become absorbed by it.

Permeability: A measure of the soil's ability to absorb water that percolates into it.

Ponding: Collecting of stormwater on the ground.

Scupper: An opening on a roof or parapet that allows excess water to empty into the surface drain system.

Secondary drain: A drain system that acts as a backup to the primary drain system. It is also called an emergency drain.

Siphonic roof drain system: An engineered drainage system that blocks air from entering the drain system during the drainage process, resulting in greater flow volume at full bore than can be achieved by conventional drainage systems.

Correct sizing of storm drainage systems depends upon two factors. The first is the maximum expected rainfall during a given period of time. The second is the designed rate of removal. Local codes provide guidance on how to size and install storm drainage systems. Storm drainage systems are required for the following:

- Building roofs
- Paved areas, such as parking lots and streets
- Residential lawns and yards

For one- and two-family homes, channel the stormwater away from the house and onto lawns or roads. Many codes require cleanouts on storm drainage systems.

Stormwater can be disposed of right away, or it can be stored and then released slowly. Consult your local code to learn how stormwater disposal is handled in your area. Designers can choose one of three methods for retaining and disposing of stormwater. Select the method that is appropriate for the type of structure and the anticipated amount of drainage:

- Above-grade drainage
- Below-grade drainage
- Roof drainage

Depending on your local code, connect storm drains to a combined sewer line or to a separate storm sewer. Never connect storm drainage to a sewage-only line. If the storm drain connects to a combined sewer line, install a trap. Traps can be installed in two ways. One option is to trap all drain branches that serve each conductor or leader. A conductor is an interior spout or vertical pipe that drains stormwater into the main drain line; leaders are exterior spouts or vertical pipes that perform the same function. The other option is to install a trap in the main drain immediately upstream of its connection with the sewer. Do not reduce the size of a drainage pipe in the direction of flow.

Building engineers usually design storm drainage systems. They must ensure that the building is strong enough to handle the weight of water. Follow the design closely when installing the system so that you will know that it works efficiently and effectively. Always refer to your local code.

 02306-14 Sizing DWV and Storm Systems

2.1.0 Calculating Rainfall Conversions

Consider the annual amount of rainfall in your region when sizing storm drains. Also consider the potential for flash floods. Unlike most other estimates that you will make, rainfall calculations are based on probabilities, not facts. Nevertheless, rainfall probabilities are a reliable tool for sizing storm drainage systems. Your local code will have copies of rainfall maps for your area. Most model and local codes get their climate information from the National Oceanic and Atmospheric Administration (NOAA). You can find NOAA online at **www.noaa.gov**. NOAA's Precipitation Frequency Data Server (PFDS), an online database that collects rainfall data for the entire United States, is a useful reference for identifying rainfall trends in your region. Data in the PFDS is presented in both table and chart form (see *Figure 8*).

The local 100-year average is the most accurate indicator of rainfall. It takes both wet and dry years into consideration. The system must also be designed to handle the most severe storm that is likely to occur. Estimates for this kind of storm are calculated for 10-, 25-, 50-, and 100-year periods (see *Figure 9*). Of course, this does not tell you when during that 100-year period the storm will happen. It may happen in the first year, it may happen in the hundredth year, or it may not happen at all. However, the system should be able to handle even this rare storm.

To size a drainage system, find the area of the surface and the volume of water it can hold. For example, you have to drain a roof that is 50 feet wide by 100 feet long. You know that its total area is 5,000 square feet (50 feet × 100 feet). Now imagine that the design calls for the roof to hold 6 inches of water. The roof will hold up to 2,500 cubic feet of water (50 feet × 100 feet × 0.5 foot). Converted to gallons, that equals 18,750 gallons (2,500 cubic feet × 7.5 gallons per cubic foot).

The building engineer has to make sure that a roof or other storage area can hold the weight of water. In the example above, 18,750 gallons of water weigh more than 156,000 pounds (18,750 gallons × 8.33 pounds per gallon). That is more than 78 tons of water. You can see why it is important to size storage and drainage systems correctly.

Just 2 inches of rainfall in an hour weighs more than 10 pounds (see *Table 7*). The same amount of rainfall equals 1.25 gallons (see *Table 8*). Use *Tables 7* and *8* to calculate the weight and amounts of rain that a drainage system will have to handle.

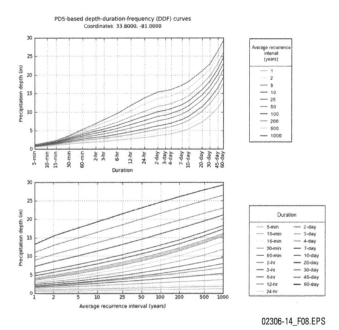

Figure 8 NOAA PFDS charts showing point precipitation frequency estimates for Columbia, South Carolina.

02306-14_F08.EPS

Figure 9 Sample map showing 100-year, 1-hour rainfall in inches.

02306-14_F09.EPS

Table 7 Weight of Rainwater

Rainfall (in/hr)	Weight (lb/sq ft)
1	5.202
2	10.404
3	15.606
4	20.808
5	26.010
6	31.212

Table 8 Rainfall Conversion Data for Calculating Water Volume

Rainfall (in/hr)	Gallons per Minute (per sq ft)	Gallons per Hour (per sq ft)
1	0.0104	0.623
2	0.0208	1.247
3	0.0312	1.870
4	0.0416	2.493
5	0.0520	3.117
6	0.0624	3.740

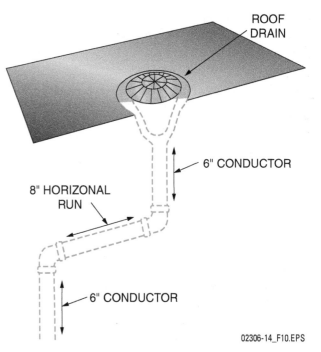

Figure 10 Typical roof drain in a conventional drainage system.

2.2.0 Sizing Roof Storage and Drainage Systems

Stormwater can be drained immediately, or it can be stored and released over a longer period. Systems that drain immediately are called conventional roof drainage systems. Systems that store water and release it slowly are called controlled-flow roof drainage systems. In this section, you will learn how to size both types of drainage systems.

2.2.1 Sizing Conventional Roof Drainage Systems

The conventional way to drain stormwater from a roof is to allow the water to drain as soon as it falls. This type of drainage system must be able to handle the largest expected rainfall in the area. Drainage systems consist of vertical conductors and/or leaders and horizontal drains (see *Figure 10*). Most codes require that a roof area of up to 10,000 square feet have a minimum of two drains. In some installations, a vertical wall diverts rainwater to the roof. If so, add half of the wall's area to the projected roof area.

Review the rainfall tables for your local area. Then convert that figure into gallons per minute per square foot (refer to *Table 8*). Size vertical conductors and leaders and horizontal drains and branches using tables in your local code (see *Tables 9* and *10*). Plumbers usually install horizontal branches with a 1-percent slope. This equals ⅛ inch of slope for every 12 inches of pipe. Increasing the slope will increase the amount of water drained per minute without increasing the pipe size. Refer to your local code before increasing the slope.

To find the total amount of water that a single drain can handle, multiply the number of gallons per square foot per minute by the number of square feet. To calculate roof drain sizes, use the following procedure:

Step 1 Determine the square footage of the roof area. Divide the roof area by 10,000 square feet. This is the number of drains required for the roof.

Step 2 Determine the local rainfall rate and convert to gallons per minute.

Step 3 Multiply the roof area by the number in Step 2. This is the number of gallons per minute that will land on the roof.

Step 4 If parapets are installed in adjacent vertical walls, add ½ the vertical area of the wall to the total square footage.

Step 5 Divide the number in Step 4 (or Step 3, if parapets are not used) by the number of drains. This is how much stormwater each individual drain must handle.

Step 6 Refer to your local code to find the size of the vertical leader and the slope of the horizontal drain.

Step 7 Determine the sizes of the vertical and horizontal pipes into which the individual drains will drain.

The illustration in *Figure 11* shows how to use these steps to size drains for a roof that is 200 feet by 300 feet. In this example, the roof drains have to handle 416 gallons per hour. Based on the informa-

Table 9 Sample Table for Sizing Vertical Leaders

Diameter of Leader (inches)[a]	Horizontally Projected Roof Area (square feet)											
	Rainfall rate (inches per hour)											
	1	2	3	4	5	6	7	8	9	10	11	12
2	2,880	1,440	960	720	575	480	410	360	320	290	260	240
3	8,800	4,400	2,930	2,200	1,760	1,470	1,260	1,100	980	880	800	730
4	18,400	9,200	6,130	4,600	3,680	3,070	2,630	2,300	2,045	1,840	1,675	1,530
5	34,600	17,300	11,530	8,650	6,920	5,765	4,945	4,325	3,845	3,460	3,145	2,880
6	54,000	27,000	17,995	13,500	10,800	9,000	7,715	6,750	6,000	5,400	4,910	4,500
8	116,000	58,000	38,660	29,000	23,200	19,315	16,570	14,500	12,890	11,600	10,545	9,600

a. Sizes indicated are the diameter of circular piping. This table is applicable to piping of other shapes, provided the cross-sectional shape fully encloses a circle of the diameter indicated in this table. Interpolation is permitted for pipe sizes that fall between those listed in this table.

Table 10 Sample Table for Sizing Horizontal Storm Drains

Size of Horizontal Piping (inches)	Horizontally Projected Roof Area (square feet)					
	Rainfall rate (inches per hour)					
	1	2	3	4	5	6
⅛ unit vertical in 12 units horizontal (1 percent slope)						
3	3,288	1,644	1,096	822	657	548
4	7,520	3,760	2,506	1,800	1,504	1,253
5	13,360	6,680	4,453	3,340	2,672	2,227
6	21,400	10,700	7,133	5,350	4,280	3,566
8	46,000	23,000	15,330	11,500	9,200	7,600
10	82,800	41,400	27,600	20,700	16,580	13,800
12	133,200	66,600	44,400	33,300	26,650	22,200
15	218,000	109,000	72,800	59,500	47,600	39,650
¼ unit vertical in 12 units horizontal (2 percent slope)						
3	4,640	2,320	1,546	1,160	928	773
4	10,600	5,300	3,533	2,650	2,120	1,766
5	18,880	9,440	6,293	4,720	3,776	3,146
6	30,200	15,100	10,066	7,550	6,040	5,033
8	65,200	32,600	21,733	16,300	13,040	10,866
10	116,800	58,400	38,950	29,200	23,350	19,450
12	188,000	94,000	62,600	47,000	37,600	31,350
15	336,000	168,000	112,000	84,000	67,250	56,000
½ unit vertical in 12 units horizontal (4 percent slope)						
3	6,576	3,288	2,295	1,644	1,310	1,096
4	15,040	7,520	5,010	3,760	3,010	2,500
5	26,720	13,360	8,900	6,680	5,320	4,450
6	42,800	21,400	13,700	10,700	8,580	7,140
8	92,000	46,000	30,650	23,000	18,400	15,320
10	171,600	85,800	55,200	41,400	33,150	27,600
12	266,400	133,200	88,800	66,600	53,200	44,400
15	476,000	238,000	158,800	119,000	95,300	79,250

tion provided, use a 6-inch vertical leader and an 8-inch horizontal drain sloped at ¼ inch per foot.

2.2.2 Sizing Siphonic Roof Drain Systems

Siphonic roof drain systems are engineered drainage systems that operate by suction rather than by gravity. Siphonic drains are fitted with an air baffle plate in the inlet to prevent air from entering the drain system during drainage (see *Figure 12*). This allows for a greater flow volume when the system is running at full bore. Siphonic drains are connected to horizontal drainage manifolds that discharge into a vertical stack (see *Figure 13*). Gravity pulls the water down through the

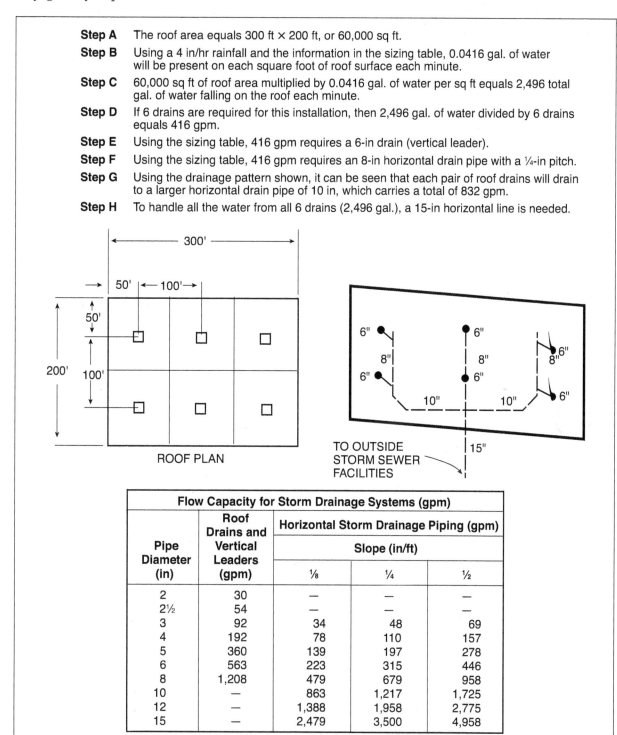

Step A The roof area equals 300 ft × 200 ft, or 60,000 sq ft.

Step B Using a 4 in/hr rainfall and the information in the sizing table, 0.0416 gal. of water will be present on each square foot of roof surface each minute.

Step C 60,000 sq ft of roof area multiplied by 0.0416 gal. of water per sq ft equals 2,496 total gal. of water falling on the roof each minute.

Step D If 6 drains are required for this installation, then 2,496 gal. of water divided by 6 drains equals 416 gpm.

Step E Using the sizing table, 416 gpm requires a 6-in drain (vertical leader).

Step F Using the sizing table, 416 gpm requires an 8-in horizontal drain pipe with a ¼-in pitch.

Step G Using the drainage pattern shown, it can be seen that each pair of roof drains will drain to a larger horizontal drain pipe of 10 in, which carries a total of 832 gpm.

Step H To handle all the water from all 6 drains (2,496 gal.), a 15-in horizontal line is needed.

Flow Capacity for Storm Drainage Systems (gpm)				
Pipe Diameter (in)	Roof Drains and Vertical Leaders (gpm)	Horizontal Storm Drainage Piping (gpm)		
		Slope (in/ft)		
		⅛	¼	½
2	30	—	—	—
2½	54	—	—	—
3	92	34	48	69
4	192	78	110	157
5	360	139	197	278
6	563	223	315	446
8	1,208	479	679	958
10	—	863	1,217	1,725
12	—	1,388	1,958	2,775
15	—	2,479	3,500	4,958

02306-14_F11.EPS

Figure 11 How to size a conventional roof drainage system.

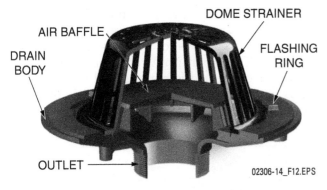

Figure 12 Cut-away view of a siphonic roof drain.

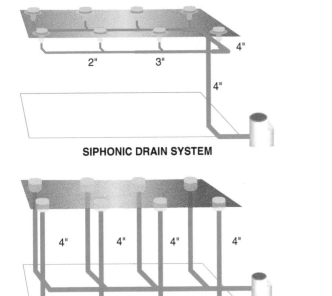

NOTE: ALL PIPE SIZES ARE NOMINAL DIMENSIONS.

02306-14_F13.EPS

Figure 13 Comparison of siphonic and gravity roof drainage systems.

vertical stack, rather than relying on the equalization of pressure as in a typical drainage system.

Unlike other types of drainage systems, siphonic systems are designed to operate at peak efficiency with the drainage piping completely filled with water. At ground level, the water is discharged into the storm system at normal atmospheric pressure and velocity. This allows for a higher rate of discharge and also improved self-cleaning of the drainage system. Because siphonic systems use smaller-diameter pipes and require less piping overall, they can reduce material and labor costs on larger projects.

Because siphonic roof drain systems are engineered for the particular installation, installers should follow the manufacturer's instructions closely when installing the system. The manufacturer's instructions will specify the permitted pipe sizes and allowable tolerances based on the engineer's calculations of roof area, volumetric flow, and depth profile. Many manufacturers provide online calculators to enable engineers to calculate the number and placement of drains, pipe size, and other factors when designing the system.

Never join vertical or horizontal pipes to the system that are larger in diameter than the sizes specified in the design. An increase in diameter will create voids that allow air into the system, thereby reducing the efficiency of the drainage. Installing sloping pipes may prevent siphonic action from forming in the system. All pipe sections in a siphonic system should be perfectly vertical or horizontal. To ensure a balanced flow in all parts of the system, the engineer may require the installation of "doglegs" from the drainpipe into the collector pipe (see *Figure 14*).

Siphonic systems must be tested once installed to ensure that they are working as designed. Most codes require tests to be performed in accordance with American Society of Plumbing Engineers (ASPE) 45, *Siphonic Roof Drainage*. ASPE 45 re-

Combined Sewer Overflows

Some sewer systems are intentionally designed to overflow. The wastewater from combined sewer systems is made up of stormwater, domestic sewage, and industrial wastewater. Approximately 900 cities in the US use combined sewer systems. During periods of heavy rainfall or snowmelt, the flow can exceed the system's capacity. Engineers design the system to drain into a nearby body of water when that happens. The result is called a combined sewer overflow, or CSO.

The Environmental Protection Agency (EPA) issued a control policy for CSOs in 1994. CSOs are classified as urban wet-weather discharges. That means CSOs are considered similar to other sanitary sewer overflows and stormwater discharges. The EPA's policy on CSOs helps communities meet the discharge and pollutant goals of the Clean Water Act. Take time to review the CSO policy and the Clean Water Act requirements on the EPA's website at **www.epa.gov**.

quires tests to be conducted with a positive pressure with a 30-foot water column, or 13 pounds per square inch gauge (psig).

2.2.3 Sizing Secondary Roof Drains

Sometimes the main storm drain can become overloaded or blocked. Secondary drains provide a backup in such cases (see *Figure 15*). Secondary drains are also called emergency drains. The *International Plumbing Code*® (IPC) requires that secondary drains be completely separate from the main roof drain. The IPC also requires the drains to be located above grade and where occupants can see them. If your local code is based on the IPC, you can size the secondary drains using the same rainfall rates as for the main roof. Do not consider the flow through the main drain when sizing the secondary drain. Refer to your local code for standards that apply in your area.

Another form of overflow protection is the installation of scuppers in the walls (see *Figure 16*) or parapets around the roof. These are openings that allow excess water to drain from the roof into the surface drainage system. Refer to your local code for the sizing and placement of scuppers.

2.3.0 Sizing Above-Grade and Below-Grade Drainage Systems

In some areas, stormwater can be allowed to drain directly onto the ground. This action is called ponding. The water then percolates, or seeps, into the soil. Ponding is permitted only on ground that has been set aside for that purpose. To construct effective aboveground storage, you need to know how well the soil can diffuse the water that flows onto it. The ability of soil to absorb water is called permeability. Consult your local agricultural extension service to learn about the permeability of soil in your area.

Your local code will list the different types of pipe that can be used in aboveground storm drainage systems. The IPC allows the following pipe materials:

- Acrylonitrile butadiene styrene (ABS) plastic
- Brass
- Cast iron
- Copper and copper alloy
- Types K, L, M, and DWV copper and copper-alloy tubing
- Galvanized steel
- Glass
- Polyolefin
- Type DWV polyvinyl chloride (PVC) plastic

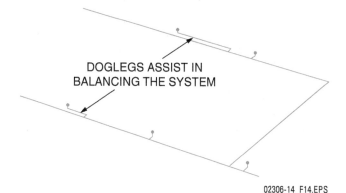

02306-14_F14.EPS

Figure 14 Doglegs in a hypothetical siphonic roof drainage system installation.

02306-14_F15.EPS

Figure 15 Primary and secondary roof drains used in a conventional drainage system.

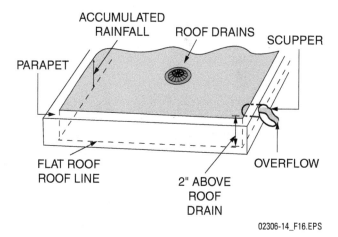

02306-14_F16.EPS

Figure 16 Scupper installation in a conventional drainage system.

A Cast-Iron Chronology

In 1562, a fountain in Langensalza, Germany, was fitted with cast-iron water supply pipe. This is the earliest recorded use of cast-iron pipe. The famous water gardens of Versailles, France, built in 1664, also used cast-iron pipe. These pipes, which still work, carry water from a river 15 miles away. London received its first cast-iron-pipe water distribution system in 1746. Sir Thomas Simpson invented the bell-and-spigot joint in 1785. This was an improvement over bolted flanges and butt joints wrapped with metal bands. Bell-and-spigot joints are still widely used in cast-iron pipe.

Around the beginning of the 19th century, cities in Pennsylvania and other eastern states began using cast-iron pipe. At the time, iron ore mined in the United States was shipped to foundries in England, and the completed pipes were shipped back across the ocean. The first successful US foundries were not established until 1819 in New Jersey. Beginning in the 1880s, foundries moved outward to the South and the Midwest, and their pipe plumbed the country's rapidly growing cities. Cast-iron pipe production grew to an industrial scale during that time. Production of cast-iron pipe in the United States peaked at 280,000 net tons in 1916.

Additional Resources

ASPE 45, *Siphonic Roof Drainage*, Latest Edition. Rosemont, IL: American Society of Plumbing Engineers.
Cast Iron Soil Pipe and Fittings Handbook. 2006. Atlanta, GA: Cast Iron Soil Pipe Institute.
http://www.cispi.org/cispi/handbook/
International Plumbing Code®, Latest Edition. Falls Church, VA: International Code Council.

2.0.0 Section Review

1. When connecting a storm drain to a combined sewer line, install a _____.

   a. trap
   b. valve
   c. sump
   d. filter

2. ASPE 45, *Siphonic Roof Drainage*, requires tests of siphonic drainage systems to be conducted with a positive pressure with a water column of _____.

   a. 30 feet
   b. 25 feet
   c. 20 feet
   d. 15 feet

3. A plastic pipe material that is *not* intended for use in aboveground storm drainage systems is _____.

   a. ABS
   b. PEX
   c. PVC
   d. polyolefin

Summary

DWV systems and storm drainage systems provide safe and efficient drainage for residences and commercial buildings. Plumbers size both types of system, and codes vary with regard to drainage requirements.

Plumbers size DWV systems by calculating the DFU load in each branch. DFUs measure the waste discharge of fixtures. Local codes provide tables for using DFU loads to size building drains and sewers. Vents are also sized by calculating the DFUs in the system. Plumbers use the same techniques to size combination drain-and-vent systems and reduced-size vent systems.

This module also discussed methods to size storm drainage systems. Plumbers consider the maximum expected rainfall and the rate of removal. Local codes provide rainfall tables, which are used to estimate the system's capacity. Above- and below-grade systems allow stormwater to percolate into the soil.

The type of roof drainage system affects its sizing. Conventional systems allow stormwater to run off as soon as it falls. Siphonic systems rely on suction, not the equalization of pressure, to draw water off the roof, allowing for smaller piping. Local codes provide sizing tables for both types of roof drainage systems.

1. Drainage fixture units measure the _____.
   a. number of fixtures in a structure
   b. amount of waste discharged by an individual fixture
   c. amount of waste discharged by all units in a structure
   d. size of the fixture's drain

2. If a structure includes a plumbing system, the system must have at least one main _____.
   a. trap
   b. cleanout
   c. vent
   d. drain

3. The sizing technique that begins with the farthest branch line and ends with the fixture nearest the main vent is referred to as _____.
   a. following the flow
   b. flow-based sizing
   c. going with the flow
   d. counterflow sizing

4. If drainage pipes run underground or beneath a cellar or basement, many codes require that they have a diameter of at least _____.
   a. 4 inches
   b. 2½ inches
   c. 2 inches
   d. 1¼ inches

5. When a drainage stack must be offset by an angle greater than 45 degrees, the offset section of the stack must be _____.
   a. larger than the vertical section
   b. sized as a vent
   c. smaller than the vertical section
   d. sized as a horizontal pipe

6. To permit a building drain or sewer to handle a larger number of DFUs, codes sometimes allow increasing the drain's _____.
   a. diameter
   b. length
   c. slope
   d. capacity rating

7. When enlarging a vent terminal to prevent frost buildup, the enlarged section should extend below the roof's surface by at least _____.
   a. 18 inches
   b. 12 inches
   c. 8 inches
   d. 6 inches

8. In calculating the sizing for a vent, the distance between the stack and the farthest drainage connection is the _____.
   a. developed length
   b. total system length
   c. drain length
   d. maximum length

9. A wet vent should *not* be installed on a _____.
   a. shower stall
   b. wall-hung water closet
   c. floor-outlet water closet
   d. lavatory

10. Counting from the top floor down, a relief vent should be installed for every _____.
    a. third branch
    b. fifth branch
    c. tenth branch
    d. twelfth branch

11. Storm drainage should never be connected to a _____.
    a. road ditch
    b. combined sewer line
    c. natural watercourse
    d. sewage-only line

12. One gallon of water weighs _____.
    a. 8.33 pounds
    b. 7.65 pounds
    c. 5.5 pounds
    d. 3.28 pounds

13. The most accurate indicator of local rainfall is the _____.

    a. 10-year average
    b. 30-year average
    c. 50-year average
    d. 100-year average

14. Storm drainage systems that drain immediately are called _____.

    a. rapid-flow drainage systems
    b. normal roof drainage systems
    c. conventional roof drainage systems
    d. pitched-roof drainage systems

15. Ponding is an action in which stormwater is allowed to flow _____.

    a. into a natural body of water
    b. onto a designated ground area
    c. onto a paved area, such as a parking lot
    d. into a cistern

Trade Terms Quiz

Fill in the blank with the correct term that you learned from your study of this module.

1. A(n) _____ is a roof drainage system that drains stormwater as soon as it falls.

2. The collecting of stormwater on the ground is called _____.

3. A(n) _____ is a drain system that acts as a backup to the primary drain system.

4. An engineered drainage system that blocks air from entering the drain system during the drainage process, resulting in greater flow volume at full bore, is called a(n) _____.

5. A(n) _____ is an opening on a roof or parapet that allows excess water to empty into the surface drain system.

6. The measure of the waste discharge of a fixture in gallons per minute is called the _____.

7. A(n) _____ is a roof drainage system that retains stormwater for an average of 12 hours before draining it.

8. The measure of the soil's ability to absorb water that percolates into it is called _____.

9. A(n) _____ is an interior vertical spout or pipe that drains stormwater into a main drain.

10. To seep into the soil and become absorbed by it is to _____.

11. A(n) _____ is an exterior vertical spout or pipe that drains stormwater into a main drain.

Trade Terms

Conductor
Controlled-flow roof drainage system
Conventional roof drainage system
Drainage fixture unit (DFU)
Leader
Percolate

Permeability
Ponding
Scupper
Secondary drain
Siphonic roof drain system

Trade Terms Introduced in This Module

Conductor: An interior vertical spout or pipe that drains stormwater into a main drain.

Controlled-flow roof drainage system: A roof drainage system that retains stormwater for an average of 12 hours before draining it. Controlled-flow systems help prevent overloads in the storm-sewer system during heavy storms.

Conventional roof drainage system: A roof drainage system that drains stormwater as soon as it falls.

Drainage fixture unit (DFU): A measure of the waste discharge of a fixture in gallons per minute. DFUs are calculated from the fixture's rate of discharge, the duration of a single discharge, and the average time between discharges.

Leader: An exterior vertical spout or pipe that drains stormwater into a main drain.

Percolate: To seep into the soil and become absorbed by it.

Permeability: A measure of the soil's ability to absorb water that percolates into it.

Ponding: Collecting of stormwater on the ground.

Scupper: An opening on a roof or parapet that allows excess water to empty into the surface drain system.

Secondary drain: A drain system that acts as a backup to the primary drain system. It is also called an emergency drain.

Siphonic roof drain system: An engineered drainage system that blocks air from entering the drain system during the drainage process, resulting in greater flow volume at full bore than can be achieved by conventional drainage systems.

Additional Resources

This module presents thorough resources for task training. The following resource material is suggested for further study.

ASPE 45, *Siphonic Roof Drainage*, Latest Edition. Rosemont, IL: American Society of Plumbing Engineers.

Cast Iron Soil Pipe and Fittings Handbook. 2006. Atlanta, GA: Cast Iron Soil Pipe Institute.
   http://www.cispi.org/cispi/handbook/

Handbook of Materials Selection. 2002. Myer Kutz, ed. New York: J. Wiley.

International Plumbing Code®, Latest Edition. Falls Church, VA: International Code Council.

Introduction to Drain, Waste, and Vent (DWV) Systems, Fourth Edition. 2012. Upper Saddle River, NJ: Pearson.

Water, Sanitary, and Waste Services for Buildings. 2002. Alan F. E. Wise and J. A. Swaffield. Boston, MA: Butterworth-Heinemann.

Figure Credits

Answer	Section Reference	Objective
Section One		
1. d	1.1.0	1a
2. c	1.2.1	1b
Section Two		
1. a	2.0.0	2
2. a	2.2.2	2b
3. b	2.3.0	2c

NCCER CURRICULA — USER UPDATE

NCCER makes every effort to keep its textbooks up-to-date and free of technical errors. We appreciate your help in this process. If you find an error, a typographical mistake, or an inaccuracy in NCCER's curricula, please fill out this form (or a photocopy), or complete the online form at **www.nccer.org/olf**. Be sure to include the exact module ID number, page number, a detailed description, and your recommended correction. Your input will be brought to the attention of the Authoring Team. Thank you for your assistance.

Instructors – If you have an idea for improving this textbook, or have found that additional materials were necessary to teach this module effectively, please let us know so that we may present your suggestions to the Authoring Team.

NCCER Product Development and Revision
13614 Progress Blvd., Alachua, FL 32615

Email: curriculum@nccer.org
Online: www.nccer.org/olf

❏ Trainee Guide ❏ Lesson Plans ❏ Exam ❏ PowerPoints Other _____

Craft / Level: _____ Copyright Date: _____

Module ID Number / Title: _____

Section Number(s): _____

Description: _____

_____

_____

_____

Recommended Correction: _____

_____

_____

_____

Your Name: _____

Address: _____

_____

Email: _____ Phone: _____

02307-14

Sewage Pumps and Sump Pumps

OVERVIEW

Building designs often require plumbers to locate a drain below the sewer line. Because these drains, called subdrains, cannot discharge into the sewer by gravity, the wastewater must be pumped upward to the higher-level drainage lines. Subdrains empty into temporary holding pits called sumps, where pumps (ejectors) provide the lift needed. These sewage removal systems, together with stormwater removal systems, perform a critical role in plumbing installations and in maintaining public health.

Module Six

Trainees with successful module completions may be eligible for credentialing through the NCCER Registry. To learn more, go to **www.nccer.org** or contact us at **1.888.622.3720**. Our website has information on the latest product releases and training, as well as online versions of our *Cornerstone* magazine and Pearson's product catalog.

Your feedback is welcome. You may email your comments to **curriculum@nccer.org,** send general comments and inquiries to **info@nccer.org,** or fill in the User Update form at the back of this module.

This information is general in nature and intended for training purposes only. Actual performance of activities described in this manual requires compliance with all applicable operating, service, maintenance, and safety procedures under the direction of qualified personnel. References in this manual to patented or proprietary devices do not constitute a recommendation of their use.

Objectives

When you have completed this module, you will be able to do the following:

1. Describe the components of sewage and stormwater removal systems and explain how to size and install them.
   a. Describe sewage pumps and sumps, and explain how to size and install them.
   b. Describe stormwater pumps and sumps, and explain how to size and install them.
   c. Describe sewage and stormwater pump controls, and explain how they work.
2. Explain how to troubleshoot and repair sewage and stormwater removal systems.
   a. Explain how to troubleshoot electrical problems.
   b. Explain how to troubleshoot mechanical problems.
   c. Explain how to replace pumps.

Performance Tasks

Under the supervision of your instructor, you should be able to do the following:

1. Using a detailed drawing provided by the instructor, identify system components.
2. Install a sump pump.
3. Install and adjust sensors, switches, and alarms in sewage and sump pumps.
4. Troubleshoot sewage and sump pumps.

Trade Terms

Air lock	Float switch	Radar	Stormwater removal
Analog switch	Impeller	Retention pond	system
Centrifugal pump	Lift station	Retention tank	Subdrain
Digital switch	Mercury float switch	Reverse-flow pump	Sump
Duplex pump	Pneumatic ejector	Run dry	Sump pump
Ejector	Pressure float switch	Sewage pump	
Field effect	Probe switch	Sewage removal system	

Industry-Recognized Credentials

If you're training through an NCCER-accredited sponsor, you may be eligible for credentials from NCCER's Registry. The ID number for this module is 02307-14. Note that this module may have been used in other NCCER curricula and may apply to other level completions. Contact NCCER's Registry at 888.622.3720 or go to **www.nccer.org** for more information.

Code Note

Codes vary among jurisdictions. Because of the variations in code, consult the applicable code whenever regulations are in question. Referring to an incorrect set of codes can cause as much trouble as failing to reference codes altogether. Obtain, review, and familiarize yourself with your local adopted code.

Contents

Topics to be presented in this module include:

Figures and Tables

1.0.0 SIZING AND INSTALLING SEWAGE AND STORMWATER REMOVAL SYSTEMS

Objective

Describe the components of sewage and stormwater removal systems and explain how to size and install them.

a. Describe sewage pumps and sumps, and explain how to size and install them.
b. Describe stormwater pumps and sumps, and explain how to size and install them.
c. Describe sewage and stormwater pump controls, and explain how they work.

Performance Tasks

Using a detailed drawing provided by the instructor, identify system components.

Install a sump pump.

Install and adjust sensors, switches, and alarms in sewage and sump pumps.

Trade Terms

Analog switch: A type of switch that is operated mechanically or electrically.

Centrifugal pump: A device that uses an impeller powered by an electric motor to draw wastewater out of a sump and discharge it into a drain line.

Digital switch: A type of switch that is operated electronically.

Duplex pump: Two centrifugal pumps and their related equipment installed in parallel in a sump.

Ejector: A common alternative name for a sewage pump.

Field effect: A change in an electrical state that occurs when a conductive material such as water passes through an electric field.

Float switch: A container filled with a gas or liquid that measures wastewater height in a sump and activates the pump when the float reaches a specified height.

Impeller: A set of vanes or blades that draws wastewater from a sump and ejects it into the discharge line.

Lift station: A popular name for a sewage removal system. The term is also used to refer to a pumping station on a main sewer line.

Mercury float switch: A float switch filled with mercury. It activates the pump when it reaches a specified height in a basin.

Pneumatic ejector: A pump that uses compressed air to force wastewater from a sump into the drain line.

Pressure float switch: An adjustable float switch that can either be allowed to rise with wastewater or be clipped to the sump wall.

Probe switch: An electrical switch used in pneumatic ejectors that closes on contact with wastewater.

Radar: A system that determines distance by bouncing radio waves off an object and measuring the time it takes for the waves to return. Radar is an acronym for "radio detection and ranging."

Retention pond: A large man-made basin into which stormwater runoff drains until sediment and contaminants have settled out.

Retention tank: An enclosed container that stores stormwater runoff until sediment and contaminants have settled out.

Reverse-flow pump: A duplex pump designed to discharge solid waste without allowing it to come into contact with the pump's impellers.

Sewage pump: A device that draws wastewater from a sump and pumps it to the sewer line. It may be either a centrifugal pump or a pneumatic ejector.

Sewage removal system: An installation consisting of a sump, pump, and related controls that stores sewage wastewater from subdrains and lifts it into the main drain line.

Stormwater removal system: An installation consisting of a sump, pump, and related controls that stores stormwater runoff from a subdrain and lifts it into the main drain line.

Subdrain: A building drain located below the sewer line.

Sump: A container that collects and stores wastewater from a subdrain until it can be pumped into the main drain line.

Sump pump: A common name for a pump used in a stormwater removal system.

Plumbing systems dispose of wastewater through drains. Drains prevent contamination by sewage wastes, and prevent flooding from runoff water. Often, the design of a building makes it necessary for a plumber to locate a drain below the sewer line. These drains are called subdrains. Such drains can't discharge into the sewer by gravity. Instead, subdrains must empty into temporary holding pits. Pumps installed in the pits provide the lift to move wastewater up out of the pit and into higher drainage lines. Pits and pumps that handle sewage wastes are called sewage removal systems, or lift stations. Pits and pumps that handle clear-water runoff are called stormwater removal systems.

Roofs, paved areas, and yards all require drains. The water from these areas drains into a separate storm sewer system or into special flood-control areas, such as ponds and tanks. Many codes do not allow stormwater to be discharged into sewage lines. Install a storm sewer line to drain clear-water runoff. If the collection point for runoff must be deeper than the sewer drain line, install a stormwater removal system to lift the wastewater.

1.1.0 Sizing and Installing Sewage Pumps and Sumps

A complete sewage removal system consists of sewage pumps, sumps, and controls. A sewage pump creates a partial vacuum that draws the waste from the pit into the pump and forces it out with sufficient pressure to reach the sewer above. Sewage pumps are often called ejectors. A sump collects sewage from the subdrain. Sumps are also called wet wells or sump pits. Each installation has controls that operate the pump. Controls measure the amount of wastewater in the sump, turn the pump on and off, and provide backup in case of pump or power failure.

There are two distinct types of sewage removal systems: centrifugal and pneumatic. The names refer to the type of pump used to remove the waste. Each of these systems and their components are discussed in more detail in the following sections.

1.1.1 Sewage Pumps

Two types of pumps are used in sewage removal systems: centrifugal pumps and pneumatic ejectors. Both types of pump are designed for specific applications. Be sure to review the design of the building's waste system. This will help you select the most appropriate type of pump. Always follow the manufacturer's instructions when installing pumps and ejectors. Join pumps to the inlet and outlet piping with either a union or a flange. This permits the pump to be easily removed for maintenance, repair, or replacement. The following installation steps are common for most sewage and stormwater pumps:

Step 1 Before installing the pump in the basin, ensure the pump is appropriate for the sump.

Step 2 Check that the voltage and phase from the electrical supply matches the power requirements shown on the pump's nameplate.

Step 3 Thread the sump's discharge pipe into the pump's discharge connection.

Step 4 Install a check or backwater valve horizontally if one is required. If this is not possible, install it at a 45-degree angle with the pivot at the vertical, to prevent solids from clogging the valve.

Step 5 Drill a ³⁄₁₆-inch (or recommended size) hole in the discharge pipe about 2 inches above the pump discharge connection to prevent air from locking the pump.

Step 6 If required, install a gate valve in the system after the check valve to permit removal of the pump for servicing.

Step 7 Install a union above the high-water line between the check valve and the pump, to allow the pump to be removed without disturbing the piping.

Centrifugal pumps are frequently used in both stormwater and sewage removal systems. Centrifugal pumps consist of the pump, an electric motor to run the pump, a plate on which to mount the pump, and controls to turn the pump on and off at the proper time. Many pumps come with motors that are designed to be submersible, which means the entire pump assembly can be placed inside the sump (see *Figure 1*). Others have a motor that must be installed above the sump, with the pump body connected to the motor by a sealed shaft (*Figure 2*). Centrifugal pumps are widely used in sewage removal systems because of their mechanical simplicity and low cost. Complete centrifugal pump systems are available as self-contained units with pump, controls, and basin. They may also be fabricated on site from individual components.

Inside the body of a centrifugal pump is a set of spinning vanes called an impeller. An impeller removes wastewater from a sump in two steps. First, it draws wastewater up from the sump by creating a partial vacuum. It then ejects the swirling water out of the pump body into the discharge line. Centrifugal pumps are water lubricated, which means that they must always be in contact with water or they will suffer damage. If this happens, repair or replace the pump. Ensure that centrifugal pumps are primed before switching them on.

Centrifugal pumps can be designed with non-clogging and self-grinding features that efficiently handle solid wastes up to a certain size.

02307-14_F01.EPS

Figure 1 Submersible centrifugal sewage pump.

The *International Plumbing Code*® (IPC), for example, requires that pumps receiving wastes from water closets must be able to handle solids up to 2 inches in diameter. Other pumps must be able to pump solids up to 1 inch in diameter. Because sand and grit in the wastewater can cause significant wear on pump components, install sediment interceptors in the drain line. Centrifugal pumps may require maintenance to replace moving parts, such as the impeller. Install pumps so that they can be removed easily from sumps.

Sometimes a design may call for a backup capability in case a pump fails, or a design may require extra pumping capacity during periods of peak flow. In such cases, install a duplex pump to fulfill the requirement. A duplex pump consists of two complete centrifugal-pump assemblies, including floor plates, controls, and high-water alarms, in-

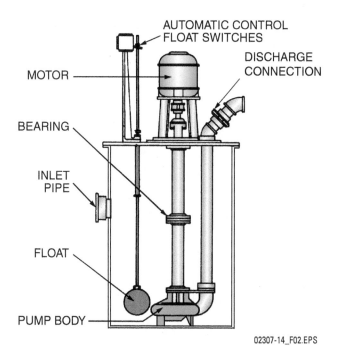

02307-14_F02.EPS

Figure 2 Centrifugal-pump installation with aboveground motor.

stalled in parallel in a sump (see *Figure 3*). Duplex pumps can be designed so that they operate on an alternating basis, reducing the wear on a single pump. Use an electrical pump alternator to switch between pumps in a duplex arrangement. Remember to size the sump to hold both pumps. Sump covers must be specially ordered to accommodate a duplex installation.

Rags and stringy solids may clog centrifugal pumps. Floating wastes and scum are out of the reach of a pump inlet that is located at the bottom of a sump. For these types of sewage wastes, use a reverse-flow pump arrangement. Reverse-flow pumps consist of a duplex pump installed with strainers, cutoff valves, and check valves (see *Figure 4*). Reverse-flow pumps are designed to discharge solid wastes without passing them through the impeller housing. Because reverse-flow pumps consist of two separate pumps, the system offers a safety backup in case one pump fails.

As wastewater enters a sump through one inlet, a strainer traps solids. When the water reaches a predetermined height, the pump switches on. The pump forces the trapped solids out of the strainer and through the discharge pipe. To prevent back pressure into the soil line when the pump is operating, install a full-opening, positive-seating, noncorroding check valve. Locate the check valve in the horizontal drain line (refer to *Figure 4*). The duplex arrangement allows wastewater to flow into the basin through one of the inlets while the

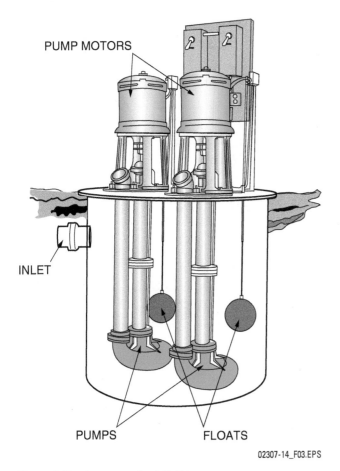

PUMP MOTORS

INLET

PUMPS   **FLOATS**

02307-14_F03.EPS

Figure 3 Duplex-pump installation.

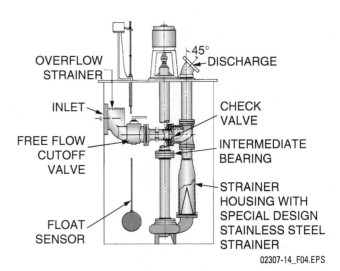

OVERFLOW STRAINER

INLET

FREE FLOW CUTOFF VALVE

FLOAT SENSOR

45°
DISCHARGE

CHECK VALVE

INTERMEDIATE BEARING

STRAINER HOUSING WITH SPECIAL DESIGN STAINLESS STEEL STRAINER

02307-14_F04.EPS

Figure 4 Reverse-flow pump installation.

other is closed during the pump discharge process. Install a pump alternator to switch between the two pumps. Occasionally, excess amounts of wastewater may surge into the basin, requiring both pumps to activate at the same time. To allow for this, install overflow strainers near the inlets for each pump. Overflow strainers allow waste-

water to enter the sump when both check valves are closed.

A pneumatic ejector is an alternative to a centrifugal pump. Plumbers often select pneumatic ejectors for large-capacity applications, such as housing subdivisions and municipal facilities. Unlike centrifugal pumps, pneumatic ejectors do not have moving parts. This means they require less maintenance than centrifugal pumps. However, they are more mechanically complex than centrifugal systems.

Sewage enters a pneumatic ejector through an inlet. When the wastes reach a predetermined height, a sensor triggers a compressor. The compressor forces air into the basin. The increased pressure causes the sewage to be moved into the discharge line and out of the basin. Pneumatic ejectors are equipped with check valves on the inlet and discharge lines. During waste ejection, the valves prevent back pressure. After ejection, they

History of the Pump

One of the earliest known pumps was the shaduf, developed in Egypt around 1550 BC. The shaduf was a lever that lifted water from a well. About 400 years later in Persia, people pumped water using a chain of pots tied to a long loop of rope. The loop was wrapped around a wheel. As the wheel turned, the pots scooped water from a well and brought it to the top. About 650 BC, Romans in Egypt pumped water with a similar device called a noria. This was a wheel with pots attached to the rim. When the wheel was lowered partly into the water and spun, the buckets dipped into the water. In 230 BC, the Greek inventor Archimedes designed a pump made out of a coil of pipe wrapped around a shaft. By dipping one end of the pipe into the water and turning the shaft, the water corkscrewed its way up the pipe coil and came out the top end. Water wheels and Archimedean screws are still widely used throughout the world.

The Byzantines used a piston and cylinder to lift water using the force of air pressure in the first century BC. Air pressure is also used in the centrifugal pump, which was invented by Leonardo da Vinci in the 1500s. It employs a set of spinning blades to force water into and out of a chamber. An engineer named Denis Papin designed the first practical centrifugal pump in 1688. The centrifugal pump is widely used in modern sewage and stormwater removal systems.

keep the waste from backflowing. When the ejector is empty, a diaphragm exhaust valve releases the compressed air and allows the chamber to refill with sewage (see *Figure 5*).

Pneumatic ejectors are designed as a unit. This means that basins, pumps, and controls are all part of a single manufactured package. Be sure to install the right-size pneumatic ejector for the anticipated load. Consult the construction drawings, and check your local code for requirements governing pneumatic-ejector installation. Install a union above the high-water line between the check valve and the pump to allow the pump to be removed for maintenance without disturbing the piping.

1.1.2 Sewage Sumps

A sump is a container that holds wastewater until it can be pumped into the sewer line. Plumbers are responsible for installing sumps. Installation includes all of the components within the sump, including pumps, controls, and inlet and outlet piping. Sumps can made of plastic, fiberglass (see *Figure 6*), and precast or site-built concrete (see *Figure 7*). Basin size and materials are specified in your local code. On pneumatic systems, install check valves on sump inlet piping, and install gate and check valves on discharge piping. Ensure that the valves are located outside the sump and are easily accessible. Residential installations may require only a check valve; refer to your local code. Also refer to your local applicable code before installing pumps in elevator shafts, as some codes prohibit this.

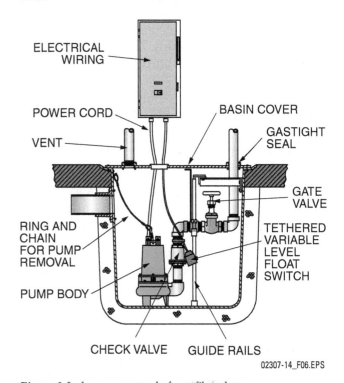

Figure 6 Indoor sump made from fiberglass.

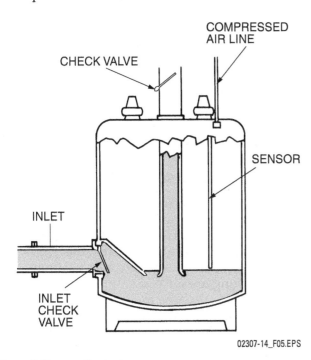

Figure 5 Pneumatic ejector.

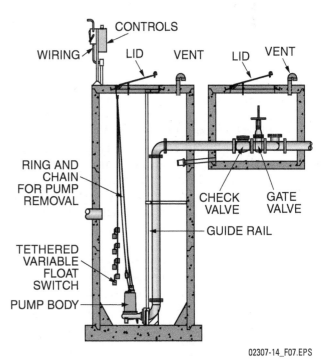

Figure 7 Outdoor sump basin made from concrete.

When designing sumps, remember to consider the drain flow rate from all of the fixtures draining into the subdrain. Calculate the flow in terms of the drainage fixture unit. The drainage fixture unit is a measure of a fixture's discharge in gallons per minute divided by 7.5 (the number of gallons in a cubic foot). Drainage-fixture-unit calculations take several factors into consideration:

- Rate of discharge
- Duration of discharge
- Average time between discharges

Table 1 shows drainage fixture units for fixture drains or traps, and *Table 2* provides drainage fixture units for fixtures in commercial and private installations.

Install airtight covers and vents on sewage sumps. Otherwise, sewer gases will escape and may pose a threat to health and safety. Consult local codes for specific venting requirements. Design the sewage removal system so that wastes are not retained for more than 12 hours. The maximum allowable height of wastewater in sumps is established in local codes. Ensure that the pump does not cycle more than six times per hour because excessive cycling could damage the pump.

1.2.0 Sizing and Installing Stormwater Pumps and Sumps

Stormwater removal systems are mainly used for flood control. The procedures for sizing and installing stormwater removal systems share some similarities with sewage removal systems. However, the applications of the two types of systems are very different. Stormwater pumps and basins are designed for clear-water wastes rather than wastes with solid matter. Local codes provide guidance on how to design stormwater removal systems.

Table 1 Sample Drainage Fixture Units for Fixture Drains or Traps

Fixture Drain or Trap Size (inches)	Drainage Fixture Unit Value
1¼	1
1½	2
2	3
2½	4
3	5
4	6

1.2.1 Stormwater Pumps

Pumps used in stormwater removal systems are often called sump pumps or bilge pumps. Install centrifugal sump pumps to lift stormwater from sumps to the drainage lines (see *Figure 8*). Like centrifugal sewage pumps, the motor of a centrifugal sump pump may be mounted either atop or inside the sump. Install pumps made of chlorinated polyvinyl chloride (CPVC) where the runoff contains corrosive chemicals.

Many local codes require gate and check valves on the discharge line (see *Figure 9*). Check valves prevent backflow into the basin. Aluminum-flapper check valves are often specified because of their quieter operation. In a single or duplex pump, the check valve should be installed in a horizontal section of the discharge line. Gate valves, which allow the plumber to shut off the discharge for maintenance, may be placed in either a horizontal or a vertical run of the discharge line.

Size the pump according to the anticipated amount of drainage. You will learn how to calculate rainfall rates later in this module. Provide a safety factor by ensuring that the pump has a capacity of 1½ times the maximum calculated flow. An additional pump can serve as a backup in case the primary pump fails. Power failures often occur during storms, which is when sump pumps are most needed. Consider installing a backup power source to keep the sump pumps running.

1.2.2 Stormwater Sumps

Stormwater sumps collect runoff from subdrains and store it until it can be pumped to storm sewers or to flood-control areas. Construct sumps from cast iron, fiberglass, or concrete. Apply waterproofing to the inside of concrete sumps. Provide a settling area inside sumps. This will allow sand and other coarse material to settle without being sucked up into the pump. Sediment can damage the pump. Sump covers do not need to be airtight; however, a secure lid will prevent objects from falling into the sump.

Table 2 Sample Drainage Fixture Units for Fixtures and Groups

Fixture Type	Drainage Fixture Unit Value as Load Factors	Minimum Size of Trap (inches)
Automatic clothes washers, commercial	3	2
Automatic clothes washers, residential	2	2
Bathroom group as defined in Section 202 (1.6 gpf water closet)[d]	5	—
Bathroom group as defined in Section 202 (water closet flushing greater than 1.6 gpf)[d]	6	—
Bathtub[a] (with or without overhead shower or whirpool attachments)	2	1½
Bidet	1	1¼
Combination sink and tray	2	1½
Dental lavatory	1	1¼
Dental unit or cuspidor	1	1¼
Dishwashing machine, domestic	2	1½
Drinking fountain	½	1¼
Emergency floor drain	0	2
Floor drains	2	2
Floor sinks		2
Kitchen sink, domestic	2	1½
Kitchen sink, domestic with food waste grinder and/or dishwasher	2	1½
Laundry tray (1 or 2 compartments)	2	1½
Lavatory	1	1¼

Shower (based on the total flow rate through showerheads and body sprays)
 Flow rate:
   5.7 gpm or less
   Greater than 5.7 gpm to 12.3 gpm
   Greater than 12.3 gpm to 25.8 gpm
   Greater than 25.8 gpm to 55.6 gpm
	
	2
	3
	5
	6
		
		1½
		2
		3
		4

Service sink	2	1½
Sink	2	1½
Urinal	4	Note [b]
Urinal, 1 gallon per flush or less	2[c]	Note [b]
Urinal, nonwater supplied	½	Note [b]
Wash sink (circular or multiple) each set of faucets	2	1½
Water closet, flushometer tank, public or private	4[c]	Note [b]
Water closet, private (1.6 gpf)	3[c]	Note [b]
Water closet, private (flushing greater than 1.6 gpf)	4[c]	Note [b]
Water closet, public (1.6 gpf)	4[c]	Note [b]
Water closet, public (flushing greater than 1.6 gpf)	6[c]	Note [b]

a. A showerhead over a bathtub or whirlpool bathtub attachment does not increase the drainage fixture unit value.
b. Trap size shall be consistent with the fixture outlet size.
c. For the purpose of computing loads on building drains and sewers, water closets and urinals shall not be rated at a lower drainage fixture unit unless the lower values are confirmed by testing.
d. For fixtures added to a bathroom group, add the dfu value of those additional fixtures to the bathroom group fixture count.

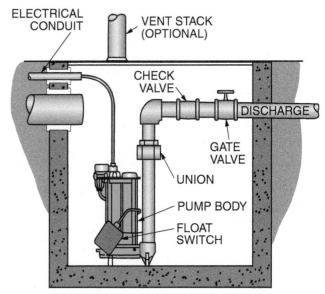

Figure 9 Check and gate valves in a centrifugal-pump installation.

02307-14_F08.EPS

Figure 8 Submersible centrifugal sump pump.

Sizing sumps properly will ensure that they perform their function efficiently. Sizing a storm-water sump is similar to the process for sizing a sewage sump. Keep two objectives in mind when sizing a sump:

- The basin must be large enough to receive the anticipated amount of stormwater.
- The basin must be large enough to eliminate frequent cycling of the pump motor.

Ensure that the basin's high-water mark is at least 3 inches below the subdrain inlet. The low-level mark must be no less than 6 inches above the bottom of the basin. This will ensure that the pump's suction end is always underwater.

Size the sump so that the pump motor does not activate more than once every five minutes. For example, if wastewater flows into a sump at 60 gallons per minute (gpm) and if the pump cycles every five minutes, then the basin has to hold at least 300 gallons (5 × 60) of wastewater.

How big will a sump have to be to handle 300 gallons of wastewater? You can size basins by using the formula for finding the volume of a cylinder. The formula is:

$$V = (\pi r^2)h$$

In other words, the volume is the product of the area of the basin's floor and its height. When converting the volume to gallons, remember that there are 7.5 gallons in a cubic foot. Using the formula, a 4-foot-diameter cylinder would have to be 3.18 feet, or 3 feet 2 inches tall, to hold 300 gallons. Adding the required 3-inch minimum at the top and 6-inch minimum at the bottom, the sump should be at least 3 feet 11 inches from the bottom of the inlet pipe to the base of the sump.

Local codes will provide the rate of wastewater collection and area rainfall rates to use when calculating the size of the removal system. Unless otherwise specified, use a collection rate of 2 gpm for every 100 square feet of sandy soil and 1 gpm for clay soils. Flow rates for roof and driveway runoff will vary considerably. For roofs and driveways, use a collection rate of 1 gpm for every 24 square feet of surface area. Local codes include specific guidelines for determining roof and driveway runoff rates.

To calculate the average amount of rain in the area, consult a table or map of local area rainfall. Tables and maps are printed in model and local codes, civil engineering handbooks, and landscape architecture manuals. You can also get information from the US Department of Agriculture or the National Weather Service of the National Oceanic and Atmospheric Administration (NOAA). The National Weather Service issues 100-year 1-hour rainfall maps for the United States. These maps show the average recorded rainfall in 1 hour for a given area over the previous century. See the *Appendix* for 100-year 1-hour rainfall maps for different parts of the country.

Design the removal system to discharge runoff into the building storm sewer. Local code may also allow the waste to discharge into a street gutter or curb drain. Some older installations connect storm drains with sewage waste lines. Many codes now prohibit this type of connection. Many cities and counties maintain retention ponds and retention tanks for storm drainage. They are used for flood control. Retention ponds are large man-made basins where sediment and contaminants settle out of the wastewater before the water returns to rivers, streams, and lakes. Retention tanks are similar to retention ponds, except they are enclosed. Install backwater valves on the drainage line to prevent contamination of the sump in case a retention pond or tank overflows.

1.3.0 Installing Sewage and Stormwater-Pump Controls

Sewage removal systems perform a critical role in plumbing installations. Proper sizing and installation will help ensure that a system will not suffer from malfunctions and breakdowns. When possible, provide a backup pump in case the lead pump fails. For pneumatic-ejector installations, provide a backup compressor. Power failures will prevent centrifugal pumps and pneumatic ejectors from working. Your local applicable code may require the installation of power-off alarms for power failures and similar conditions. If possible, provide a backup power supply. Remember that plumbers are responsible for helping to maintain public safety through the fast and efficient removal of wastes. These extra steps will allow the system to continue operating while plumbers are called in to make repairs.

1.3.1 Switches

When wastewater in a sump reaches a preselected height, a switch in the basin activates the pump. Switches are also called sensors. Sewage removal systems use analog switches or digital switches. An analog switch is operated mechanically or electrically. A digital switch is operated electronically. This section discusses both types of switches used in sump systems.

There are four types of analog switches:

- Float switch
- Pressure float switch
- Mercury float switch
- Probe switch

A float switch works a lot like the float in a water-closet tank. As the wastewater level rises, the float rises within the sump. An arm attached to the float activates the pump when the float reaches a specified level. Set the float to activate the pump at the highest water level possible inside the sump. This will prevent excessive cycling of the pump. Consider installing a standby pump or backup power system for emergency situations. An alarm will provide occupants with a warning signal during an emergency condition like a pump failure. Review the installation requirements and construction drawings to see if these measures are appropriate.

A pressure float switch (*Figure 10*) is an adjustable type of float switch. Install pressure float switches to rise with wastewater or clamp them to the basin wall.

Mercury float switches are a specialized type of float sensor that plumbers can install at various heights within the basin. Although most manufacturers have discontinued the manufacture of mercury switches due to the risk that mercury poses to the environment and to health, they can still be found in many existing installations. It is important for plumbers to know how to service them when encountered in the field. Usually two mercury float switches are required, and a third switch can serve as a high-water alarm if desired. If more than one pump is used in the basin, four mercury float switches will typically be installed. Always follow

02307-14_F10.EPS

Figure 10 Pressure float switches in a sewage sump basin.

---

the manufacturer's instructions when working with mercury switches, and refer to your local applicable code for proper disposal of old mercury switches. Wear appropriate personal protective equipment when handling or working with mercury switches.

Probe switches, also called electrodes, are used in pneumatic ejectors. They work similarly to thermostatic probes installed on water heaters. When wastewater in the sump reaches a certain height, the probe triggers the compressor. When the waste is ejected, the switch resets.

> **WARNING!**
>
> Review the manufacturer's instructions before attempting to repair or replace a pump. Ensure that valves are closed and that electrical connections to the pump motor and controls have been turned off and tagged out. Wear appropriate personal protective equipment. Do not work in a walk-in stormwater sump without prior knowledge of safe confined-space procedures.

In recent years, manufacturers have introduced digital technologies for monitoring wastewater levels in sumps. These include lasers, field effect sensors, and even radar.

A laser system works by beaming a concentrated light beam at a target across the open area of the sump. When rising water interrupts the beam, the pump is triggered to drain the wastewater.

Field-effect sensors are electronic detectors that are designed to respond to the presence of materials that can conduct electricity, such as water. When a field-effect sensor is electrically charged, it emits a weak electrical field. When a conductive substance passes through the field—for example, water rising in a sump basin—the sensor sends a signal to the pump to turn it on.

A radar system broadcasts a continuous series of radio-wave pulses from the top of the tank down toward the surface of the wastewater, and measures the amount of time it takes for the signal to bounce back to the transmitter. As the water level in the sump rises, the time between the transmission and the return echo decreases. When the time decreases to a preset point, the pump is turned on until the radar

detects that the wastewater level has returned to a preset height. Radar level-measurement systems generally require less maintenance because they are less prone to failure from clogging, solid-waste buildup, and foaming in the wastewater.

1.3.2 Other Controls

Project specifications may call for the installation of battery backup pumps in sewage and stormwater systems (see *Figure 11*). As their name suggests, these pumps are battery powered and are designed to operate during a power failure or when the capacity of the main pump is exceeded.

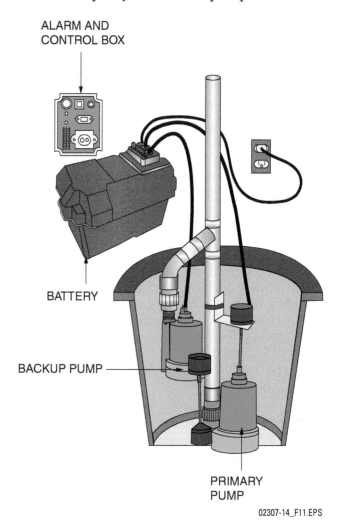

ALARM AND
CONTROL BOX

BATTERY

BACKUP PUMP

PRIMARY
PUMP

02307-14_F11.EPS

Figure 11 Battery backup pump with charger.

Battery backup pumps are fitted with their own controls and many come with an alarm as well. Many battery backup pumps are designed to use rechargeable batteries. These can be connected to the main electrical service to ensure they are fully charged when needed.

Manufacturers offer pre-assembled sewage and stormwater sump systems that include the pump, controls, alarms, piping, and installation hardware in a single package (see *Figure 12*). Pre-assembled systems are available in a variety of sizes and are designed for ease of installation. The sump basins are typically made from a durable plastic. Because they are self-contained, pre-assembled systems are typically rated for both indoor and outdoor use.

02307-14_F12.EPS

Figure 12 Cut-away of a pre-assembled sewage system.

Pumps in History

Pumps have played a vital supporting role in some of history's greatest inventions and scientific discoveries. For example, in 1660, English chemist Robert Boyle (1627–1691) published a book called *The Spring and Weight of the Air* in which he used a hand pump of his own design to remove the air from a large sealed chamber. He placed a bell in the chamber and showed how the sound of the bell faded as more air was pumped out. He also placed a burning candle in the chamber and showed how it snuffed out as air was pumped out. His experiments with air pressure shattered many widely held beliefs about the nature of air. His work encouraged others to take up scientific experiments. Boyle's use of pumps to discover the nature of air is considered one of the most important examples of how empirical, or experimental, science can lead to new discoveries.

As a young man, the Scottish inventor and engineer James Watt (1736–1819) became interested in steam-powered pumps. Miners used primitive and inefficient steam-powered piston pumps to suck water out of England's coal mines. Watt did research to find ways to improve how the steam pumps worked. His research led him to discover the properties of steam, which in turn led to an improved steam-engine design that eventually powered the Industrial Revolution in the nineteenth century.

Additional Resources

Building Services Engineering: A Review of Its Development. 1982. Neville S. Billington. New York: Pergamon Press.

International Plumbing Code®, Latest Edition. Falls Church, VA: International Code Council.

Pipefitters Handbook. 1967. Forrest R. Lindsey. New York: Industrial Press Inc.

1.0.0 Section Review

1. Centrifugal pumps are lubricated by _____.
   a. hydraulic fluid
   b. air
   c. oil
   d. water

2. The height of the low-level mark in the bottom of a stormwater sump must be no less than _____.
   a. 6 inches
   b. 8 inches
   c. 10 inches
   d. 12 inches

3. Probe switches are also called _____.
   a. mercury switches
   b. sensors
   c. electrodes
   d. float switches

2.0.0 TROUBLESHOOTING AND REPAIRING SEWAGE AND STORMWATER REMOVAL SYSTEMS

Objective

Explain how to troubleshoot and repair sewage and stormwater removal systems.
a. Explain how to troubleshoot electrical problems.
b. Explain how to troubleshoot mechanical problems.
c. Explain how to replace pumps.

Performance Task

Troubleshoot sewage and sump pumps.

Trade Terms

Air lock: A pump malfunction caused by air trapped in the pump body, interrupting wastewater flow.

Run dry: To operate a pump without water.

Occasionally, pumps break down. When they do, they must be repaired or replaced. A malfunctioning pump usually means an entire drainage system must be shut down until the pump can be repaired. Quick and efficient troubleshooting and repair or replacement is essential. Plumbers can perform most pump repairs. Pump problems can usually be traced to one or both of the following causes:

- Electrical, involving the pump motor and wiring
- Mechanical, involving the pump's moving parts

The most common types of problems and their solutions are discussed in the following sections.

2.1.0 Troubleshooting Electrical Problems

Most problems with sewage and sump pumps can be traced to failure of the pump's electric motor. Power failures often happen during storms, when sump pumps are needed most. The loss of electrical power causes the pump to stop working, which puts the structure at risk of flooding. If the water level rises to the level of the pump motor, remove the electrical cord from the power source. If the motor is in danger of getting wet, it may be necessary to remove it from the pump. If a motor gets wet, ensure that it is thoroughly dry before connecting it to a power source. Consider installing a battery backup pump to operate the removal system during power outages or to assist the main pump if its capacity is exceeded.

If a pump is not running, follow these steps to diagnose the problem:

Step 1 Check the power source to see if the voltage level is the same as the pump's required voltage. This is specified on the pump's nameplate.

Step 2 Check all fuses, circuit breakers, and electrical connections for breaks or loose electrical connections.

Step 3 Inspect the power cord. If the power-cord insulation is damaged, replace the cord with a new one of the same rating. If the power cord is open or grounded, you will need to contact an electrician to check the resistance between the cord's hot and neutral leads.

Occasionally, the impeller will lock. Have an electrician check the amps drawn by the pump motor. If the motor is drawing more amps than permitted by the manufacturer, the impeller may be blocked, the bearings may be frozen, or the impeller shaft may be bent. Remove the pump for inspection, and repair or replace the damaged part. If a motor's overload protection has tripped, contact an electrician to inspect the motor.

> **WARNING!**
> Electrical pump motors that have been exposed to water may cause electrical shock when run. Wear appropriate personal protective equipment, and ensure that electric motors are thoroughly dry before testing.

Pumps sometimes fail to deliver their rated capacity because of electrical problems. If the impeller appears to be operating below normal speed, check the voltage supply against the motor's required voltage. Sometimes backflow can cause the impeller to rotate in the wrong direction. For single-phase electrical motors, simply shut off the power and allow the impeller to stop rotating. Then turn the pump on again; the problem should correct itself. For three-phase electrical motors, contact an electrician to inspect the motor.

2.2.0 Troubleshooting Mechanical Problems

Nonelectrical, mechanical problems with pumps include worn, damaged, or broken controls or pump components. Float sensors are a common culprit in mechanical failures. Carefully inspect the floats for corrosion and sediment buildup, which can cause floats to stick. To test if a float switch is broken, bypass the switch. If the pump operates, the switch will need to be replaced. Ensure the float is properly adjusted and that it has not slipped from its desired location. In any case, when repairing a waste removal system, always recalibrate and test the floats for proper operation.

Pneumatic ejectors use a rubber diaphragm exhaust valve to allow compressed air to escape and wastewater to resume filling the sump. If an ejector will not shut off, inspect the diaphragm and its switch. If the switch is broken or the rubber diaphragm is weak, replace the part immediately. Clean the areas around the diaphragm to remove any sediment lodged between the retainer ring and the diaphragm.

Broken pipes can damage pumps. If a pipe feeding the pump breaks and drains, the pump will continue to run. Pumps that are designed to be primed should not be allowed to run dry as a result of a broken pipe. This will damage the pump and it will have to be repaired or even replaced as a result. If there is evidence that a pump has been damaged from running dry, examine the length of the pipe supplying the pump for leaks, and repair them using the installation and repair techniques you have already learned.

Plugged vent pipes can also prevent a pump from shutting off. Inspect the vent pipe and clear it out if necessary. The pump may have an air lock, which means that air trapped in the pump body is interfering with the flow of wastewater. Turn off the pump for about one minute, then restart it. Repeat this several times until the air is cleared from the pump. If the removal system has a check valve, ensure that a ³⁄₁₆-inch hole has been drilled in the discharge pipe about 2 inches above the connection. The wastewater inflow may match the pump's capacity. If this is the case, recalculate the amount of wastewater entering the system and install a suitable pump.

> **WARNING!**
> Stormwater runoff may contain corrosive chemicals such as oil and grease. Wear appropriate personal protective equipment when performing maintenance on stormwater sumps.

If the pump runs but won't discharge wastewater, the check valves may have been installed backward. Inspect the flow-indicating arrow on the check valves. If the check valve is still suspect, remove it and test to make sure it is not stuck or plugged. Check the pump's rating to ensure the lift is not too high for the pump. Inspect the inlet to the impeller and remove any solids that may be blocking it. Turn the pump on and off several times to remove an air lock.

When a pump is not delivering its rated capacity, check to see if the impeller is rotating in the wrong direction. Wastewater can drain back into a sump from a long discharge pipe if a check valve is not installed. Reverse flow will cause the impeller to rotate backward. Turn the pump off until the impeller stops rotating. Install a check valve on the discharge line to prevent the problem from recurring. Inspect the impeller for wear due to abrasives or corrosion, and replace a worn impeller immediately. Then turn the pump back on. Finally, check the pump rating to ensure the lift is not too high for the pump.

If the pump cycles continually, the problem could again be wastewater draining back into the basin from the discharge line. Install a check valve in the discharge line. If a check valve is already installed, inspect it for leakage. Repair or replace leaking valves. The sump may be too small for the wastewater inflow. Recalculate the inflow, and if necessary, install a larger basin.

Odors associated with a sewage removal system may mean that the sump lid is not properly sealed. Check the cover to see if its seal is intact. Inspect the vent to ensure the sump has open-air access. Some commercial structures, such as self-service laundries, use special lint traps. If the sewer lines are backed up, check to make sure all the lint traps are clean.

2.3.0 Replacing Pumps

If a pump cannot be repaired, it will have to be replaced. When replacing a sewage or stormwater pump, always remember to install a properly sized replacement. If a pump repeatedly breaks down, it may be the wrong size or type. Review the construction drawings, plumbing installation design, and removal-system specifications. If necessary, have an engineer appraise the design and installation. These steps will ensure that the pump is matched with the job it is supposed to perform.

Many pumps are water lubricated. This means that they must be immersed in water before they can be tested or run. Pumps are often self-priming. This means they automatically fill with water before operating. Failure to run the pump in water could burn out the pump's bearings quickly. Follow manufacturer's specifications closely when installing and testing sump and sewage pumps.

> **WARNING!**
> When removing a pump for servicing, ensure the wires have been resealed before returning the pump to service. Exposure to gaseous waste could damage the pump controls. A spark could cause an explosion.

> **WARNING!**
> Review Occupational Safety and Health Administration (OSHA) confined-space safety protocols before working in a walk-in sump. Wear appropriate personal protective equipment.

Plumbing Code Requirements for Sewage and Stormwater Removal Systems

Local plumbing codes govern the installation and operation of sewage and stormwater removal systems. Many local codes are based on one of the model plumbing codes. Always refer to your local code before installing sewage and stormwater removal systems. Here are some general guidelines from two model codes: the *International Plumbing Code®* (IPC) and the *Uniform Plumbing Code™* (UPC).

The UPC requires sewage removal pumps in single dwelling units to be able to pass a 1½-inch ball and have a discharge capacity of at least 20 gallons per minute (gpm). Lines serving sewage removal pumps are required to be fitted with an accessible backwater or swing check valve and a gate valve.

According to the UPC, stormwater removal systems require sumps if the drainage cannot be conveyed by gravity. The minimum dimensions of a sump are 15 inches in diameter and 18 inches deep, with a minimum discharge capacity of 15 gpm. The discharge pipe must be at least 1½ inches in diameter and fitted with a quick-disconnect fitting to ensure the pump is accessible for servicing.

The IPC requires sump pits to be covered with a gastight lid and be vented. Dimensions of the basin must be at least 18 inches in diameter and 24 inches deep. Effluent (the system's outlet) must be kept at least 2 inches below the gravity drain system's inlet. Accessible check valves and full-open valves are required on the discharge piping between the pump and the drainage system.

Additionally, the IPC mandates minimum ejector capacities: for discharge pipes 2 inches in diameter, the capacity must be 21 gpm; for discharge pipes 2½ inches in diameter, the capacity must be 30 gpm; and for discharge pipes 3 inches in diameter, the capacity must be 46 gpm.

The IPC's stormwater removal systems require subdrains to discharge into a sump serviced by an appropriately sized pump. Sumps must be at least 18 inches in diameter and 24 inches deep. Electrical service outlets must conform to National Fire Protection Association (NFPA) standards. Discharge pipes require a gate valve and full-flow check valve.

The First Modern Sanitary Engineer

New York engineer Julius W. Adams laid the foundations for Brooklyn's sewer system in 1857. He ended up inventing a system that defined modern sanitary engineering. His design provided sewage drains for all of Brooklyn's 20 square miles. Adams used technology already in use by other cities—reservoirs feeding into underground mains that branched off to connect commercial and residential buildings. However, Adams did something that no one had done before: he distilled his experience into a sewer design manual, which he then published. For the first time, cities throughout the country could consult a standard reference for sewer design.

Additional Resources

Handbook of Materials Selection. 2002. Myer Kutz, ed. New York: J. Wiley.
Introduction to Drain, Waste, and Vent (DWV) Systems, Fourth Edition. 2012. Upper Saddle River, NJ: Pearson.
Water, Sanitary, and Waste Services for Buildings. 2002. Alan F. E. Wise and J. A. Swaffield. Boston, MA: Butterworth-Heinemann.

2.0.0 Section Review

1. An impeller can be forced to rotate in the wrong direction due to _____.

    a. low pressure
    b. an electrical short
    c. alternated current
    d. backflow

2. In a pneumatic ejector, compressed air is allowed to escape through a _____.

    a. brass gate valve
    b. plastic butterfly valve
    c. rubber diaphragm exhaust valve
    d. latex diaphragm inlet check valve

3. Pumps that automatically fill with water before operating are called _____.

    a. self-priming
    b. auto-priming
    c. dry-running
    d. self-running

SUMMARY

Sewage and stormwater removal systems are important components of plumbing installation. They pump wastewater in subdrains into the sewer line, where they can discharge into a public sewer, retention pond, or retention tank. Pneumatic ejectors and centrifugal pumps are used to lift wastewater from basins into the discharge line. Sumps hold the wastewater until the pump discharges it. Switches installed on the inside of the basin trigger the pump when the wastewater reaches a preselected height.

Plumbers are responsible for maintaining, repairing, and replacing pumps. Electrical and mechanical problems can be diagnosed by inspecting various components. Always remember to take appropriate precautions when working with electrical pumps near water. Consult with engineers and electricians if the problem requires more than the repair or replacement of a part or a whole pump. To protect yourself against harmful solids and liquids in wastewater, always wear appropriate personal protective equipment when working with sewage and sump pumps. Because sewage and stormwater removal systems help maintain public health and safety, respond to pump problems quickly and professionally.

1. Pumps and pits that move sewage up to a higher-level sewer line are called _____.
   a. uplifts
   b. elevation units
   c. lift stations
   d. boosters

2. Stormwater runoff from paved areas can be directed into storm sewers or into _____.
   a. tempering basins
   b. ponds or tanks
   c. relief reservoirs
   d. overflow structures

3. Another term for a sewage pump is _____.
   a. ejector
   b. centrifuge
   c. turbine
   d. injector

4. To prevent air from locking a pump, drill a 3/16-inch-diameter hole _____.
   a. ½ inch above the pump discharge connection
   b. ½ inch above the pump inlet connection
   c. 2 inches above the pump discharge connection
   d. 2 inches above the pump inlet connection

5. The motors used on many centrifugal sewage pumps are _____.
   a. air primed
   b. rated for industrial applications
   c. fitted with more than one impeller stage
   d. submersible

6. To capture sand and grit in wastewater before it enters the pump, a drain line can be fitted with a _____.
   a. sand trap
   b. sediment interceptor
   c. grit filter
   d. sediment scrubber

7. As specified in the *International Plumbing Code*®, a pump that receives waste from a water closet must be able to handle solids with a diameter up to _____.
   a. 2 inches
   b. 1½ inches
   c. 1 inch
   d. ¾ inch

8. A pneumatic injector requires less maintenance than a centrifugal pump because it _____.
   a. is more ruggedly constructed
   b. has fewer moving parts
   c. has no moving parts
   d. has a built-in sediment filter

9. Sewage sumps may be made of concrete, plastic, or _____.
   a. steel
   b. fiberglass
   c. brick or block
   d. aluminum

10. A sewage removal system should be designed to retain wastes for not longer than _____.
    a. 4 hours
    b. 8 hours
    c. 12 hours
    d. 24 hours

11. Analog switches are classified into _____.
    a. three types
    b. four types
    c. six types
    d. eight types

12. In areas subject to power outages, the sump pump should be supplemented by installing a _____.
    a. larger sump
    b. hand-operated pump
    c. compressor
    d. battery backup pump

13. If a pump impeller is rotating at a slower-than-normal speed, the problem may be caused by _____.

    a. sediment clogging
    b. insufficient voltage
    c. a leaking or broken input line
    d. an electrical short

14. Mechanical failure in a pump can often be traced to _____.

    a. fluid lock
    b. a reversed check valve
    c. a float sensor
    d. sediment buildup

15. If a pump breaks down frequently, it may _____.

    a. be the wrong size or type
    b. need different controls
    c. have a manufacturing defect
    d. be improperly installed

Trade Terms Quiz

Fill in the blank with the correct term that you learned from your study of this module.

1. A(n) _____ is an enclosed container that stores stormwater runoff until sediment and contaminants have settled out.

2. A building drain located below the sewer line is called a(n) _____.

3. A(n) _____ is a switch that is operated mechanically or electrically.

4. An adjustable float switch that can either be allowed to rise with wastewater or be clipped to the sump wall is called a(n) _____.

5. A(n) _____, commonly known by the alternative name _____, is a device that draws wastewater from a sump and pumps it to the sewer line.

6. The system that determines distance by bouncing radio waves off an object and measuring the time it takes for the waves to return is called _____.

7. A(n) _____ is a set of vanes or blades that draws wastewater from a sump and ejects it into the discharge line.

8. A pump malfunction caused by air trapped in the pump body, interrupting wastewater flow, is called a(n) _____.

9. A(n) _____, also known as a(n) _____, is an installation consisting of a sump, pump, and related controls that stores sewage wastewater from subdrains and lifts it into the main drain line.

10. A container that collects and stores wastewater from a subdrain until it can be pumped into the main drain line is called a(n) _____.

11. To _____ is to operate a pump without water.

12. A common name for a pump used in a stormwater removal system is _____.

13. A(n) _____ is a pump that uses compressed air to force wastewater from a sump into the drain line.

14. The change in an electrical state that occurs when a conductive material such as water passes through an electric field is called the _____.

15. A(n) _____ is an electrical switch used in pneumatic ejectors that closes on contact with wastewater.

16. An installation consisting of a sump, pump, and related controls that stores stormwater runoff from a subdrain and lifts it into the main drain line is called a(n) _____.

17. A(n) _____ is a device that uses an impeller powered by an electric motor to draw wastewater out of a sump and discharge it into a drain line.

18. A switch that is operated electronically is called a(n) _____.

19. A(n) _____ is a float switch filled with mercury that activates the pump when it reaches a specified height in a basin.

20. A duplex pump designed to discharge solid waste without allowing it to come into contact with the pump's impellers is called a(n) _____.

21. A(n) _____ is a container filled with a gas or liquid that measures wastewater height in a sump and activates the pump when the float reaches a specified height.

22. An arrangement of two centrifugal pumps and their related equipment installed in parallel in a sump is called a(n) _____.

23. A(n) _____ is a large man-made basin into which stormwater runoff drains until sediment and contaminants have settled out.

Trade Terms

Air lock	Lift station	Run dry
Analog switch	Mercury float switch	Sewage pump
Centrifugal pump	Pneumatic ejector	Sewage removal system
Digital switch	Pressure float switch	Stormwater removal system
Duplex pump	Probe switch	Subdrain
Ejector	Radar	Sump
Field effect	Retention pond	Sump pump
Float switch	Retention tank	
Impeller	Reverse-flow pump	

Trade Terms Introduced in This Module

Air lock: A pump malfunction caused by air trapped in the pump body, interrupting wastewater flow.

Analog switch: A type of switch that is operated mechanically or electrically.

Centrifugal pump: A device that uses an impeller powered by an electric motor to draw wastewater out of a sump and discharge it into a drain line.

Digital switch: A type of switch that is operated electronically.

Duplex pump: Two centrifugal pumps and their related equipment installed in parallel in a sump.

Ejector: A common alternative name for a sewage pump.

Field effect: A change in an electrical state that occurs when a conductive material such as water passes through an electric field.

Float switch: A container filled with a gas or liquid that measures wastewater height in a sump and activates the pump when the float reaches a specified height.

Impeller: A set of vanes or blades that draws wastewater from a sump and ejects it into the discharge line.

Lift station: A popular name for a sewage removal system. The term is also used to refer to a pumping station on a main sewer line.

Mercury float switch: A float switch filled with mercury. It activates the pump when it reaches a specified height in a basin.

Pneumatic ejector: A pump that uses compressed air to force wastewater from a sump into the drain line.

Pressure float switch: An adjustable float switch that can either be allowed to rise with wastewater or be clipped to the sump wall.

Probe switch: An electrical switch used in pneumatic ejectors that closes on contact with wastewater.

Radar: A system that determines distance by bouncing radio waves off an object and measuring the time it takes for the waves to return. Radar is an acronym for "radio detection and ranging."

Retention pond: A large man-made basin into which stormwater runoff drains until sediment and contaminants have settled out.

Retention tank: An enclosed container that stores stormwater runoff until sediment and contaminants have settled out.

Reverse-flow pump: A duplex pump designed to discharge solid waste without allowing it to come into contact with the pump's impellers.

Run dry: To operate a pump without water.

Sewage pump: A device that draws wastewater from a sump and pumps it to the sewer line. It may be either a centrifugal pump or a pneumatic ejector.

Sewage removal system: An installation consisting of a sump, pump, and related controls that stores sewage wastewater from subdrains and lifts it into the main drain line.

Stormwater removal system: An installation consisting of a sump, pump, and related controls that stores stormwater runoff from a subdrain and lifts it into the main drain line.

Subdrain: A building drain located below the sewer line.

Sump: A container that collects and stores wastewater from a subdrain until it can be pumped into the main drain line.

Sump pump: A common name for a pump used in a stormwater removal system.

100-YEAR, 1-HOUR RAINFALL IN INCHES

02307-14_A01.EPS

Figure A1 Western United States.

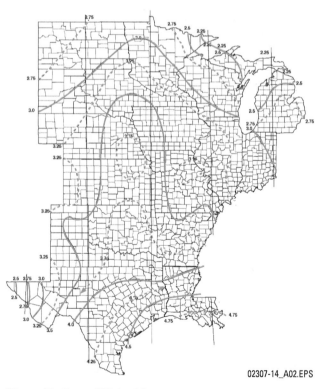

02307-14_A02.EPS

Figure A2 Central United States.

02307-14_A03.EPS

Figure A3 Eastern United States.

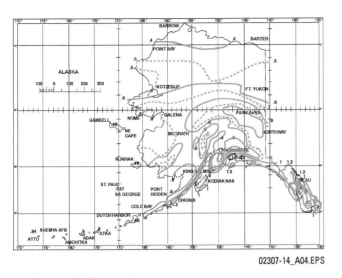

Figure A4 Alaska.

02307-14_A04.EPS

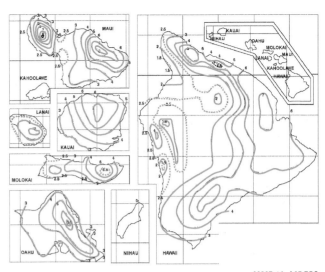

Figure A5 Hawaii.

02307-14_A05.EPS

Additional Resources

This module presents thorough resources for task training. The following resource material is suggested for further study.

Building Services Engineering: A Review of Its Development. 1982. Neville S. Billington. New York: Pergamon Press.

Handbook of Materials Selection. 2002. Myer Kutz, ed. New York: J. Wiley.

International Plumbing Code®, Latest Edition. Falls Church, VA: International Code Council.

Introduction to Drain, Waste, and Vent (DWV) Systems, Fourth Edition. 2012. Upper Saddle River, NJ: Pearson.

Pipefitters Handbook. 1967. Forrest R. Lindsey. New York: Industrial Press Inc.

Water, Sanitary, and Waste Services for Buildings. 2002. Alan F. E. Wise and J. A. Swaffield. Boston, MA: Butterworth-Heinemann.

Figure Credits

Answer	Section Reference	Objective
Section One		
1. d	1.1.1	1a
2. a	1.2.2	1b
3. c	1.3.1	1c
Section Two		
1. d	2.1.0	2a
2. c	2.2.0	2b
3. a	2.3.0	2c

NCCER CURRICULA — USER UPDATE

NCCER makes every effort to keep its textbooks up-to-date and free of technical errors. We appreciate your help in this process. If you find an error, a typographical mistake, or an inaccuracy in NCCER's curricula, please fill out this form (or a photocopy), or complete the online form at **www.nccer.org/olf**. Be sure to include the exact module ID number, page number, a detailed description, and your recommended correction. Your input will be brought to the attention of the Authoring Team. Thank you for your assistance.

Instructors – If you have an idea for improving this textbook, or have found that additional materials were necessary to teach this module effectively, please let us know so that we may present your suggestions to the Authoring Team.

NCCER Product Development and Revision
13614 Progress Blvd., Alachua, FL 32615

Email: curriculum@nccer.org
Online: www.nccer.org/olf

❏ Trainee Guide   ❏ Lesson Plans   ❏ Exam   ❏ PowerPoints   Other _____

Craft / Level: _____   Copyright Date: _____

Module ID Number / Title: _____

Section Number(s): _____

Description: _____

_____

_____

_____

Recommended Correction: _____

_____

_____

_____

Your Name: _____

Address: _____

_____

Email: _____   Phone: _____

02308-14

Corrosive-Resistant Waste Piping

OVERVIEW

Unlike household chemical wastes such as bleaches and detergents, which are not harsh enough to damage drainage pipes, corrosive wastes created by industrial, commercial, and laboratory processes require a separate removal system. The codes and standards governing these systems are very strict. Plumbers are responsible for installing waste piping that safely discharges corrosive wastes.

Module Seven

Trainees with successful module completions may be eligible for credentialing through the NCCER Registry. To learn more, go to **www.nccer.org** or contact us at **1.888.622.3720**. Our website has information on the latest product releases and training, as well as online versions of our *Cornerstone* magazine and Pearson's product catalog.

Your feedback is welcome. You may email your comments to **curriculum@nccer.org**, send general comments and inquiries to **info@nccer.org**, or fill in the User Update form at the back of this module.

This information is general in nature and intended for training purposes only. Actual performance of activities described in this manual requires compliance with all applicable operating, service, maintenance, and safety procedures under the direction of qualified personnel. References in this manual to patented or proprietary devices do not constitute a recommendation of their use.

Objectives

When you have completed this module, you will be able to do the following:

1. Identify corrosive wastes and handle them safely.
   a. Identify the different types of corrosive waste.
   b. Identify the safety issues and hazard communication systems related to working with corrosive-resistant waste.
2. Explain how to join and install different types of corrosive-resistant waste piping.
   a. Explain how to join and install borosilicate glass pipe.
   b. Explain how to join and install plastic pipe.
   c. Explain how to join and install silicon cast-iron pipe.
   d. Explain how to join and install stainless steel pipe.
   e. Explain how to install acid dilution and neutralization sumps.

Performance Tasks

Under the supervision of your instructor, you should be able to do the following:

1. Identify the circumstances in which corrosive-resistant waste piping should be installed.
2. Identify the neutralizing agents of an acid.
3. Connect three different types of corrosive-resistant waste piping using proper techniques and materials.

Trade Terms

Adhesive
Austenitic stainless steel
Beaded end
Biohazard
Borosilicate glass
Corrosive waste

Coupling
Duplex stainless steel
Ferritic stainless steel
Furan
Fusion coil
Hazard communication

Martensitic stainless steel
Thermoplastic
Thermoset
Uniform Laboratory Hazard
  Signage

Industry-Recognized Credentials

If you're training through an NCCER-accredited sponsor, you may be eligible for credentials from NCCER's Registry. The ID number for this module is 02308-14. Note that this module may have been used in other NCCER curricula and may apply to other level completions. Contact NCCER's Registry at 888.622.3720 or go to **www.nccer.org** for more information.

Code Note

Codes vary among jurisdictions. Because of the variations in code, consult the applicable code whenever regulations are in question. Referring to an incorrect set of codes can cause as much trouble as failing to reference codes altogether. Obtain, review, and familiarize yourself with your local adopted code.

Contents

Topics to be presented in this module include:

Figures

1.0.0 IDENTIFYING AND HANDLING CORROSIVE WASTE

Objective

Identify corrosive wastes and handle them safely.
 a. Identify the different types of corrosive waste.
 b. Identify the safety issues and hazard communication systems related to working with corrosive-resistant waste.

Performance Task

Identify the circumstances in which corrosive-resistant waste piping should be installed.

Trade Terms

Biohazard: An organism or chemical that is dangerous to the health of living creatures.

Corrosive waste: Waste products that contain harsh chemicals that can damage pipe and cause illness, injury, or even death.

Hazard communication: A workplace safety program that informs and educates employees about the risks from chemicals.

Uniform Laboratory Hazard Signage: Signs placed on laboratory walls that use pictographs to indicate hazards.

Plumbers install drain, waste, and vent (DWV) systems to safely handle a wide variety of wastes. You have learned about DWV systems that handle sewage and stormwater. You may also install systems designed to handle corrosive wastes. Corrosive wastes, also called corrosives, contain harsh chemicals that can damage the inside of ordinary drainage and vent pipes. Some corrosive wastes contain residue that could restrict the flow in the drainage system. Corrosive wastes can damage sewage treatment systems. They can also cause illness, injury, and death if people are exposed to them.

Corrosive wastes must be handled differently from other types of waste. They require the installation of a separate drainage system (see *Figure 1*). Corrosive wastes must also be treated before they can be discharged. Drainage pipes must be made of special materials that can resist corrosive wastes. In this module, you will learn about the

different types of corrosive-resistant pipes and how to install them. You will also learn how to identify and label dangerous chemicals in the workplace. Study each section carefully. The rules and guidelines are there to protect you and your customers from harm.

> **NOTE**
> Plumbers use the term *corrosive-resistant waste piping* to refer to pipe that is designed specifically to handle corrosive wastes. The term does not imply that such pipe is designed to resist other types of corrosion—for example, electrolysis, galvanic corrosion, scale, or rust.

> **NOTE**
> Most jurisdictions require plumbers to be certified or to be manufacturer-trained in order to install corrosive-resistant waste piping. Refer to your local applicable code for the standards that apply in your area.

1.1.0 Identifying Types of Corrosive Waste

The US Environmental Protection Agency (EPA) defines corrosive waste as liquid that has a pH value of 2 or lower (extremely acidic) or 12 or higher (extremely alkaline). Corrosive waste can eat through steel in a relatively short time. It can erode the insides of many types of metal and plastic waste pipes. It can create toxic or noxious fumes that can spread throughout a building. In other

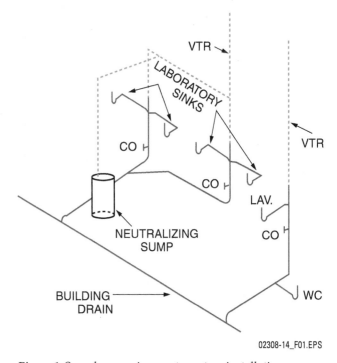

02308-14_F01.EPS

Figure 1 Sample corrosive-waste system installation.

words, corrosive waste is hazardous not only to sanitation but to health.

Many common corrosive wastes are created in the home. Even though some of them could damage a waste system, they usually are not treated before they are discharged. Most wastes created in the home are less harsh than corrosive wastes created by industry. As a result, standard DWV piping can handle them. The following common home products create corrosive wastes:

- Scalding water
- Bleaches
- Detergents
- Drain cleaners
- Scouring powders
- Other household chemicals

Almost any acid-resistant material can be used for DWV piping in the home. Concrete, cast iron, plastic, steel, and copper can all be used. Some home septic systems are equipped with a separate tank to handle laundry wastes. Otherwise, little extra protection is needed from household corrosive wastes.

This module focuses on corrosive wastes created by industrial, commercial, and laboratory processes. These wastes are harsher than household chemical wastes. They must be treated before they can be discharged into the drainage system, so a separate corrosive-waste system must be installed. Many types of industrial or manufacturing businesses create corrosive wastes:

- Auto painting and detailing shops
- Construction companies
- Laboratories
- Metal manufacturers
- Paint manufacturers
- Pharmaceutical manufacturers
- Photograph-processing facilities
- Pulp and paper mills
- Vehicle maintenance facilities

Corrosive wastes that you are likely to find on the job include rust removers, cleaning fluids, solvents, battery acid, degreasers, paint wastes, printing ink, pesticides, and agricultural chemicals. Plumbers are responsible for installing waste piping that safely discharges corrosive wastes.

1.2.0 Identifying Safety Issues and Hazard Communication Systems Related to Corrosive Waste

A recent government survey found that 650,000 hazardous chemicals are commonly used in workplaces all around the country. Each of them can pose health and safety risks if not handled properly. Workers have a right to know what the risks are. Employers develop hazard communication (HazCom) plans to educate and protect their employees. HazCom programs are required by law. As part of a HazCom program, an employer must do the following:

- Post lists of hazardous chemicals in the workplace.
- Create a written safety plan.
- Put labels or tags on all chemical containers.
- Keep safety data sheets (SDSs) on each of the hazardous chemicals that are used or stored.
- Train workers in proper safety and handling techniques.

HazCom programs vary depending on the types of chemicals that are used and stored. Be sure to find out about your company's HazCom program. Your safety and that of your co-workers depend on it.

1.2.1 HazCom Labels

The United Nations Globally Harmonized System of Classification and Labeling of Chemicals (GHS) is a new international standard for safety data sheets (SDSs, formerly called material safety data sheets, or MSDSs). The purpose of the GHS system is to define and communicate the health, physical, and environmental hazards of chemicals as well as the necessary protective measures on labels and SDSs (see *Figure 2*). In the United States, the GHS system has replaced the labeling system developed by the National Fire Protection Association (NFPA), known as the NFPA diamond. The standardized label elements that are included in a typical GHS label include:

- Hazard symbols, called pictograms, that convey visual information about the health, physical, and environmental hazards associated with the labeled product.

Corrosive Chemicals

Corrosive chemicals did not appear in large quantities until the nineteenth century. As a result of the Industrial Revolution, European and American companies began to use increasing amounts of chemicals. People soon noticed that chemicals were damaging waste pipes when they were discarded. Engineers began to develop new metal alloys that could resist these harsh wastes, marking the beginning of the modern corrosive-resistant waste piping industry.

TOLUENE
CAS # 108-88-3
Annex VI Index

DANGER

Highly flammable liquid and vapor. May be fatal if swallowed and enters airways. Causes skin irritation. May cause drowsiness or dizziness. Suspected of damaging unborn children. May cause severe damage to organs through prolonged or repeated exposure.

First Aid Statements

Keep away from heat/sparks/open flames/hot surfaces. No smoking. Ground/bond container and receiving equipment. Wear protective gloves and eye/face protection. Do not breathe dust/fumes/gas/mist/vapor spray. IF ON SKIN: Wash with plenty of soap and water. If skin irritation occurs, get medical advise/attention. IF INHALED: Remove to fresh air and keep at rest in a position comfortable for breathing. Call a POISON CENTER or doctor/physician if you feel unwell. IF SWALLOWED: Immediately call a POISON CENTER or doctor/physician. Do NOT induce vomiting. Store in a well-ventilated place. Keep cool. Dispose of contents/container to special waste disposal facility.

Name and Address of Company:
Treatment Services, Inc.
4258 N. 21st Avenue
Hohot, NE 55555
Telephone Number:
(555) 555-1212

02308-14_F02.EPS

Figure 2 The elements of a GHS label.

- Signal words that indicate the severity of the hazard. The signal word "Danger" is used to indicate the most severe hazards. The signal word "Warning" is used to indicate hazards that are less severe.
- Hazard statements that describe the nature of the hazard using standard phrases.

The symbols used in the GHS system, called pictograms, are grouped into categories that identify the various classes of hazards, hazards during transport, and oral toxicity (that is, the level of toxicity if ingested). *Figures 3, 4,* and *5* illustrate the pictograms in each of these categories.

GHS labels should be posted near all doors leading into an area where chemicals are stored

 • OXIDIZERS

- FLAMMABLES
- SELF REACTIVES
- PYROPHORICS
- SELF-HEATING
- EMITS FLAMMABLE GAS
- ORGANIC PEROXIDES

- EXPLOSIVES
- SELF REACTIVES
- ORGANIC PEROXIDES

 • ACUTE TOXICITY (SEVERE)

 • CORROSIVES

 • GASES UNDER PRESSURE

- CARCINOGEN
- RESPIRATORY SENSITIZER
- REPRODUCTIVE TOXICITY
- TARGET ORGAN TOXICITY
- MUTAGENICITY
- ASPIRATION TOXICITY

 • ENVIRONMENTAL TOXICITY

- IRRITANT
- DERMAL SENSITIZER
- ACUTE TOXICITY (HARMFUL)
- NARCOTIC EFFECTS
- RESPIRATORY TRACT
- IRRITATION

02308-14_F03.EPS

Figure 3 GHS pictograms indicating hazard classes.

 02308-14 Corrosive-Resistant Waste Piping

• FLAMMABLE LIQUID
• FLAMMABLE GAS
• FLAMMABLE AEROSOL

 • FLAMMABLE SOLID SELF-REACTIVE SUBSTANCES

 • PYROPHORICS (SPONTANEOUSLY COMBUSTIBLE) SELF-HEATING SUBSTANCES

 • SUBSTANCES WHEN IN CONTACT WITH WATER EMIT FLAMMABLE GASES (DANGEROUS WHEN WET)

 • OXIDIZING GASES
• OXIDIZING LIQUIDS
• OXIDIZING SOLIDS

 • EXPLOSIVE DIVISIONS 1.1, 1.2, 1.3

 • EXPLOSIVE DIVISION 1.4

 • EXPLOSIVE DIVISION 1.5

 • EXPLOSIVE DIVISION 1.6

 • COMPRESSED GASES

 • ACUTE TOXICITY (POISON): ORAL, DERMAL, INHALATION

 • CORROSIVE

 • MARINE POLLUTANT

 • ORGANIC PEROXIDES

02308-14_F04.EPS

Figure 4 GHS pictograms indicating transport hazards.

ACUTE ORAL TOXICITY - ANNEX 1					
	Category 1	Category 2	Category 3	Category 4	Category 5
Ld_{50}	5 mg/kg	> 5 < 50 mg/kg	50 < 300 mg/kg	300 < 2000 mg/kg	2000 < 5000 mg/kg
Pictogram					No symbol
Signal word	Danger	Danger	Danger	Warning	Warning
Hazard statement	Fatal if swallowed	Fatal if swallowed	Toxic if swallowed	Harmful if swallowed	May be harmful if swallowed

02308-14_F05.EPS

Figure 5 GHS pictograms indicating oral toxicity.

NCCER – *Plumbing Level Three* 02308-14

or used. Ensure that labels are posted by outside entrances where rescue workers can see them. The GHS labeling standard is designed to protect workers and the public. It is also intended to help response teams react swiftly in an emergency. The labels do not include detailed information about the types of chemicals being stored.

Laboratories use special signs to indicate potential hazards. The Uniform Laboratory Hazard Signage (ULHS) system is widely used in many different types of laboratories. ULHS signs are standards that are used around the world. They indicate the type of risk by using special graphics called pictographs (see *Figure 6*). Signs warn against oxidizers, electrical hazards, cancer hazards, biohazards, and other risks. A biohazard is an organism or chemical that poses a health risk to humans, animals, or plants. Refer to your local code before posting laboratory signs.

1.2.2 Safety Data Sheets

SDSs are bulletins that contain a wide range of useful safety information. Manufacturers are required to create an SDS for each of their chemical products. Companies must keep SDSs on file for all of the chemicals that are used or stored on site. The SDSs should be kept where they can be found easily. All employees must know where to find SDSs.

An SDS lists the chemical and common names of a product as well as the health and safety hazards associated with it. If the chemical is a compound, the names of dangerous ingredients are listed. The SDS also describes the product's boiling and freezing points, density and specific gravity, appearance, and odor. If the product could ignite or explode, the SDS will list the conditions that would present a danger. Chemicals that react with the product are also listed.

The SDS provides instructions for safe handling and use of the product. It also describes what to do in an emergency. Always consult the SDS before you work with a chemical product. Never dispose of a chemical product without consulting the SDS or your safety officer. Refer to your workplace health code for more information.

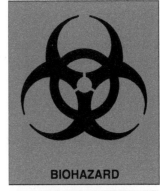

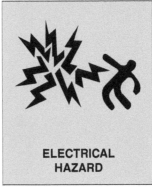
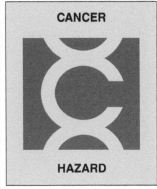

02308-14_F06.EPS

Figure 6 Examples of laboratory hazard signs.

1.2.3 Worker Training

Training programs are an effective way to ensure that employees understand the company's HazCom program. Train new employees when they join the company. Conduct annual reviews for all employees who work with hazardous chemicals. The training should cover the following topics:

- Where to find information such as SDSs and the written safety plan
- How to detect the release of chemicals
- How to identify health and physical hazards
- How to protect yourself and others in the event of a chemical release

All employees must know the safety regulations and where dangerous chemicals are stored and used. Study your company's HazCom plan. Don't be afraid to ask questions about the details of the plan. In an emergency, this knowledge could save your life and the lives of your co-workers.

Additional Resources

Hazard Communication Made Easy: A Checklist Approach to OSHA Compliance. 2000. Sean M. Nelson and John R. Grubbs. Rockville, MD: ABS Group, Government Institutes.

1.0.0 Section Review

1. The EPA defines corrosive wastes as liquids that have a pH value of _____.
   a. 5 or lower
   b. 6 or higher
   c. between 2 and 12
   d. 2 or lower or 12 or higher

2. The signal words used in the GHS are _____.
   a. "Caution" and "Warning"
   b. "Caution" and "Danger"
   c. "Danger" and "Warning"
   d. "Danger" and "Toxic"

2.0.0 JOINING AND INSTALLING CORROSIVE-RESISTANT WASTE PIPING

Objective

Explain how to join and install different types of corrosive-resistant waste piping.

   a. Explain how to join and install borosilicate glass pipe.

   b. Explain how to join and install plastic pipe.

   c. Explain how to join and install silicon cast-iron pipe.

   d. Explain how to join and install stainless steel pipe.

   e. Explain how to install acid dilution and neutralization sumps.

Performance Tasks

Identify the neutralizing agents of an acid.

Connect three different types of corrosive-resistant waste piping using proper techniques and materials.

Trade Terms

Adhesive: A chemical used to bond thermoset pipes and fittings without fusing them together.

Austenitic stainless steel: A stainless steel alloy that contains nickel and stays strong at very low temperatures. Austenitic steel is the most popular type of steel for corrosive-waste pipes.

Beaded end: A raised lip around the rim of borosilicate glass pipe.

Borosilicate glass: A pipe material made from specially reinforced glass that does not expand or contract much with changes in temperature.

Coupling: Removable metal clamps with plastic seals, used to connect borosilicate glass pipe.

Duplex stainless steel: A stainless steel alloy mix of austenitic and ferritic stainless steels that offers very good corrosion resistance.

Ferritic stainless steel: A stainless steel alloy that is very resistant to rust and has a very low carbon content.

Furan: Any resin that is highly resistant to solvents. Furans are used to make thermoset pipe.

Fusion coil: A device that uses heat to join thermoplastic pipe.

Martensitic stainless steel: A stainless steel alloy that is very strong but is mostly used to handle mild corrosive wastes.

Thermoplastic: A type of plastic pipe that can be softened and reshaped when it is exposed to heat.

Thermoset: A type of plastic pipe that is hardened into shape when manufactured.

Each waste system requires special pipe and fittings that are suited for the type of waste being created. The codes and standards that govern corrosive-waste systems are very strict. Consult your local code to find out what pipe materials may be used in your area. Many manufacturers provide tables that list the chemicals that their pipes can safely handle. Corrosive-resistant pipe and fittings are more expensive than regular waste pipe and fittings.

The following are the most common materials used to make pipe and fittings for corrosive-waste systems:

- Borosilicate glass
- Thermoplastic and thermoset plastics
- Silicon cast-iron
- Stainless steel

Each of these materials has unique corrosive-resistant properties. They are suitable for a wide variety of waste systems. The properties and uses of each type of pipe material are discussed in more detail in the following sections. Review each section carefully. Talk with an experienced plumber to learn more about the different types of pipe. Remember that the goal is to select the pipe that will provide years of safe and trouble-free operation.

You can install corrosive-resistant waste systems using techniques that you have already learned, but some of the specific requirements are different from those for other waste systems. Ensure that the system is completely separate from other drainage systems in the building. Install a trap on every fixture in the system. The wastes must be diluted or neutralized before they can be discharged. Refer to your local code for installation requirements in your area.

Engineers design corrosive-resistant waste systems. Refer to the project plans for the materials to be used in the system. If you have any questions about the design, ask the engineer. Install the corrosive-resistant waste system so that it can be expanded or changed in the future. Remember that these installation instructions are for general reference only. Refer to your local code for specific guidelines. Test the system in accordance with local codes when the installation is complete.

2.1.0 Joining and Installing Borosilicate Glass Pipe

Many corrosive wastes are discharged at extreme temperatures. Use borosilicate glass pipe (see *Figure 7*) for these wastes. Borosilicate glass pipe is made from a specially reinforced transparent glass. Changes in temperature cause very little expansion or contraction. As a result, borosilicate glass pipes can safely handle liquids ranging from freezing to 212°F. However, sudden extreme changes in temperature may cause the pipe to crack.

> **CAUTION**
>
> Pay careful attention to the pipe lengths when preparing material takeoffs for borosilicate glass pipe. Incorrect lengths could lead to damage to the drainage system.

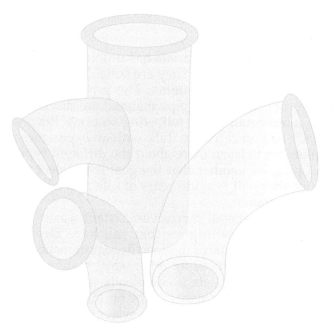

02308-14_F07.EPS

Figure 7 Borosilicate glass pipe and fittings.

2.1.1 Characteristics of Borosilicate Glass Pipe

Borosilicate glass pipe comes with either plain or beaded ends. A beaded end is a lip around the pipe rim. The pipe and fittings in a glass drainage system can be reused. Glass pipe cutters (*Figure 8*) and internal-pressure pipe cutters (*Figure 9*) can be used to cut the pipe to new lengths as needed. To use a standard glass pipe cutter, insert the cutter to the desired depth, adjust the cutting head manually until it makes contact with the inner surface of the pipe, and then turn the shaft in a circular motion to score and cut the pipe. Internal-pressure pipe cutters use pressure supplied from a pump to force knife blades connected to a piston into contact with the inner surface of the pipe.

Use couplings to join pipes and fittings. Couplings are removable clamps with plastic seals on the inner rim (see *Figure 10*). Couplings provide leak-free joints even when the pipe is slightly deflected. The seals on the inside of the coupling are made from tetrafluoroethylene (TFE) plastic. TFE resists chemicals and acids better than any other type of plastic. The outer shell of the coupling is made of stainless steel.

2.1.2 Joining and Installing Borosilicate Glass Pipe

Glass pipes and fittings are usually packed in Styrofoam and delivered to the job site in rugged shipping cartons. To avoid damaging them, unpack them only when they are needed. Couplings join glass pipe by clamping the two ends together. Pipes can be joined beaded end to beaded end, beaded end to plain end, and metal or plastic to plain end (see *Figure 11*). Use the following steps to join pipes using a beaded-end-to-beaded-end coupling (see *Figure 12*):

Step 1 Dip the coupling in water or wet the inside with a damp cloth.

Step 2 Push and rotate the coupling to snap it over the pipe bead. Start with the side of the coupling that is opposite the bolt.

CUTTING POINT   PLAIN OR BEADED-END GLASS PIPE   TENSION ADJUSTMENT KNOB

02308-14_F08.EPS

Figure 8 Glass pipe cutter.

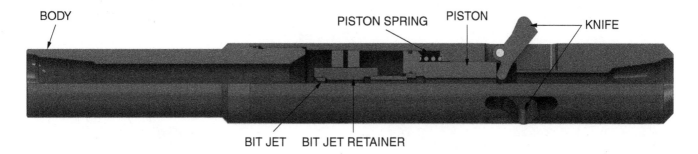

BODY   PISTON SPRING   PISTON   KNIFE

BIT JET   BIT JET RETAINER

02308-14_F09.EPS

Figure 9 Internal-pressure pipe cutter.

02308-14_F10.EPS

Figure 10 Coupling used to join glass pipe.

When the coupling has been snapped onto the first pipe, insert the other section of pipe into the other side of the coupling, using the same technique.

Step 3 Tighten the coupling bolt using a ratchet wrench. Do not exceed the manufacturer's torque specifications when tightening the bolt.

The procedure for joining a beaded-end pipe to a plain-end glass, metal, or rigid plastic pipe is similar:

Step 1 Dip the coupling in water or wet the inside with a damp cloth.

Step 2 Snap the coupling over the beaded end.

Step 3 Insert the plain-end pipe into the opposite side of the coupling, ensuring that the plain end is fully seated. Do not force the pipe into the coupling.

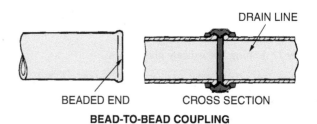

BEADED END   CROSS SECTION   DRAIN LINE

BEAD-TO-BEAD COUPLING

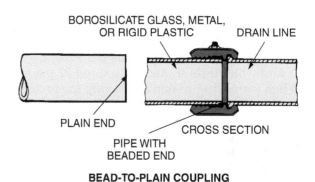

BOROSILICATE GLASS, METAL, OR RIGID PLASTIC   DRAIN LINE

PLAIN END   CROSS SECTION

PIPE WITH BEADED END

BEAD-TO-PLAIN COUPLING

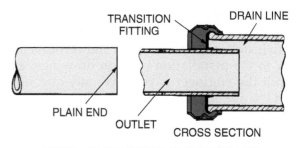

TRANSITION FITTING   DRAIN LINE

PLAIN END   OUTLET   CROSS SECTION

METAL- OR PLASTIC-TO-PLAIN COUPLING

02308-14_F11.EPS

Figure 11 Three types of couplings.

Step 4 Tighten the coupling bolt using a ratchet wrench. Do not exceed the manufacturer's torque specifications when tightening the bolt.

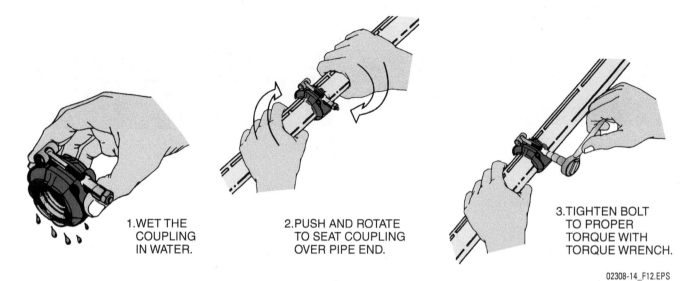

1. WET THE COUPLING IN WATER.

2. PUSH AND ROTATE TO SEAT COUPLING OVER PIPE END.

3. TIGHTEN BOLT TO PROPER TORQUE WITH TORQUE WRENCH.

02308-14_F12.EPS

Figure 12 Joining borosilicate glass pipes.

Though glass pipe looks very different from other types of pipe, it is installed the same way. It can be installed in aboveground and underground systems. Remember that glass pipe can break if it is strained. Reduce strain by allowing vertical and horizontal installations to have some limited movement. After you install borosilicate glass pipe, test it according to your local code. Air tests should not exceed 5 pounds per square inch (psi), and water tests should not exceed 22 psi.

When preparing material takeoffs for projects that require the use of borosilicate glass pipe, be sure to consider the pipe lengths that will be required. Refer to the manufacturer's product catalog to ensure that you order the correct type and number of couplings for the job based on the number, type, and lengths of pipes ordered.

Glass piping is available in 5-foot and 10-foot lengths. Diameters range from 1½ to 6 inches. It can be installed to almost any length or height and can be installed behind walls. Glass pipe is safe to install in both horizontal and vertical runs. Refer to the installation's design drawings, and consult the engineer if you have any questions about the use of glass pipe in an installation.

Use padded hangers for horizontal runs of glass pipe (see *Figure 13*). The padding will help prevent the metal hangers from scratching the glass. Space hangers 8 to 10 feet apart. If the pipe has three or more couplings within an 8- to 10-foot span, install an extra hanger. Do not force glass pipe into a hanger. Always move the hanger to the glass pipe instead. Tighten each coupling as each joint is made. Ensure that horizontal runs are free to move sideways. To change the pitch of a run of glass pipe, loosen the coupling bolt, tilt

the pipe to the pitch required, and retighten the coupling bolt.

For horizontal runs of glass pipe behind interior walls, use pipe sleeves with a diameter that is at least 2 inches greater than the outside diameter of the pipe. Pack the space between the pipe and the sleeve with fiberglass or glass wool. For pipes that pass through exterior fire-, water-, or explosionproof walls, pack the sleeve with appropriate caulking material. Space the couplings no more than 6 inches apart on either side of the wall.

Use riser clamps for vertical installation of glass pipe. The clamps must be lined with ¼ inch of neoprene rubber (see *Figure 14*). Refer to your local code for the proper number and spacing of riser clamps, and to the manufacturer's instructions for additional support requirements. Install riser clamps below couplings, not above them. Install a clamp below the bottom coupling in a glass pipe stack. Where possible, install clamps below couplings on every third floor. Do not support a

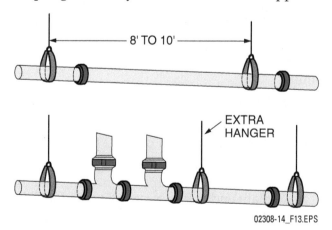

8' TO 10'

EXTRA HANGER

02308-14_F13.EPS

Figure 13 Installing a horizontal pipe run using padded hangers.

vertical glass pipe stack with a horizontal pipe run. Locate the first hanger on a horizontal line 6 to 8 feet from where a vertical stack connects to it.

Fit pipe sleeves on all vertical stacks that pass through a floor or slab. The sleeve's diameter should be at least 2 inches greater than the diameter of the pipe. Pack the space between the pipe and sleeve with fiberglass or glass wool. Install a coupling within 6 inches of the floor or slab. The coupling allows the pipe to have some flexibility of motion. Some installations will require you to pass pipe vertically through fire-, water-, or explosionproof floors. Pack the sleeve with appropriate caulking material or water plug cement. Follow your local code carefully when you install pipe through fireproof floors or walls.

Glass pipe can be used to vent a stack through the roof. Install vent flashing to prevent rainwater from running into the building through the hole cut for the vent. Depending on the location, job, and contract, this may be a job for the plumber, carpenter, or roofer. Check with your supervisor to find out if installing the vent flashing is part of your job. Wrap the vent pipe with tape or insulating material according to your local code and the building plans. Use either seamless lead roof flashing with a caulked counterflashing sleeve or seamless lead or copper roof flashing.

To connect the glass pipe stack to the floor drain, install a glass-to-steel coupling to connect the glass pipe to a stainless steel pipe. Connect the metal pipe to the drain according to the manufac-

turer's instructions. Another option is to insert the glass pipe into the floor drain using the following steps:

Step 1 Cut the glass pipe to the desired length. Smooth the sharp outer edges of the pipe using the proper finishing tool.

Step 2 Insert the pipe into the outlet of the floor drain.

Step 3 Seal the glass pipe in place using asbestos rope and acidproof cement. Allow the cement to cure for the time listed in the manufacturer's instructions. Be sure to use appropriate personal protective equipment when you are using asbestos and cement.

Glass pipe and fittings can be used for underground drainage systems. Use glass pipe that is fitted with a protective casing. Expanded polystyrene is commonly used as a casing material. Most codes specify 5-foot lengths of pipe for underground installations. Wrap the fittings in thin polyvinyl film or other recommended material. Remove the casing before you cut a pipe. Cut the casing 2 inches shorter than the new length of pipe. Replace the casing so that 1 inch of pipe is exposed at each end. The gap provides enough space to install a coupling.

To lay glass pipe underground, begin by excavating the pipe trench to a width of at least 24 inches at the bottom. If the soil is sandy, dig the trench 1 to 2 inches below the final grade. For rocky soil or clay, dig the trench 4 to 6 inches below the final grade (see *Figure 15*). Ensure that the trench bedding is firm enough to support the glass pipe uniformly along the full length of the run.

Assemble several glass pipe joints to form a section. Tighten the couplings firmly, and lower the section into the trench. Bury the pipe using thin layers of rock-free sand or soil until it is 12 inches above the glass pipe. Tamp the sand firmly by hand. Then backfill the trench with available soil using a backhoe or other mechanical means (refer to *Figure 15*).

2.2.0 Joining and Installing Plastic Pipe

Codes allow plastic pipe to be used for corrosive-waste drainage. Plastic pipe has the following advantages over other pipe materials:

- Resistance to chemical and galvanic corrosion
- Low friction loss
- High handling temperatures
- Flexibility
- Inflammability

NEOPRENE

RISER CLAMP   GLASS PIPE

02308-14_F14.EPS

Figure 14 Installing a riser clamp on a waste pipe.

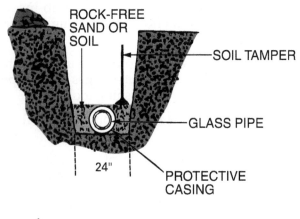

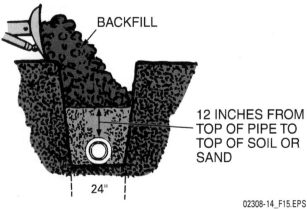

02308-14_F15.EPS

Figure 15 Laying borosilicate glass pipe underground.

2.2.1 Characteristics of Plastic Pipe

Refer to the manufacturer's catalog when selecting plastic pipe. The catalog will list the properties of each type of pipe. It will help you select the pipe material that is best suited for the type of corrosive waste in a given system.

Thermoplastic pipe and thermoset pipe can be used to drain corrosive wastes. Both types of plastic pipe are composed of a resin base with chemical additives. The difference is that thermoplastic pipe can be softened and reshaped when exposed to heat. Thermoset pipe, on the other hand, is hardened into shape when manufactured. Suitable thermoplastic pipe materials include the following:

- Polyvinyl chloride (PVC)
- Chlorinated polyvinyl chloride (CPVC)
- High-density polyethylene (HDPE)
- Polyvinylidene fluoride (PVDF)
- Polypropylene (PP)

You are already familiar with PVC and CPVC pipe. HDPE has an excellent strength-to-weight ratio and is resistant to a wide range of solvents; also, it is designed to tear rather than shatter in the event of an accident. HDPE pipe used for corrosive waste is designed with separate inner and outer layers for added protection. PVDF is more resistant to corrosion and abrasion than other types of plastic pipe. PP is a light, chemically inert plastic. PP is safe for use in kitchens and other food preparation areas. Thermoplastic and thermoset pipe can handle acetic acids and chemical compounds containing aluminum, ammonia, barium, calcium, sodium, and mild sulfuric acids. Resistance varies depending on the type of pipe.

> **WARNING!**
>
> Corrosive chemicals can cause injury or death if they are not handled properly. Read all warning labels before you handle any chemical. Wear appropriate personal protective equipment when you work with chemicals, and always follow required safety guidelines.

Thermoset pipe is made from epoxies, polyesters, vinyl esters, or furans. Furans are resins that are highly resistant to solvents. The pipe is wrapped in fiberglass or graphite fibers for reinforcement. The fiber wrapping makes the pipe rigid and strong. Thermoset plastic pipe is highly resistant to corrosion. Because it is more rigid than other types of pipe, hangers must be closely spaced to provide adequate support.

Thermoset pipe can handle many of the same corrosive chemicals as other types of plastic pipe. Properties of specific types of pipe vary depending on the manufacturer. Always refer to the specifications for each pipe before you select it. Refer to your local code to determine whether the pipe is suitable for the type of installation.

Some pipes are not suitable for certain types of corrosive wastes. Pipes made out of furan resins, for example, should not be used with oxidized wastes. PVDF is unsuitable for certain strong alkalies or acids.

> **NOTE**
>
> Some codes do not allow PVC waste pipe to be used aboveground for corrosive-waste disposal. Fire-retardant PVC waste pipe may be required for aboveground corrosive-waste disposal installations, particularly in plenums. Refer to your local applicable code.

2.2.2 Joining and Installing Plastic Pipe

Thermoplastic and thermoset pipe can be joined and installed in several ways. Use the following to join plastic pipe:

- Fusion method
- Solvent cement
- Mechanical joints
- Pipe threading

The fusion method uses heat to join PVDF and PP pipe, as well as outdoor applications of HDPE pipe. Insert a pipe end into a fusion fitting, which contains a fusion coil and electrical terminals. Clamp the fitting according to the manufacturer's instructions, and connect electrical leads from the control box to the coil. Ensure that the control box has been calibrated and certified prior to using it. The fusion coil heats the pipe end and the fitting socket, causing them to fuse. Follow the manufacturer's directions for heating, cooling, and testing the connection. The fusion method is effective for pipe systems that handle corrosive wastes. Indoor applications of HDPE pipes can be joined with simple butt fusion.

> **NOTE**
>
> Most manufacturers require installers to be certified in their proprietary fusion systems before being allowed to join plastic pipe using the fusion method.

Older systems require the worker to position the fusion coil using clamps (see *Figure 16*). Newer systems embed heating elements in a coupling and fuse both ends of the connection at the same time (see *Figure 17*). Use the following steps to connect plastic pipe using removable fusion-coil fittings and clamps:

Step 1 Measure the full depth of the fitting socket.

Step 2 Transfer the measurement to the pipe, marking around its circumference. Ensure that the pipe is cut square and has been deburred.

Step 3 Sand the shiny surface from the area of the pipe to be fused.

Step 4 Insert the pipe completely into the fitting end that contains the fusion coil. Ensure that it meets the measured mark around the pipe when inserted.

Step 5 Position the clamp band according to the manufacturer's instructions. Tighten the clamp by hand. Note that pipe sizes of 3 inches and greater require additional tightening with a clamping tool.

Step 6 Connect the electrical leads to the coil terminals.

Step 7 Verify the settings on the control box and press the Start button on either the output head or the control box. Follow the manufacturer's instructions for cooling and testing.

Use the following steps to connect plastic pipe using couplings equipped with embedded heating elements:

Step 1 Mark the coupling depth on the pipe or fitting, following the manufacturer's instructions for specific pipe sizes.

Step 2 Sand the shiny surface from the area of the pipe to be fused.

Step 3 Insert the pipe completely into the coupling. Ensure that it meets the measured mark around the pipe when inserted.

Step 4 Repeat Step 3 with the fitting. Both ends of the coupling will fuse at the same time.

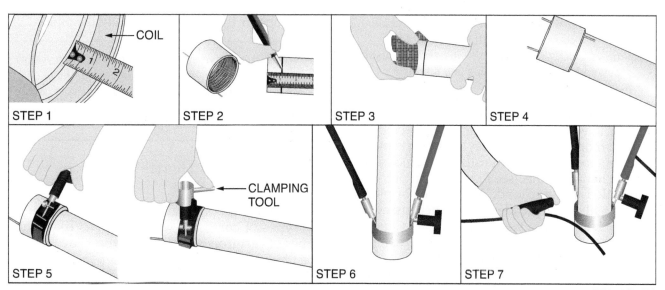

02308-14_F16.EPS

Figure 16 Using the fusion method to join plastic pipe with coil fittings and clamps.

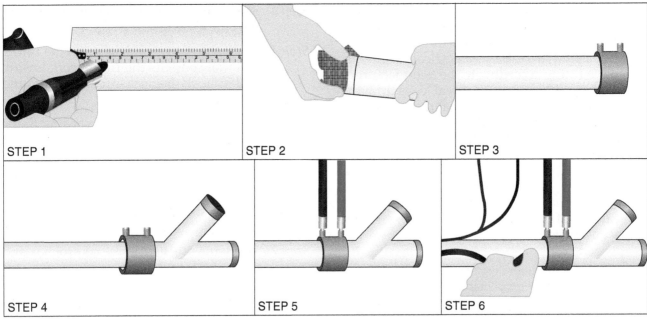

Figure 17 Using the fusion method to join plastic pipe with embedded electrofusion wire.

Step 5 Connect the electrical leads to the coil terminals.

Step 6 Enter the appropriate settings on the control box and press the Start button on either the output lead or the control box.

> **CAUTION**
>
> Fusion-seal machines require regular calibration in order to join pipe properly. Do not use a fusion-seal machine that has not been properly calibrated according to the manufacturer's instructions.

> **NOTE**
>
> Before cementing a joint, ensure that the pipe and fitting are free of dust, dirt, water, and oil. Solvent cement acts fast. It is important to move quickly and efficiently when you are using solvent cements to join pipe and fittings.

The other methods of joining plastic pipe are familiar to you, including using solvent cement to join thermoplastic pipe. Always follow the manufacturer's instructions carefully. Most codes permit mechanical joints (see *Figure 18*) on pipe that is installed underground. Some types of plastic pipe can also be threaded and joined using fittings such as elbows, tees, wyes, adapters, and reducers. Your local code will specify the preferred methods of joining thermoplastic pipe.

Thermoset pipe can be joined with threading or mechanical joints, but the most common way to join thermoset pipe is to use an epoxy adhesive. Adhesives are chemicals that bond surfaces without fusing them together. Use them on pipes that have tapered bell-and-spigot fittings. Apply adhesives the same way you would apply solvent cement. Note that most pipes and adhesives are not interchangeable among manufacturers.

When you install plastic pipe, follow your local code. Building plans will describe the proposed system. Avoid hangers that may cut or squeeze the pipe and tight clamps that prevent the pipe from moving or expanding. Support the pipe at intervals of no more than 4 feet and at branches, at changes of direction, and when you use large fittings. Do not install plastic pipe near sources of excessive heat or vibration. Avoid installing the hanger near sharp objects that could rub and cut the pipe.

> **WARNING!**
>
> Use protective eyewear and gloves when you are using solvent cement. Avoid skin or eye contact with the cement. If contact occurs, wash the affected area immediately. Always apply cement in a well-ventilated area.

2.3.0 Joining and Installing Silicon Cast-Iron Pipe

Silicon cast-iron pipe can be used to drain almost all types of corrosive waste. It looks like other types of cast-iron pipe, but the iron alloy used in the pipe is 14 percent silicon and 1 percent carbon. The silicon and carbon make the pipe and fittings highly resistant to corrosives.

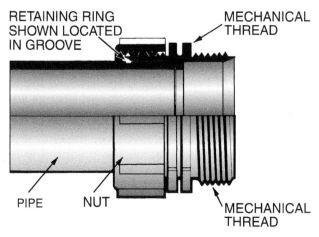

Figure 18 Mechanical joints for thermoplastic pipe.

2.3.1 Characteristics of Silicon Cast-Iron Pipe

Silicon cast-iron pipe can handle most corrosive wastes regardless of concentration or temperature. It is often used to drain sulfuric and nitric acids. Waste systems made from silicon cast-iron pipe usually last the life of the building. Silicon cast-iron is available in hub-and-spigot (see *Figure 19*) and no-hub styles, and in 5' and 10' laying lengths. Use mechanical joints to connect no-hub pipe. A typical mechanical joint has an inner sleeve made of Teflon® surrounded by neoprene. Teflon® has very high corrosion resistance, and the outer neoprene sleeve adds strength. Neoprene is resistant to heat, cold, weathering, and abrasion. Surrounding the inner and outer sleeves is a stainless steel coupling fastened by two bolts. The coupling allows the pipe to have a small amount of deflection.

Use mechanical or hydraulic spring-loaded pipe cutters to cut silicon cast-iron pipe; cold chis-

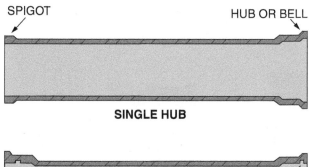

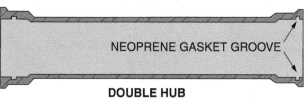

Figure 19 Hub-and-spigot silicon cast-iron pipe.

els may also be used. Support the pipe on a sandbag to absorb the shock. A properly cut silicon cast-iron pipe has a smooth edge. If the edge is smooth, the pipe can be installed with no further preparation.

Do not use silicon cast-iron pipe in a pressurized corrosive-resistant waste system. Your local code will specify the type of pipe to be used in a pressurized system. Silicon cast-iron pipe can be painted.

> **CAUTION**
>
> Remember that cast-iron pipe must be labeled with the name of the manufacturer, the trademark of the Cast Iron Soil Pipe Institute, and the pipe diameter. Most codes will not allow you to install unlabeled pipe. Refer to your local code.

2.3.2 Joining and Installing Silicon Cast-Iron Pipe

As mentioned in the previous section, silicon cast-iron pipe comes in hub-and-spigot and no-hub styles. Consult the waste system's design drawings to find out which type is preferred for the installation. Hub-and-spigot silicon cast-iron pipe is joined using neoprene gaskets inserted into the hub (see *Figure 20*). The materials are specially designed to resist corrosion. Always refer to your local applicable code before installing silicon cast-iron pipe, to ensure that it can be used and that you are joining and installing it according to the code requirements of your jurisdiction. Use the following steps to join hub-and-spigot pipe:

Step 1 Clean the hub and spigot to ensure they are free of dirt and other contaminants.

Step 2 Use a file or a ball peen hammer to smooth the pipe edges.

Step 3 Insert the gasket completely into the hub until only the flange remains outside the hub.

Step 4 Use a rag or paintbrush to lubricate the gasket according to the manufacturer's instructions. Pipe with a diameter larger than 5 inches should be coated with an adhesive lubricant. Apply lubricants to the inside of the gasket and the outside of the spigot.

Wood Pipes

Some heavily wooded countries still use wood for supply and waste pipes. Wood is highly resistant to the harsh effects of many chemicals and can handle some chemicals better than metal pipe.

Step 5 Align the pipe and push or pull the spigot through the gasket's sealing rings until you feel the pipe "bottom" into the hub. This action will cause the gasket to compress and completely seal the joint.

No-hub silicon cast-iron pipe is relatively easy to assemble. Use mechanical joints to assemble the pipe and fittings. Review the manufacturer's instructions to ensure that the joints are suitable for the type of pipe being installed. Use the following procedure to join pipes and fittings with mechanical joints (see *Figure 21*):

Step 1 Slide the joint sleeve over the end of the pipe until it reaches the positioning lug on the pipe.

Step 2 Slide the other pipe or fitting into the joint.

Step 3 Tighten the two bolts on the joint. The torque on the bolts should be at least 9 foot-pounds.

Hub-and-spigot and no-hub silicon cast-iron pipe can be installed underground or in concrete without special preparation. For underground installations, ensure that the first few inches of fill are free of rocks. Some mechanical joints require an asphalt mastic coating if they are installed on underground pipe. Refer to the manufacturer's instructions.

Install silicon cast-iron pipe using the same methods you would use for regular cast-iron soil pipe. Support each length of pipe in a horizontal run. If the line includes fittings, place the supports no farther than 7 feet apart. Do not completely tighten supports around the horizontal pipe because the pipe should be free to expand and contract. Install supports directly beneath fittings that connect to vertical stacks.

Support vertical runs at least every 14 feet. Refer to the manufacturer's instructions for approved hangers. Install suitable cleanout plugs on iron waste-piping systems. Clean waste systems using a sewer auger or chemical cleaner that is safe for use in silicon cast-iron pipe.

MOLDED BEAD

02308-14_F20.EPS

Figure 20 Corrosive-resistant neoprene gaskets used in hub-and-spigot silicon cast-iron piping.

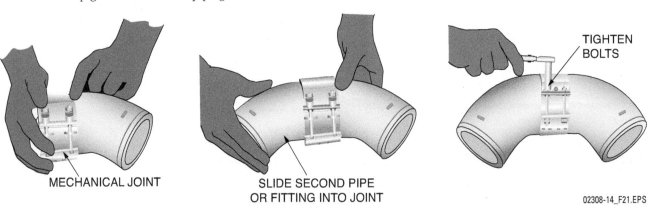

MECHANICAL JOINT

SLIDE SECOND PIPE OR FITTING INTO JOINT

TIGHTEN BOLTS

02308-14_F21.EPS

Figure 21 Installing a mechanical joint on a no-hub pipe.

16   NCCER – *Plumbing Level Three*   02308-14

2.4.0 Joining and Installing Stainless Steel Pipe

Plumbers can choose from four types of stainless steel pipe. Each type is made of a different metal alloy that offers its own special corrosion-resistance capabilities. The following are the four types of stainless steel alloys available for corrosive-resistant waste piping:

- Ferritic
- Austenitic
- Martensitic
- Duplex

2.4.1 Characteristics of Stainless Steel Pipe

Ferritic stainless steel is very resistant to rust and has an extremely low carbon content. Use ferritic steel for chloride and organic acid wastes. Ferritic stainless steel is often used in food-processing facilities.

Austenitic stainless steel contains nickel, which makes the metal more stable than ferritic steel. It stays strong at very low temperatures. Austenitic steel is the most popular type of steel for corrosive-waste pipes.

Martensitic stainless steel is very strong, but it does not resist corrosion as well as austenitic steel. It is mostly used to handle mild corrosive wastes, such as organic matter.

Duplex stainless steel is a mix of austenitic and ferritic steels. It is stronger than austenitic steel, which means that thinner, lighter pipe can be used. Duplex stainless steel offers very good corrosion resistance.

2.4.2 Joining and Installing Stainless Steel Pipe

Stainless steel corrosive-resistant waste pipe can be welded. It is usually too thin to be joined by threading. Do not attempt to weld steel pipe unless you have been trained in proper welding techniques. Always use appropriate personal protective equipment when you are working with welding tools. Ensure that the weld rod has the same alloy content as the pipe metal. Otherwise, the weld joints could be weakened through contact with the corrosive waste. Note that different welding techniques apply to the individual steel alloys.

Install stainless steel pipe as you would other types of metal piping. Ensure that hangers and supports are properly sized and located according to your local code. Provide additional support for fittings and other accessories on the line. When you are installing insulated pipe, you are required to place hangers and supports

closer together. Never spring stainless steel pipe into place. Springing causes stress that will seriously weaken the pipe. External loads on an unsupported length of pipe will also cause stress fractures. Always follow the manufacturer's directions when installing stainless steel pipe.

Many codes allow the use of press fittings (see *Figure 22*) to join corrosive-resistant stainless steel pipe. Press fittings can be used to join pipe mechanically. Press pipe, valves, and fittings are available in 304- and 316-grade stainless steel.

Grooved stainless steel pipe can be joined with couplings (see *Figure 23*). Couplings designed for use with corrosive waste are fitted with gaskets made from materials that are designed to resist damage caused by the waste. Refer to your local applicable code to determine whether these types of fittings are permitted in your jurisdiction.

02308-14_F22.EPS

Figure 22 Press fittings for stainless steel pipe.

02308-14_F23.EPS

Figure 23 Coupling for grooved stainless steel pipe.

2.5.0 Installing Acid Dilution and Neutralization Sumps

Corrosive liquids, spent acids, and toxic chemicals pose serious health risks. They cannot be discharged directly into the drainage system. They must first be diluted, chemically treated, or neutralized. Install acid dilution and neutralization sumps to treat corrosive wastes. Consult your local code for specific requirements. Many codes require that you obtain approval before you install acid dilution and neutralization sumps.

If the waste flow is intermittent, it can be diluted. Ensure that the sump is large enough to permit the waste to dilute to an acceptable level according to local code. Refer to your local code for chemicals that can be treated through dilution. Sumps used for dilution must be vented (see *Figure 24*). Chemical waste and sanitary vents must extend through the roof. Some dilution tanks may be vented separately; refer to the manufacturer's instructions and to your local applicable code. Never vent a dilution tank into the sanitary system.

Some wastes must pass through a neutralizing agent before they can be discharged (see *Figure 25*). Neutralizing agents react chemically with corrosive wastes. The result is a waste product that can be discharged safely. One of the most commonly used neutralizing agents is appropriately sized limestone. Stones that are too small will restrict the flow. Install vents and gastight lids on sumps with neutralizing agents. Ensure that the sump contains enough of the neutralizing agent to treat all of the special waste that will flow through it.

Install the sump where it is easily accessible. Change the neutralizing agent regularly. The manufacturer's instructions will specify how often it should be changed. Allow only trained professionals to service a sump. Take proper precautions when you open a sealed sump lid, and seal the lid tightly when you are finished. Refer to your local code for specific guidelines on servicing sumps.

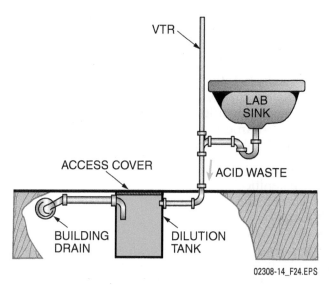

02308-14_F24.EPS

Figure 24 Venting a dilution tank through the roof.

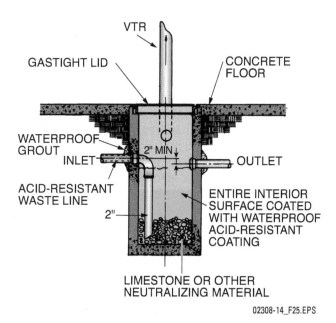

02308-14_F25.EPS

Figure 25 Neutralizing sump with neutralizing agent.

Gastight Seals

When you are dealing with contaminated stainless steel, you can use a lightweight crimping tool to create gastight seals. These seals can be used in contaminated stainless steel tubing up to 1 inch in diameter. The crimper removes the contaminated section and provides a uniform seal across the crimped surface.

Additional Resources

Corrosion-Resistant Piping Systems. 1994. Philip A. Schweitzer. New York: Marcel Dekker, Inc.
Water, Sanitary, and Waste Services for Buildings. 2002. Alan F. E. Wise and J. A. Swaffield. Boston, MA: Butterworth-Heinemann.

2.0.0 Section Review

1. Borosilicate glass pipe is joined by _____.
   a. fusion
   b. couplings
   c. adhesive
   d. threaded fittings

2. In the event of an accident, HDPE pipe is designed to _____.
   a. tear
   b. shatter
   c. bend
   d. thicken

3. When installing a horizontal run of silicon cast-iron pipe with fittings, place the supports no farther than _____.
   a. 7 feet apart
   b. 10 feet apart
   c. 13 feet apart
   d. 16 feet apart

4. When preparing to install stainless steel drainage pipe that is designed to handle chloride and organic acid waste, the type of steel that should be used is _____.
   a. duplex
   b. ferritic
   c. martensitic
   d. austenitic

5. One of the most commonly used neutralizing agents in acid dilution and neutralization sumps is _____.
   a. zeolite
   b. chloride
   c. sodium
   d. limestone

SUMMARY

Plumbers install DWV systems that are designed to handle corrosive wastes. Corrosive wastes are extremely acidic or alkaline chemicals. They can damage pipes and cause illness, injury, or death. Corrosive wastes must be disposed of through a separate drainage system after being treated. This module described how to install corrosive-resistant waste piping in commercial facilities.

Hazard communication plans help prevent chemical accidents. Employers must post a list of all chemicals that they use or store and create a written safety plan in case of an emergency. They must label or tag all chemical containers and keep SDSs for all chemicals on site.

The most common materials used in corrosive-waste pipe and fittings are borosilicate glass, thermoplastic and thermoset plastics, silicon cast-iron, and stainless steel. Each of these materials is used to handle different types of corrosive chemicals.

Borosilicate glass pipe is made from a specially reinforced transparent glass. It is used in systems with high-temperature wastes. Thermoplastic and thermoset pipe can handle acetic acids and chemical compounds containing aluminum, ammonia, barium, calcium, sodium, and mild sulfuric acids. Resistance will vary depending on the type of pipe. Silicon cast-iron pipe can handle almost any type of corrosive waste. Stainless steel pipe is available in four types of alloys for corrosive-resistant waste piping. The four alloys are ferritic, austenitic, martensitic, and duplex. Each type offers its own special corrosion-resistance capabilities.

Install corrosive-resistant waste systems using techniques that you have already learned. Specific requirements depend on the type of pipe. Glass pipe can be installed both horizontally and vertically. Use padded hangers for horizontal runs of glass pipe. Fit a protective casing on glass piping that is installed underground. Thermoplastic and thermoset pipe can be joined by fusion, solvent, mechanical joints, or threading. Do not use hangers that may cut or squeeze the pipe. Silicon cast-iron pipe comes in hub-and-spigot and no-hub styles. Install acid dilution and neutralization sumps according to local code. Stainless steel pipe may be welded. Different welding techniques apply to each steel alloy.

Review Questions

1. Some home septic systems are equipped with a separate tank to handle _____.
   a. food wastes
   b. stormwater runoff
   c. laundry wastes
   d. cleaning products

2. The term *SDS* replaces the older term _____.
   a. NFPA
   b. ULHS
   c. GHS
   d. MSDS

3. In the United States, the GHS HazCom system replaced the labeling system developed by the _____.
   a. National Fire Protection Association
   b. Underwriters Laboratories
   c. Occupational Safety and Health Administration
   d. National Safety Council

4. An organism or chemical that poses a health risk to humans, animals, or plants is classified as a(n) _____.
   a. toxic agent
   b. biohazard
   c. environmental hazard
   d. contaminant

5. For each chemical used or stored on site, a company must _____.
   a. notify the local health department
   b. file a report with OSHA
   c. keep an SDS on file
   d. prepare a disaster plan

6. The type of glass used to make corrosive-resistant waste piping is called _____.
   a. borosilicate
   b. thermoplastic
   c. silicon
   d. thermoset

7. In a corrosive-waste system, every fixture must _____.
   a. be certified for corrosive-waste handling
   b. be equipped with an antisiphon valve
   c. have a trap installed
   d. be inspected annually

8. If corrosive wastes are discharged at extreme temperatures, the piping used for the system should be _____.
   a. stainless steel
   b. CPVC
   c. silicon cast-iron
   d. borosilicate glass

9. The seals in couplings used to join borosilicate glass pipe are made from _____.
   a. neoprene
   b. TFE plastic
   c. silicone
   d. ABS plastic

10. If glass pipe is to be installed underground, most codes specify a maximum pipe length of _____.
    a. 12 feet
    b. 10 feet
    c. 5 feet
    d. 3 feet

11. Thermoset plastic piping made of furan resins should *not* be used with _____.
    a. acidic wastes
    b. food-processing wastes
    c. alkaline wastes
    d. oxidized wastes

12. To absorb the shock when cutting silicon cast-iron pipe, support the pipe _____.
    a. on a sling
    b. on a sandbag
    c. using saddles
    d. at the pipe ends

13. A vertical run of silicon cast-iron pipe should be supported every _____.

    a. 14 feet
    b. 12 feet
    c. 10 feet
    d. 8 feet

14. The type of stainless steel piping used to handle mild corrosive wastes is _____.

    a. duplex
    b. martensitic
    c. austenitic
    d. ferritic

15. Corrosive waste can be diluted in a sump if the flow is _____.

    a. no greater than 100 cubic feet per minute
    b. continuous
    c. intermittent
    d. metered

Trade Terms Quiz

Fill in the blank with the correct term that you learned from your study of this module.

1. Plastic pipe that is hardened into shape when manufactured is called _____.

2. A workplace safety program that informs and educates employees about the risks from chemicals is called a(n) _____.

3. Plastic pipe that can be softened and reshaped when it is exposed to heat is called _____.

4. Waste products that contain harsh chemicals that can damage pipe and cause illness, injury, or even death are called _____.

5. A device that uses heat to join thermoplastic pipe is called a(n) _____.

6. A(n) _____ is a chemical used to bond thermoset pipes and fittings without fusing them together.

7. The stainless steel alloy that contains nickel and stays strong at very low temperatures is called _____.

8. A(n) _____ is an organism or chemical that is dangerous to the health of living creatures.

9. The pipe material made from specially reinforced glass that does not expand or contract much with changes in temperature is called _____.

10. The _____ system consists of signs placed on laboratory walls that use pictographs to indicate hazards.

11. A resin that is highly resistant to solvents this is used to make thermoset pipe is called a(n) _____.

12. _____ is a stainless steel alloy that is very resistant to rust and has a very low carbon content.

13. A raised lip around the rim of borosilicate glass pipe is called a(n) _____.

14. _____ is a stainless steel alloy mix of austenitic and ferritic stainless steels that offers very good corrosion resistance.

15. Removable metal clamps with plastic seals, used to connect borosilicate glass pipe, are called _____.

16. _____ is a stainless steel alloy that is very strong but is mostly used to handle mild corrosive wastes.

Trade Terms

Adhesive
Austenitic stainless steel
Beaded end
Biohazard
Borosilicate glass
Corrosive waste

Coupling
Duplex stainless steel
Ferritic stainless steel
Furan
Fusion coil
Hazard communication

Martensitic stainless steel
Thermoplastic
Thermoset
Uniform Laboratory Hazard
   Signage

Trade Terms Introduced in This Module

Adhesive: A chemical used to bond thermoset pipes and fittings without fusing them together.

Austenitic stainless steel: A stainless steel alloy that contains nickel and stays strong at very low temperatures. Austenitic steel is the most popular type of steel for corrosive-waste pipes.

Beaded end: A raised lip around the rim of borosilicate glass pipe.

Biohazard: An organism or chemical that is dangerous to the health of living creatures.

Borosilicate glass: A pipe material made from specially reinforced glass that does not expand or contract much with changes in temperature.

Corrosive waste: Waste products that contain harsh chemicals that can damage pipe and cause illness, injury, or even death.

Coupling: Removable metal clamps with plastic seals, used to connect borosilicate glass pipe.

Duplex stainless steel: A stainless steel alloy mix of austenitic and ferritic stainless steels that offers very good corrosion resistance.

Ferritic stainless steel: A stainless steel alloy that is very resistant to rust and has a very low carbon content.

Furan: Any resin that is highly resistant to solvents. Furans are used to make thermoset pipe.

Fusion coil: A device that uses heat to join thermoplastic pipe.

Hazard communication: A workplace safety program that informs and educates employees about the risks from chemicals.

Martensitic stainless steel: A stainless steel alloy that is very strong but is mostly used to handle mild corrosive wastes.

Thermoplastic: A type of plastic pipe that can be softened and reshaped when it is exposed to heat.

Thermoset: A type of plastic pipe that is hardened into shape when manufactured.

Uniform Laboratory Hazard Signage: Signs placed on laboratory walls that use pictographs to indicate hazards.

Additional Resources

This module presents thorough resources for task training. The following resource material is suggested for further study.

Corrosion-Resistant Piping Systems. 1994. Philip A. Schweitzer. New York: Marcel Dekker, Inc.

Hazard Communication Made Easy: A Checklist Approach to OSHA Compliance. 2000. Sean M. Nelson and John R. Grubbs. Rockville, MD: ABS Group, Government Institutes.

Water, Sanitary, and Waste Services for Buildings. 2002. Alan F. E. Wise and J. A. Swaffield. Boston, MA: Butterworth-Heinemann.

Figure Credits

Courtesy of IRWIN Tools, CO07

U.S. Department of Labor **http://www.dol.gov**, Figures 3–5

Logan Oil Tools, Inc., Figure 9

Courtesy of Charlotte Pipe and Foundry Company, Figure 20

Courtesy of Victaulic Company, Figure 22

Grinnell Products, Tyco Mech. Products brand, Figure 23

Answer	Section Reference	Objective
Section One		
1. d	1.1.0	1a
2. c	1.2.1	1b
Section Two		
1. b	2.1.1	2a
2. a	2.2.1	2b
3. a	2.3.2	2c
4. b	2.4.1	2d
5. d	2.5.0	2e

NCCER CURRICULA — USER UPDATE

NCCER makes every effort to keep its textbooks up-to-date and free of technical errors. We appreciate your help in this process. If you find an error, a typographical mistake, or an inaccuracy in NCCER's curricula, please fill out this form (or a photocopy), or complete the online form at **www.nccer.org/olf**. Be sure to include the exact module ID number, page number, a detailed description, and your recommended correction. Your input will be brought to the attention of the Authoring Team. Thank you for your assistance.

Instructors – If you have an idea for improving this textbook, or have found that additional materials were necessary to teach this module effectively, please let us know so that we may present your suggestions to the Authoring Team.

NCCER Product Development and Revision
13614 Progress Blvd., Alachua, FL 32615

Email: curriculum@nccer.org
Online: www.nccer.org/olf

❏ Trainee Guide   ❏ Lesson Plans   ❏ Exam   ❏ PowerPoints   Other _____

Craft / Level: _____   Copyright Date: _____

Module ID Number / Title: _____

Section Number(s): _____

Description: _____

_____

_____

_____

_____

Recommended Correction: _____

_____

_____

_____

_____

Your Name: _____

Address: _____

_____

Email: _____   Phone: _____

02309-14

Compressed Air

OVERVIEW

Compressed air is used to operate a variety of tools, machines, and equipment in many industries. Compressed-air systems can be portable or they may be permanently installed in buildings such as factories to provide several machines or tools with a constant supply of compressed air. Plumbers are responsible for installing compressed-air systems.

Module Eight

Trainees with successful module completions may be eligible for credentialing through the NCCER Registry. To learn more, go to **www.nccer.org** or contact us at **1.888.622.3720**. Our website has information on the latest product releases and training, as well as online versions of our *Cornerstone* magazine and Pearson's product catalog.

Your feedback is welcome. You may email your comments to **curriculum@nccer.org,** send general comments and inquiries to **info@nccer.org**, or fill in the User Update form at the back of this module.

This information is general in nature and intended for training purposes only. Actual performance of activities described in this manual requires compliance with all applicable operating, service, maintenance, and safety procedures under the direction of qualified personnel. References in this manual to patented or proprietary devices do not constitute a recommendation of their use.

Objectives

When you have completed this module, you will be able to do the following:

1. Explain the types, functions, and capacities of different compressed-air systems.
   a. Explain the properties of air.
   b. Explain how to measure air pressure.
2. Identify the different methods of conditioning compressed air.
   a. Describe how aftercoolers are used to condition compressed air.
   b. Describe how air dryers are used to condition compressed air.
3. Identify the safety issues related to compressed-air systems.
   a. Explain the safety issues related to high temperature and high air pressure.
   b. Explain the safety issues related to installing, repairing, and servicing compressed-air systems.
4. Explain how to install a basic compressed-air system.
   a. Identify the components of compressed-air systems.
   b. Describe the steps required for installing a compressed-air system.

Performance Tasks

Under the supervision of your instructor, you should be able to do the following:

1. Identify components of compressed-air systems.
2. Design a basic compressed-air system.

Trade Terms

Actual cubic feet per minute	Constant speed control	Pneumatics	Standard cubic feet per minute
Aftercooler	Control range	Reciprocating compressor	Stop/start control
Air dryer	Coupling	Refrigeration air dryer	Throttle
Air end	Desiccant air dryer	Rotary compressor	Unloading
Air-pressure regulator	Hydraulics	Rotary screw	Use factor
Air receiver	Intercooler	Sequencing control	Water drip leg
Air throttle valve	Lubricator	Standard conditions	Whip check
Compressible	Network control		
	Part-load control		

Industry-Recognized Credentials

If you're training through an NCCER-accredited sponsor, you may be eligible for credentials from NCCER's Registry. The ID number for this module is 02309-14. Note that this module may have been used in other NCCER curricula and may apply to other level completions. Contact NCCER's Registry at 888.622.3720 or go to **www.nccer.org** for more information.

Code Note

Codes vary among jurisdictions. Because of the variations in code, consult the applicable code whenever regulations are in question. Referring to an incorrect set of codes can cause as much trouble as failing to reference codes altogether. Obtain, review, and familiarize yourself with your local adopted code.

Contents

Topics to be presented in this module include:

Figures

SECTION ONE

1.0.0 TYPES, FUNCTIONS, AND CAPACITIES OF COMPRESSED-AIR SYSTEMS

Objective

Explain the types, functions, and capacities of different compressed-air systems.
a. Explain the properties of air.
b. Explain how to measure air pressure.

Trade Terms

Actual cubic feet per minute: A measure of the air volume at the air-compressor inlet under actual temperature and pressure conditions. It is often abbreviated acfm.

Compressible: Able to be reduced in volume.

Hydraulics: The set of physical laws governing the motion of water and other liquids.

Pneumatics: The set of physical laws governing the motion of air and other gases.

Standard conditions: An air pressure of 14.7 psia, temperature of 68°F, and barometric pressure of 29.92 inches of mercury.

Standard cubic feet per minute: A measure of the air volume at the air-compressor inlet under standard conditions. Standard cubic feet per minute is often abbreviated scfm.

Much of your training has focused on systems that circulate or dispose of water and liquid wastes. You have learned about temperature, pressure, density, and friction, which are concepts related to hydraulics, the rules and principles that govern the motion of water and other liquids.

Systems that use compressed air instead of water rely on a different set of rules called pneumatics. The rules and principles of pneumatics govern the motion of air and other gases. Temperature, pressure, density, and friction play important roles in pneumatics.

Air pressure is an important principle in pneumatics. It plays a key role in many plumbing operations by:

- Equalizing pressure in drainage pipes through vents
- Absorbing shock waves inside water-hammer arresters
- Providing power for air hammers, nail guns, and other tools

There are many similarities and some important differences between water and air. Learning the basics of pneumatics will help you understand how compressed-air systems work.

1.1.0 Understanding the Properties of Air

Unlike water, air is compressible. That means it can be squeezed to fit into a smaller space than it occupies at normal air pressure. This increases its pressure. The higher the air's pressure, the more force it applies to the walls of the surrounding container. Storage tanks and piping have to be strong enough to withstand the air pressure in a system. Always check the pressure rating of materials and equipment before you install them.

When air is compressed, its temperature increases, and it gives off heat. When it expands, it absorbs heat from its surroundings. Consequently, compressed-air systems are designed to protect operators against temperature extremes. Compressors have insulation to prevent operators from touching a hot surface. Pneumatic tools are designed to protect operators from the cold. Learn how to handle compressed-air tools properly. Always wear appropriate personal protective equipment when working with compressed-air systems and tools.

1.2.0 Measuring Air Pressure

Compressed-air pressure is measured in pounds per square inch gauge (psig). Psig uses local air pressure as the zero point. At sea level, 0 psig is

---

Who Invented Compressed Air and Steam Power?

More than 1,500 years ago, a Greek mathematician and inventor named Hero described how water and compressed air could be used to power machines. He wrote a book called *The Pneumatics* that explained his ideas in detail. One of his inventions used compressed air to force a jet of water through a nozzle. Another device channeled steam from boiling water into a small sphere held by two pivots. As the steam escaped through two bent tubes on the sphere, it caused the sphere to rotate. This device is one of the first recorded descriptions of a steam engine.

14.7 pounds per square inch absolute (psia). The flow rate of compressed air is measured in cubic feet per minute (cfm).

The air-inlet capacity of an air compressor can be expressed in terms of two forms of cfm: either standard cubic feet per minute (scfm) or actual cubic feet per minute (acfm). Scfm is the volume of air measured at standard conditions. The following are considered standard conditions:

- A sea-level air pressure of 14.7 psia
- A temperature of 68°F
- A barometric pressure of 29.92 inches of mercury

On the other hand, acfm is a measure of the air volume under the actual temperature and pressure conditions at the inlet of the air compressor. The acfm of an air compressor remains constant regardless of the ambient temperature and pressure.

Where Do the Words "Hydraulic" and "Pneumatic" Come From?

The word *hydraulic* comes from the ancient Greek word *hydraulis*, which was a water-powered musical instrument. The word *pneumatic* comes from another ancient Greek word, *pnein*, which means to sneeze!

Boyle's Law

To convert measurements from standard cubic feet per minute (scfm) to actual cubic feet per minute (acfm) and vice versa, use Boyle's law. Boyle's law states that if the air pressure in a system decreases, the air volume will increase proportionally:

$$P_1V_1 = P_2V_2$$

Where:

$$P = \text{Pressure}$$

$$V = \text{Volume}$$

The calculation is performed using 29.92 inches of mercury as an absolute value for P_1.

Consider the following example: You need to find the equivalent of 35 scfm for an air compressor with an inlet vacuum level of 27 inches of mercury. The air compressor's inlet temperature is 68°F, and the compressor is located at sea level. Begin by converting the given inlet vacuum value to an absolute figure.

$$P_2 = 29.92 - 27$$

$$P_2 = 2.92$$

Now substitute the given numbers for the variables in Boyle's law and perform the math:

$$P_1V_1 = P_2V_2$$

$$29.92 \times 35 \text{ scfm} = 2.92 \times V_2 \text{ acfm}$$

$$1047.2 \text{ scfm} = 2.92 \times V_2 \text{ acfm}$$

$$V_2 \text{ acfm} = 1047.2/2.92$$

$$V_2 \text{ acfm} = 358.63$$

The equivalent air volume for the air compressor is 358.63 acfm.

Additional Resources

Compressed Air Data: Handbook of Pneumatic Engineering Practice. William Lawrence Saunders and Charles Austin Hirschberg, editors. 2002. Honolulu, HI: University Press of the Pacific.

Fact Sheet #4, "Pressure Drop and Controlling System Pressure." 1998. Compressed Air Challenge. **compressedairchallenge.org**

Hydraulics and Pneumatics: A Technician's and Engineer's Guide, Third Edition. Andrew Parr. 2011. New York: Butterworth-Heinemann.

1.0.0 Section Review

1. When air expands, what does it absorb from its surroundings?

   a. More air
   b. Heat
   c. Moisture
   d. Contaminants

2. The barometric pressure at standard conditions is _____.

   a. 29.92 inches of mercury
   b. 14.7 inches of mercury
   c. 4.34 inches of mercury
   d. 68 inches of mercury

SECTION TWO

2.0.0 CONDITIONING COMPRESSED AIR

Objective

Identify the different methods of conditioning compressed air.
  a. Describe how aftercoolers are used to condition compressed air.
  b. Describe how air dryers are used to condition compressed air.

Trade Terms

Aftercooler: A device that cools air by passing it through a heat exchanger, causing moisture to condense out of the air.

Air dryer: A device that removes all moisture from compressed air.

Desiccant air dryer: A type of air dryer in which chemical pellets absorb moisture from compressed air.

Refrigeration air dryer: A type of air dryer in which air is cooled as it passes through a heat exchanger cooled by a liquid refrigerant.

It takes 7 cubic feet of air to make 1 cubic foot of compressed air at 100 psig. However, the moisture in the air is compressed along with the air. That means that the compressed air has seven times the moisture of free air. Water vapor in the air can cause rust inside the compressed-air system and inside the machines that use the air. In addition, compressed air is hotter than free air. Excess heat could damage the system. Before compressed air can be used, therefore, it has to be dried and cooled. The following sections discuss the two code-approved methods for conditioning compressed air.

2.1.0 Using Aftercoolers to Condition Compressed Air

Compressed-air systems use aftercoolers to cool and dry the air. An aftercooler cools the air by passing it through a heat exchanger. The drop in temperature causes the moisture in the air to condense. The condensed moisture precipitates out of the air and drains away. Aftercoolers may use either air or water to cool the air (see *Figure 1*). Water cooling systems require the use of pipe materials that will not corrode when exposed to water, such as copper or stainless steel.

02309-14_F01.EPS

Figure 1 Packaged compressed-air unit designed for air or water cooling.

2.2.0 Using Air Dryers to Condition Compressed Air

Aftercoolers remove most of the moisture from compressed air, but a small amount can remain. Certain applications require air with no trace of moisture. In such cases, air dryers are installed to remove all moisture. Refrigeration air dryers and desiccant air dryers are the most common types.

Refrigeration units pass air through a heat exchanger. A liquid refrigerant is also pumped through the heat exchanger. The refrigerant chills the air to the point where all the moisture in the air precipitates and is drained away. Refrigeration air dryers are also effective at removing compressor lubrication oil from the air.

Desiccant air dryers circulate air through a container filled with chemical pellets that absorb all the moisture from the air. Silica gel is a common desiccant. The pellets should be replaced regularly. Desiccant air dryers can only be used with air compressors if an aftercooler is installed upstream of the air dryer to cool the air. Both types of air dryer should be drained daily. Refer to your local code before installing an aftercooler or air dryer.

> **WARNING!**
>
> While some types of air compressors can be used in respiration applications, none of the compressors described in this module can be used for respiration or SCBA (self-contained breathing apparatus) applications. Oil-filled air entering the lungs under pressure is a serious health hazard. Codes prohibit the installation of compressors with oil-lubricated bearings in medical or respiration applications.

Small Electric Motors and Air Compressors

Air compressors that are powered by small electric motors (under 25 horsepower) usually stop running between cycles. Install an air receiver in the system to allow the compressor to stop and start as required. When the air pressure in the air receiver reaches a preset maximum, a pressure switch turns off the motor. When the air pressure drops below a preset minimum, the switch turns on the motor again.

Additional Resources

Fact Sheet #1, "Assessing Compressed Air Needs." 1998. Compressed Air Challenge.
compressedairchallenge.org
Pipefitters Handbook. Forrest R. Lindsey. New York: Industrial Press Inc.

2.0.0 Section Review

1. In an aftercooler, air is cooled by passing it through a(n) _____.

    a. cold-water tube
    b. heat exchanger
    c. compressor
    d. expansion chamber

2. A commonly used desiccant material in desiccant air dryers is _____.

    a. chlorine
    b. sodium
    c. carbon
    d. silica

3.0.0 COMPRESSED-AIR SYSTEM SAFETY

Objective

Identify the safety issues related to compressed-air systems.

a. Explain the safety issues related to high temperature and high air pressure.
b. Explain the safety issues related to installing, repairing, and servicing compressed-air systems.

Trade Term

Coupling: A snap-fit plug connecting a tool or machine to a compressed-air system.

Working with compressed air can present a number of safety risks. As with all aspects of installing and servicing plumbing systems, you should maintain good safety practices when installing and working with compressed-air systems.

Compressed air can damage tools and equipment or cause injury if not properly handled. Before operating tools that use compressed air, always refer to the manufacturer's manual and become familiar with any specific operating procedures and safety rules. When using compressed-air and air-powered tools, consider the safety of any co-workers that are in the proximity of your work.

3.1.0 Understanding High-Temperature and High-Air-Pressure Safety

Compressed air can be dangerous because of its high temperature and high air pressure. Air compressors can become very hot and can cause serious burns, and accidental release of compressed air can cause injury or death. Remember to exercise caution in the following ways:

- Wear appropriate personal protective equipment when using any air-powered tool.
- Ensure that guards remain in place around rotary components.
- Never place a compressed-air source in or near your nose, mouth, eyes, or skin.
- Wear proper hearing protection. The very high pitch of some air-operated tools could cause hearing damage or loss.
- Keep your hands away from the rotary components of a compressed-air system.
- Ensure that all connections are secure before activating an air-powered tool or machine. A loose connection can cause a hose to whip around with tremendous force.

Take suitable safety precautions when installing, repairing, or servicing air compression systems. Always discharge the air pressure for any tool before you replace parts such as bits, blades, or discs. Always discharge the air pressure before you perform maintenance on any power tool, and never bypass the trigger lock on any power tool. Ensure that all pressurized air has been released from the system before installing or servicing any components on an air compression system.

3.2.0 Understanding Installation, Repair, and Service Safety

Use caution during installation of piping for compressed-air systems. Wear appropriate personal protective equipment when installing threaded pipe. The chips made from threading pipe are sharp and could cut you. Always use a rag to wipe off excess oil; never use your bare hands. In addition, always ensure that the inside of the pipe is clean before you install it. Debris inside the pipe will become projectiles when accelerated by the air, and can damage equipment and cause serious injury.

Finally, protect compressed-air tools and equipment from damage. When using any air-operated tool or machine, make sure that all hoses and couplings are secured. Otherwise, the pressure may be too low for the tool or machine to operate properly. The result could damage the device. Before you service a compressed-air system, shut off the system and allow the pressure to escape from all components and air lines.

Additional Resources

Fact Sheet #7, "Compressed Air System Leaks." 1998. Compressed Air Challenge.
compressedairchallenge.org
Pneumatic Handbook, Eighth Edition. 1998. Antony Barber. New York: Elsevier Science.

3.0.0 Section Review

1. When is it safe to place a compressed-air source in or near your nose, mouth, eyes, or skin?

   a. Under no circumstances
   b. When the compressed-air supply is turned off
   c. When the pressure is below 50 psig
   d. When wearing appropriate personal protective equipment

2. To remove excess oil from pipe, always use _____.

   a. your hand
   b. a rag
   c. blasts of compressed air
   d. water

4.0.0 INSTALLING COMPRESSED-AIR SYSTEMS

Objective

Explain how to install a basic compressed-air system.

a. Identify the components of compressed-air systems.
b. Describe the steps required for installing a compressed-air system.

Performance Tasks

Identify components of compressed-air systems.
Design a basic compressed-air system.

Trade Terms

Air end: A chamber at the end of the rotary screws in a rotary compressor where air is compressed.

Air-pressure regulator: A valve that keeps air pressure within the control range.

Air receiver: A storage tank for compressed air.

Air throttle valve: A spring-loaded valve used to control or shut off airflow in a compressed-air system.

Constant speed control: A control that vents excess air pressure from an air compressor.

Control range: The range between the highest- and lowest-allowable air pressures in a compressed-air system.

Intercooler: A heat exchanger that removes heat from air as it moves through a multistage air-compressor system.

Lubricator: A device that injects a fine oil mist into the air supply line, lubricating tools and machines.

Network control: A device that controls multiple compressors by linking them together into a single network controlled by a computer.

Part-load control: A device that allows an air compressor to operate at a given percentage of its normal output.

Reciprocating compressor: A type of air compressor in which air is compressed by a piston.

Rotary compressor: A type of air compressor in which rotary screws compress the air.

Rotary screw: Helical devices in a rotary compressor that compress air by passing it down the screw path.

Sequencing control: A device that adjusts air pressure in a compressed-air system by activating and deactivating individual compressors.

Stop/start control: A device that turns an air compressor off when the control range is exceeded.

Throttle: A device that controls airflow into an air compressor by adjusting the compressor's intake valve.

Unloading: The process of relieving excess air pressure from an air compressor.

Use factor: Expressed as a percentage of an average hour of use, the amount of time that a tool or machine will be operated.

Water drip leg: A vertical pipe on an air supply line that acts as a drain for condensed water in the line.

Whip check: A cable attached to the ends of an air supply hose and a tool, or to the ends of two hoses that are connected to each other, to prevent the hose from whipping uncontrollably when charged with compressed air.

A good compressed-air system provides the right amount of high-quality air at the correct pressure. Like other plumbing systems, compressed-air systems must be installed correctly to work well. Work closely with the architect, the customer, and the manufacturers. They can give you information about the desired volumes, pressures, and use factors. In this section, you will learn about the various components that make up a compressed-air system and how to install systems according to the local applicable code.

4.1.0 Identifying Compressed-Air System Components

A compressed-air system collects, prepares, and distributes air to a variety of tools and machines (see *Figure 2*). The following are the major components of a compressed-air system:

- Air compressors and related equipment
- Aftercoolers and air dryers
- Supply piping
- Controls

Each of these components is discussed in greater detail in the following sections. Review how each one works, both on its own and in combination with the other components. Your local code will have specific requirements for components that can be used in your area.

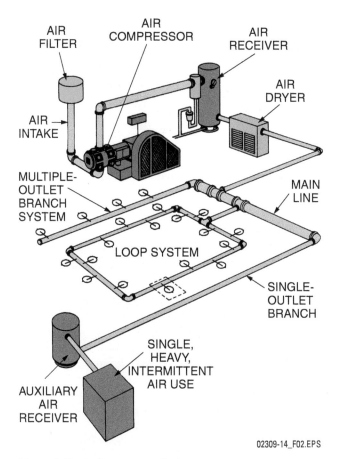

Figure 2 Typical compressed-air system.

4.1.1 Air Compressors

Air compressors condense air to the desired pressure and distribute it throughout the system. Air compressors come in a variety of sizes. If you have worked on a construction site, you have probably seen a portable air compressor like the one in *Figure 3*. Many trades use portable air compressors to operate pneumatic tools. Plumbers use portable air compressors for pressure tests on water supply systems. On the other end of the scale are compressors that supply air to an entire factory.

Air pressures vary widely. Some air compressors used in precision work produce 10 psig or less. Air compressors used in wind tunnels, by contrast, produce more than 2,500 psig. In the average building system, air compressors generally operate at pressures of 500 psig or lower. Most air compressors operate between 90 and 125 psig. Rotary compressors and reciprocating compressors are the most common types of air compressors. Centrifugal air compressors are also available, but these are large systems used in highly specialized applications. Plumbers are less likely to encounter these on the job than rotary and reciprocating types.

Rotary compressors use rotary screws to compress the air (see *Figure 4*). Rotary screws work by compressing air as it travels down the screw path.

02309-14_F03.EPS

Figure 3 Portable air compressor.

The air is directed into a chamber, called the air end, at the end of the screws. As more air collects there, it is compressed further. Rotary compressors are widely used in commercial and industrial applications.

In a reciprocating compressor, air is compressed by a piston (see *Figure 5*). The compression stroke pumps the air into the supply line. Reciprocating compressors are most widely used in residential applications and to power smaller tools such as nail guns. They are less common than rotary compressors in commercial applications.

Electric motors or gasoline engines provide the power for most air compressors. All compressors have air filters on the inlet side.

From the compressor, the air flows into an air receiver, which acts as a storage tank and also smooths out the airflow from a reciprocating com-

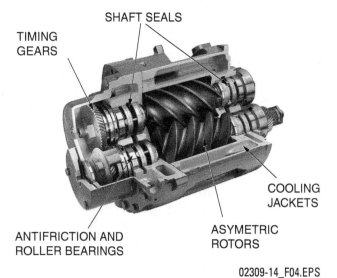

02309-14_F04.EPS

Figure 4 Rotary screw element in a rotary compressor.

02309-14_F05.EPS

Figure 5 Reciprocating compressor with piston.

02309-14_F06.EPS

Figure 6 Air-receiver relief valve.

pressor. Water that was trapped in the air while it was being compressed condenses out in the air receiver. Many air receivers have automatic drains to remove the water from the tank. Other tanks must be drained manually. Relief valves allow air receivers to vent excess pressure (see *Figure 6*). Codes require that all air receivers and relief valves meet the safety standards of the American Society of Mechanical Engineers (ASME).

Multistage air compressors have more than one compressor. Each stage is a complete compressor. Each stage passes air on to the next one to be compressed further until the desired pressure is reached. Multistage air compressors can deliver a larger volume of air than single-stage units. Multistage compressors are usually equipped with an intercooler. Intercoolers are heat exchangers that remove heat from compressed air as it moves from the lower pressures of the early stages to the higher pressures of the later stages. The number of stages depends on the volume demands of the compressed-air system.

> **WARNING!**
>
> Never use a compressed-air system to clean surfaces, tools, equipment, or yourself or others. High-pressure air can send dirt and debris flying at high speed, which can cause injury. It can also cause severe injury if pointed at soft tissue.

Air compressors, motors, and air receivers can be arranged to fit the available space. The motor and air compressor can be mounted on top of the storage tank (see *Figure 7*). Larger systems are often installed with the air storage tank separate from the air compressor and motor.

Air compressors tend to make noise and vibrate while operating. You can install a variety of devices to reduce noise and vibration. Intake silencers quiet the flow of air into the compressor. Flexible metal hoses prevent the transmission of vibrations from the compressor to the rest of the system (see *Figure 8*). Shock-absorbing mounts isolate the unit from the floor.

02309-14_F07.EPS

Figure 7 Air compressor with motor and compressor mounted on top of the air receiver.

4.1.2 Piping

When the design calls for supply pipe that is 2½ inches in diameter or larger, use standard-weight Schedule 40 black-iron pipe. Use of galvanized steel is discouraged because flakes from the gal-

02309-14_F08.EPS

Figure 8 Flexible metal hose used for absorbing vibrations in an air compressor.

vanizing layer can come loose in the airstream. These flakes could damage air and hydraulic components. Some large industrial systems call for welded steel pipe.

For systems that require pipe with a diameter smaller than 2 inches, use Type L copper tubing. When installing this type of pipe in a compressed-air system, use either cast-brass or wrought-copper solder joint fittings. Red-brass pipe with threaded cast-brass fittings can also be used.

Install piping for compressed-air systems using the methods you have already learned. Your local code will specify the correct intervals for pipe hangers and supports. Use only fasteners that are permitted by code. Never use a black-iron fitting designed for drainage systems in a compressed-air system. The interior of such a fitting is shaped for the flow of liquids, not air.

4.1.3 Hoses and Couplings

Flexible hoses are used to connect compressed-air sources to tools and other equipment (see *Figure 9*). Hoses are commonly manufactured from nylon, polyurethane, and rubber. They are available in a variety of lengths and are rated for the maximum operating pressure of the compressed-air system. To prevent hoses from rupturing when under pressure, hoses are typically manufactured with one or more layers of reinforcing material such as metal mesh. This also helps prevent the hose from kinking, or pinching, when bent sharply. Recoil hoses are designed to automatically return to a coiled position when not in use.

The air hose must be connected properly and securely. An unsecured air hose can come loose and whip around violently, causing serious injury. To prevent this, some fittings require the use of whip checks to keep them from coming loose (see *Figure 10*). A whip check is a strong steel cable with loops on either end. The loops are attached to the ends of the air supply hose and the tool, or to the ends of two hoses that are connected to each other. Occupational Safety and Health Administration (OSHA) regulations require the use of whip checks or other means of preventing air supply hoses from whipping if compressed air is flowing through them when they are disconnected.

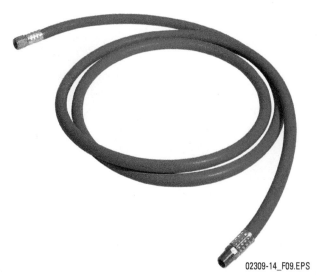

02309-14_F09.EPS

Figure 9 Coiled rubber compressed-air hose.

> **WARNING!**
>
> Whip checks are rated by the maximum working pressure of the compressed-air system. Always use a properly rated whip check. Using a whip check rated for less pressure than the system can result in failure of the whip check in the event of an accident, which can result in serious injury or even death.

Tools that require very high air pressure, such as jackhammers, use hoses with a larger diameter than those used for other types of air tools. These hoses are fitted with cam-operated twist locks secured by a removable pin (see *Figure 11*). Be sure to follow the manufacturer's instructions when using these locks to connect high-pressure hoses with tools. Always check to ensure that pressurized air has been bled completely from these hoses before disconnecting them from the tool.

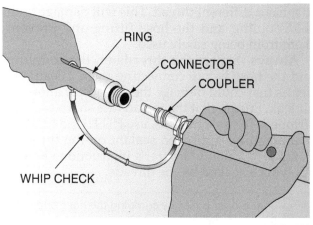

02309-14_F10.EPS

Figure 10 Whip check properly installed on a hose and a compressed-air drill.

02309-14_F11.EPS

Figure 11 Lock for a high-pressure air hose.

The connection between the compressed-air system and the tool or equipment using the air can be a location of pressure loss. Pressure is maintained through the use of properly sized couplings (see *Figure 12*). A coupling is a snap-fit plug that connects a tool or machine to the compressed-air system. A spring-loaded valve prevents airflow until the connection is secure. Couplings are available in rigid and swivel varieties. Rigid couplings do not rotate. Swivel couplings are designed to rotate independently of the hose. This makes it easier to attach and remove threaded fittings.

Couplings and hose fittings are available in a variety of threads sizes. In the United States, compressed-air connections follow American National Standards Institute (ANSI) Standard *B1.20.1, Pipe Threads, General Purpose (Inch)*, which lists the dimensions and gauges of pipe threads used in general-purpose applications. Never attempt to thread a connection onto a hose fitting that has a different thread. This will damage both the coupling and the hose fitting and prevent them from being safely used.

Always inspect the threads on the coupling and the hose before connecting them. Clean any debris that could damage the threads or allow an air leak. Ensure that the threads are sharp and not stripped. To prevent leaks, use PTFE tape or other manufacturer-approved sealant on the threads. Refer to the manufacturer's instructions before applying a sealant on the threads.

> **CAUTION**
>
> Join rubber hoses by crimping the hose ends. Never use hose clamps to join rubber hose in a compressed-air system. Hose clamps are not designed to be used in pressurized systems and will fail.

4.1.4 Controls

Compressed-air systems are designed to operate within a specified control range, the range between the highest- and lowest-allowable air pressures. If the system pressure falls below the control range, the machines will not be able to operate properly. If the air pressure is too high, the system and the machines could be damaged. Compressed-air systems use a variety of controls to keep the system pressure within the control range. The controls are installed on the air compressor and in the distribution system. The number and types of controls depend on the size and complexity of the system.

Air-Compressor Controls – A stop/start control turns the air compressor off when the system pressure exceeds the control range. The stop/start control is a simple pressure switch connected to a pressure gauge. Stop/start controls should be used only in compressed-air systems that do not cycle frequently. They are intended for light-duty systems.

Throttles control the airflow into an air compressor. The throttle widens and narrows the air compressor's intake valve. Adjusting the valve modulates the output. Throttles are used in centrifugal and rotary compressors.

Constant speed controls are more complex than stop/start and throttling controls. They allow excess air pressure to bleed out of the air compressor when the control range is exceeded. This is called unloading the air compressor. A constant speed control allows the motor to run as the air compressor unloads. The system's energy efficiency may drop as a result.

Part-load controls allow air compressors to operate at a percentage of their normal output. For example, a four-step control allows the operator to select an output of 25, 50, 75, or 100 percent.

Many large compressed-air systems use more than one compressor. Two types of controls can be used to adjust the pressure in such a system. Sequencing controls adjust the overall system pressure by turning individual compressors on or off. Sequencing controls are also called single-master controls. Network controls are more complex and efficient. Individual air-compressor controls are linked together into a single network controlled by a computer. Network controls allow for more precise control of the system pressure. Network controls are also called multi-master controls.

Distribution System Controls – Shutoff valves cut off the airflow in the supply line in an emergency. Use either ball or gate valves for this pur-

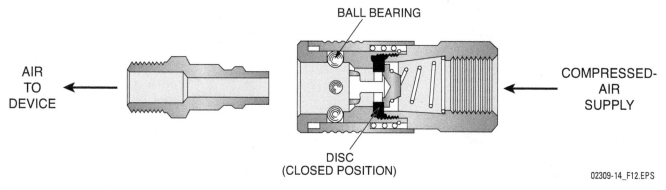

BALL BEARING

AIR TO DEVICE ←

COMPRESSED-AIR SUPPLY ←

DISC (CLOSED POSITION)

02309-14_F12.EPS

Figure 12 Cross section of a coupling used in compressed-air systems.

pose. Spring-loaded, plug-style air throttle valves are suitable for use as shutoff valves. They can be used as throttles on individual tools as well. Ensure that the pressure rating of the valve is suitable for the application.

High air pressure can damage tools and equipment. Air-pressure regulators (*Figure 13*) ensure that the air pressure remains below the high end of the control range. The regulator valve opens, allowing excess pressure to bleed into the regulator body. When the line pressure falls within the control zone again, the valve closes. Regulators are equipped with a pressure gauge. Install an air filter ahead of the regulator to trap particles that could get stuck in the regulator.

Some systems may also require a lubricator downstream from the regulator (see *Figure 14*). A lubricator injects a fine oil mist into the supply line. The oil mist lubricates the tools, equipment, and machines powered by compressed air. The manufacturer's instructions list the types of oil that are safe to use.

4.2.0 Installing a Compressed-Air System

Install the air compressor, air receiver, filters, and air dryers below the tools and machines that they serve. A basement is a good location. Like water supply systems, typical compressed-air distribution systems have the following components:

- A main line
- Risers
- Branch lines
- Feeder lines

Locate takeoffs for branch lines and risers where tools and machines will be used. Install feeder lines to connect the branch lines to individual tools and machines. Install valves on all ris-

ers, branches, and connections that are installed in anticipation of future system expansion. The takeoffs for all feeder and branch lines should be at the top of the main lines. Install mains and branches so that they slope downward in the direction the air is flowing. Refer to your local code for the proper degree of slope. Use approved hangers and install them at the proper intervals.

Aftercoolers and air dryers remove water vapor from air before it enters the distribution system. Another option is to design the distribution system so that it removes the water. Install a water drip leg at the end of the main line and in branch lines. Water drip legs are vertical pipes into which water can drain. Install either a manual blowdown valve or an automatic drain valve on each water drip leg. The valve prevents the water drip leg from disrupting the airflow. Install water drip legs at regular intervals along the distribution system. One leg every 50 to 100 feet should be enough.

If the compressed-air system is installed correctly, it will require only periodic maintenance. Low pressure is one of the most common problems in compressed-air systems. The following are common causes of low air pressure:

- Inadequate sizing of pipes and hoses
- Excessive leakage
- Insufficient air-compressor capacity

When inspecting or servicing a compressed-air system, consider each of these factors. Ensure that the piping, hoses, and air compressors are all properly sized for the needs of the system. Inspect the system for leaks, which are common at fitting connections and pipe bends. Check to see whether the couplings from the air supply to the tool or machine are the correct size and type. A careful installation will prevent unnecessary service calls in the future.

02309-14_F13.EPS

Figure 13 Air-pressure regulator.

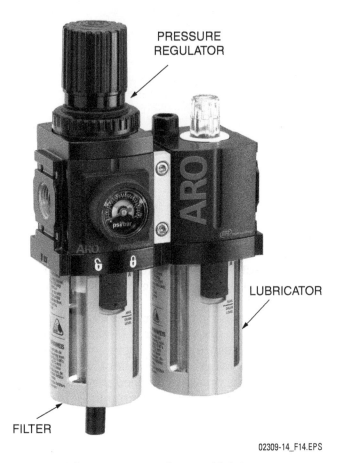

PRESSURE
REGULATOR

LUBRICATOR

FILTER

02309-14_F14.EPS

Figure 14 Filter, pressure regulator, and lubricator installation.

Additional Resources

Best Practices for Compressed Air Systems, Second Edition. William Scales. 2007. The Compressed Air Challenge.

Compressed Air Systems: A Guidebook on Energy and Cost Savings. E. M. Talbot. 1993. Lilburn, GA: Fairmont Press.

Fact Sheet #5, "Maintenance of Compressed Air Systems for Peak Performance." 1998. Compressed Air Challenge. **compressedairchallenge.org**

Fact Sheet #6, "Compressed Air System Controls." 1998. Compressed Air Challenge. **compressedairchallenge.org**

Improving Compressed Air System Performance. Vienna, VA: Resource Dynamics Corporation for Compressed Air Challenge. **compressedairchallenge.org**

4.0.0 Section Review

1. In a typical building system, the operating pressure of an air compressor is generally no more than _____.

   a. 200 psig
   b. 300 psig
   c. 400 psig
   d. 500 psig

2. One of the most common problems in compressed-air systems is _____.

   a. low temperature
   b. high temperature
   c. low pressure
   d. high pressure

SUMMARY

Compressed-air systems provide air for a variety of tools, machines, and equipment. In this module, you learned how to size and install large compressed-air systems designed to serve an entire building. Compressed-air systems operate under the principles of pneumatics. Temperature, pressure, density, and friction are each an important aspect of pneumatics.

Air pressure plays a key role in many plumbing operations. Compressed-air pressure is measured in psig. The flow rate of compressed air is measured in cfm. Air-compressor inlet capacity is measured in acfm, which measures the actual inlet conditions. Another common measure of air pressure is scfm, which measures air volume at standard conditions.

This module also described the components of compressed-air systems. These systems consist of air compressors, aftercoolers and air dryers, supply piping, and controls. Air compressors condense free air to a specific pressure and distribute it. Most systems use either reciprocating or rotary compressors. Aftercoolers remove a large amount of moisture from compressed air, and refrigeration and desiccant air dryers remove the remaining moisture. Schedule 40 black iron is the most common type of pipe used in a compressed-air system. A variety of compressor and distribution controls maintain system pressure within the control range.

Compressed-air systems are designed to meet the customer's volume, pressure, and conditioning needs. The skills and techniques required for installing compressed-air systems are the same as those required for installing water supply systems.

Air compressors, air receivers, filters, and dryers are usually installed in the basement. The main line, risers, branch lines, and feeder lines are installed according to local code. Water drip legs can be installed to allow water to drain out of the supply pipes. After installation, inspect the system for leaks. If compressed-air systems are correctly installed and regularly maintained, they will provide years of reliable service.

1. Measured at sea level, air pressure expressed in psig is _____.
   a. 0
   b. 14.7
   c. 29.92
   d. 98.6

2. As compressed air passes through an after-cooler's heat exchanger, the moisture in the air _____.
   a. evaporates
   b. increases
   c. condenses
   d. sublimates

3. If all moisture must be removed from compressed air, it is passed through a(n) _____.
   a. evaporator
   b. air dryer
   c. precipitator
   d. air heater

4. The chemical pellets used to remove moisture from compressed air are called _____.
   a. absorbents
   b. evaporants
   c. precipitants
   d. dessicants

5. Compressed air is dangerous because of its high air pressure and its high _____.
   a. flow rate
   b. temperature
   c. noise level
   d. toxicity

6. Plumbers use portable air compressors to _____.
   a. clear lines of foreign matter
   b. pressure-test water supply systems
   c. operate impact wrenches
   d. clean and dry tools after use

7. The type of compressor most commonly used to power small pneumatic tools such as nail guns is the _____.
   a. centrifugal compressor
   b. rotary compressor
   c. axial compressor
   d. reciprocating compressor

8. All air receivers and relief valves are required by code to meet the safety standards set by the _____.
   a. Occupational Safety and Health Administration
   b. Underwriters Laboratories
   c. American Society of Mechanical Engineers
   d. National Safety Council

9. Multistage compressors are usually equipped with a(n) _____.
   a. intercooler
   b. heat sink
   c. interlock
   d. evaporator

10. To eliminate the transmission of vibrations from the compressor to the rest of the system, install _____.
    a. shock-absorbing mounts
    b. silicon rubber hoses
    c. vibration dampers
    d. flexible metal hoses

11. If the design of a compressed-air system requires supply pipe with a diameter of 2½ inches or greater, use _____.
    a. CPVC pipe
    b. Schedule 40 black-iron pipe
    c. Type L copper tubing
    d. galvanized steel pipe

12. Flexible hoses used to connect compressed-air sources to tools are reinforced to prevent them from _____.

    a. coiling
    b. collapsing
    c. rupturing
    d. disconnecting

13. Avoid pressure loss at connections between tools and the compressed-air source by using _____.

    a. pressure reducers
    b. reinforced hoses
    c. flexible sealants
    d. properly sized couplings

14. The process of removing excess air pressure from a compressor when the control range is exceeded is called _____.

    a. draining
    b. unloading
    c. bleeding
    d. decompressing

15. When installing a compressed-air system, locate the takeoffs for branch and feeder lines _____.

    a. at the top of the main line
    b. at the sides of the main line
    c. at the end of the main line
    d. at the bottom of the main line

Trade Terms Quiz

Fill in the blank with the correct term that you learned from your study of this module.

1. A control that vents excess air pressure from an air compressor is called a(n) _____.

2. A(n) _____ is a cable attached to the ends of an air supply hose and a tool, or to the ends of two hoses that are connected to each other, to prevent the hose from whipping uncontrollably when charged with compressed air.

3. A vertical pipe on an air supply line that acts as a drain for condensed water in the line is called a(n) _____.

4. _____ is the set of physical laws governing the motion of air and other gases.

5. The process of relieving excess air pressure from an air compressor is called _____.

6. A(n) _____ is a heat exchanger that removes heat from air as it moves through a multistage air-compressor system.

7. A device that cools air by passing it through a heat exchanger, causing moisture to condense out of the air is called a(n) _____.

8. A(n) _____ is a type of air compressor in which air is compressed by a piston.

9. A device that adjusts air pressure in a compressed-air system by activating and deactivating individual compressors is called a(n) _____.

10. A(n) _____ is a type of air compressor in which rotary screws compress the air.

11. The set of physical laws governing the motion of water and other liquids is called _____.

12. The _____, expressed as a percentage of an average hour of use, is the amount of time that a tool or machine will be operated.

13. A device that controls multiple compressors by linking them together into a single network controlled by a computer is called a(n) _____.

14. The _____ is the range between the highest- and lowest-allowable air pressures in a compressed-air system.

15. A device that turns an air compressor off when the control range is exceeded is called a(n) _____.

16. A(n) _____ is a snap-fit plug connecting a tool or machine to a compressed-air system.

17. A device that injects a fine oil mist into the air supply line, lubricating tools and machines, is called a(n) _____.

18. A(n) _____ is a device that allows an air compressor to operate at a given percentage of its normal output.

19. To be able to be reduced in volume is to be _____.

20. A(n) _____ is a spring-loaded valve used to control or shut off airflow in a compressed-air system.

21. A device that removes all moisture from compressed air is called a(n) _____.

22. _____ is a measure of the air volume at the air-compressor inlet under actual temperature and pressure conditions.

23. A helical device in a rotary compressor that compress air by passing it down a screw-shaped path is called a(n) _____.

24. A(n) _____ is a device that controls airflow into an air compressor by adjusting the compressor's intake valve.

25. A chamber at the end of the rotary screws in a rotary compressor where air is compressed is called a(n) _____.

26. A(n) _____ is a storage tank for compressed air.

27. A valve that keeps air pressure within the control range is called a(n) _____.

28. _____ is the measure of the air volume at the air-compressor inlet under standard conditions.

29. An air pressure of 14.7 psia, temperature of 68°F, and barometric pressure of 29.92 inches of mercury are considered _____.

30. A(n) _____ is a type of air dryer in which chemical pellets absorb moisture from compressed air.

31. An air dryer in which air is cooled as it passes through a heat exchanger cooled by a liquid refrigerant is called a(n) _____.

Trade Terms

Actual cubic feet per minute
Aftercooler
Air dryer
Air end
Air-pressure regulator
Air receiver

Air throttle valve
Compressible
Constant speed control
Control range
Coupling
Desiccant air dryer
Hydraulics

Intercooler
Lubricator
Network control
Part-load control
Pneumatics
Reciprocating compressor

Refrigeration air dryer
Rotary compressor
Rotary screw
Sequencing control
Standard conditions
Standard cubic feet per minute

Stop/start control
Throttle
Unloading
Use factor
Water drip leg
Whip check

Trade Terms Introduced in This Module

Actual cubic feet per minute: A measure of the air volume at the air-compressor inlet under actual temperature and pressure conditions. It is often abbreviated acfm.

Aftercooler: A device that cools air by passing it through a heat exchanger, causing moisture to condense out of the air.

Air dryer: A device that removes all moisture from compressed air.

Air end: A chamber at the end of the rotary screws in a rotary compressor where air is compressed.

Air-pressure regulator: A valve that keeps air pressure within the control range.

Air receiver: A storage tank for compressed air.

Air throttle valve: A spring-loaded valve used to control or shut off airflow in a compressed-air system.

Compressible: Able to be reduced in volume.

Constant speed control: A control that vents excess air pressure from an air compressor.

Control range: The range between the highest- and lowest-allowable air pressures in a compressed-air system.

Coupling: A snap-fit plug connecting a tool or machine to a compressed-air system.

Desiccant air dryer: A type of air dryer in which chemical pellets absorb moisture from compressed air.

Hydraulics: The set of physical laws governing the motion of water and other liquids.

Intercooler: A heat exchanger that removes heat from air as it moves through a multistage air-compressor system.

Lubricator: A device that injects a fine oil mist into the air supply line, lubricating tools and machines.

Network control: A device that controls multiple compressors by linking them together into a single network controlled by a computer.

Part-load control: A device that allows an air compressor to operate at a given percentage of its normal output.

Pneumatics: The set of physical laws governing the motion of air and other gases.

Reciprocating compressor: A type of air compressor in which air is compressed by a piston.

Refrigeration air dryer: A type of air dryer in which air is cooled as it passes through a heat exchanger cooled by a liquid refrigerant.

Rotary compressor: A type of air compressor in which rotary screws compress the air.

Rotary screw: Helical devices in a rotary compressor that compress air by passing it down the screw path.

Sequencing control: A device that adjusts air pressure in a compressed-air system by activating and deactivating individual compressors.

Standard conditions: An air pressure of 14.7 psia, temperature of 68°F, and barometric pressure of 29.92 inches of mercury.

Standard cubic feet per minute: A measure of the air volume at the air-compressor inlet under standard conditions. Standard cubic feet per minute is often abbreviated scfm.

Stop/start control: A device that turns an air compressor off when the control range is exceeded.

Throttle: A device that controls airflow into an air compressor by adjusting the compressor's intake valve.

Unloading: The process of relieving excess air pressure from an air compressor.

Use factor: Expressed as a percentage of an average hour of use, the amount of time that a tool or machine will be operated.

Water drip leg: A vertical pipe on an air supply line that acts as a drain for condensed water in the line.

Whip check: A cable attached to the ends of an air supply hose and a tool, or to the ends of two hoses that are connected to each other, to prevent the hose from whipping uncontrollably when charged with compressed air.

Additional Resources

This module presents thorough resources for task training. The following resource material is suggested for further study.

Best Practices for Compressed Air Systems, Second Edition. William Scales. 2007. The Compressed Air Challenge.

Compressed Air Data: Handbook of Pneumatic Engineering Practice. William Lawrence Saunders and Charles Austin Hirschberg, editors. 2002. Honolulu, HI: University Press of the Pacific.

Compressed Air Systems: A Guidebook on Energy and Cost Savings. E. M. Talbot. 1993. Lilburn, GA: Fairmont Press.

Fact Sheet #1, "Assessing Compressed Air Needs." 1998. Compressed Air Challenge. **compressedairchallenge.org/library**

Fact Sheet #4, "Pressure Drop and Controlling System Pressure." 1998. Compressed Air Challenge. **compressedairchallenge.org/library**

Fact Sheet #5, "Maintenance of Compressed Air Systems for Peak Performance." 1998. Compressed Air Challenge. **compressedairchallenge.org/library**

Fact Sheet #6, "Compressed Air System Controls." 1998. Compressed Air Challenge. **compressedairchallenge.org/library**

Fact Sheet #7, "Compressed Air System Leaks." 1998. Compressed Air Challenge. **compressedairchallenge.org/library**

Hydraulics and Pneumatics: A Technician's and Engineer's Guide, Third Edition. Andrew Parr. 2011. New York: Butterworth-Heinemann.

Improving Compressed Air System Performance. Vienna, VA: Resource Dynamics Corporation for Compressed Air Challenge. **compressedairchallenge.org/library**

Pipefitters Handbook. Forrest R. Lindsey. New York: Industrial Press Inc.

Pneumatic Handbook, Eighth edition. 1998. Antony Barber. New York: Elsevier Science.

Figure Credits

Answer	Section Reference	Objective
Section One		
1. b	1.1.0	1a
2. a	1.2.0	1b
Section Two		
1. b	2.1.0	2a
2. d	2.2.0	2b
Section Three		
1. a	3.1.0	3a
2. b	3.2.0	3b
Section Four		
1. d	4.1.1	4a
2. c	4.2.0	4b

NCCER CURRICULA — USER UPDATE

NCCER makes every effort to keep its textbooks up-to-date and free of technical errors. We appreciate your help in this process. If you find an error, a typographical mistake, or an inaccuracy in NCCER's curricula, please fill out this form (or a photocopy), or complete the online form at **www.nccer.org/olf**. Be sure to include the exact module ID number, page number, a detailed description, and your recommended correction. Your input will be brought to the attention of the Authoring Team. Thank you for your assistance.

Instructors – If you have an idea for improving this textbook, or have found that additional materials were necessary to teach this module effectively, please let us know so that we may present your suggestions to the Authoring Team.

NCCER Product Development and Revision
13614 Progress Blvd., Alachua, FL 32615

Email: curriculum@nccer.org
Online: www.nccer.org/olf

❏ Trainee Guide   ❏ Lesson Plans   ❏ Exam   ❏ PowerPoints   Other _____

Craft / Level: _____ Copyright Date: _____

Module ID Number / Title: _____

Section Number(s): _____

Description: _____

_____

_____

_____

_____

Recommended Correction: _____

_____

_____

_____

_____

Your Name: _____

Address: _____

_____

Email: _____ Phone: _____

02311-14

Service Plumbing

OVERVIEW

Plumbers must diagnose and repair plumbing problems quickly and correctly. Most service calls are for routine, nonemergency repairs; however, sometimes a potentially dangerous situation requires immediate safety measures and fast, effective troubleshooting. Servicing water supply systems may involve locating and repairing leaks and frozen pipes or correcting problems caused by low or high water pressure. Ensuring that a building's water service meets quality standards may require diagnosing and correcting problems such as hard water or small particles or organisms in the water. When servicing DWV systems, plumbers most often remove blockages in drains, traps, and vents, and diagnose and correct the causes of unpleasant odors.

Module Nine

Trainees with successful module completions may be eligible for credentialing through the NCCER Registry. To learn more, go to **www.nccer.org** or contact us at **1.888.622.3720**. Our website has information on the latest product releases and training, as well as online versions of our *Cornerstone* magazine and Pearson's product catalog.

Your feedback is welcome. You may email your comments to **curriculum@nccer.org,** send general comments and inquiries to **info@nccer.org**, or fill in the User Update form at the back of this module.

This information is general in nature and intended for training purposes only. Actual performance of activities described in this manual requires compliance with all applicable operating, service, maintenance, and safety procedures under the direction of qualified personnel. References in this manual to patented or proprietary devices do not constitute a recommendation of their use.

Objectives

When you have completed this module, you will be able to do the following:

1. Recognize and observe standards of safety and etiquette when making service calls to residential and commercial facilities.
   a. Observe safety on service calls.
   b. Observe etiquette on service calls.
2. Explain how to troubleshoot and repair problems with water supply systems.
   a. Explain how to troubleshoot and repair leaks.
   b. Explain how to troubleshoot and repair frozen pipes.
   c. Explain how to troubleshoot and repair water pressure problems.
   d. Explain how to troubleshoot and repair water quality problems.
   e. Explain how to troubleshoot and repair water flow-rate problems.
   f. Explain how to troubleshoot and repair underground pipes.
   g. Explain how to diagnose cross-connections.
3. Explain how to troubleshoot and repair problems with fixtures and appliances.
   a. Identify common repair and maintenance requirements and procedures for fixtures, valves, and faucets.
   b. Explain how to troubleshoot and repair water heaters.
   c. Explain how to troubleshoot, repair, and replace tubs, showers, and water closets.
   d. Explain how to troubleshoot and repair other fixtures and appliances.
4. Explain how to troubleshoot and repair problems with DWV systems.
   a. Explain how to troubleshoot and address blockages.
   b. Explain how to troubleshoot and correct odor problems.

Performance Tasks

Under the supervision of your instructor, you should be able to do the following:

1. Troubleshoot and repair problems with water supply systems.
2. Troubleshoot and repair problems with fixtures, valves, and faucets using the proper tools and replacement parts.
3. Use manufacturer's instructions to disassemble and re-assemble a valve.
4. Troubleshoot and repair problems with water heaters.
5. Troubleshoot and repair problems with DWV systems.

Trade Terms

Activated carbon filter	Dynamic water pressure	Reseating tool	Tap
Auger	Electric heat tape	Reverse-osmosis	Twist packing
Condensation	Etiquette	method	Wall hydrant
Corrosion	Hard water	Rodder	Water filter
Deionizer	Jetter	Schrader valve	Water softener
Drum machine	Packing extractor	Screw extractor	Yard hydrant
Dry well	Removable seat wrench	Static water pressure	Zeolite system

Industry Recognized Credentials

If you're training through an NCCER-accredited sponsor you may be eligible for credentials from NCCER's Registry. The ID number for this module is 02311-14. Note that this module may have been used in other NCCER curricula and may apply to other level completions. Contact NCCER's Registry at 888.622.3720 or go to **www.nccer.org** for more information.

Code Note

Codes vary among jurisdictions. Because of the variations in code, consult the applicable code whenever regulations are in question. Referring to an incorrect set of codes can cause as much trouble as failing to reference codes altogether. Obtain, review, and familiarize yourself with your local adopted code.

Contents

Topics to be presented in this module include:

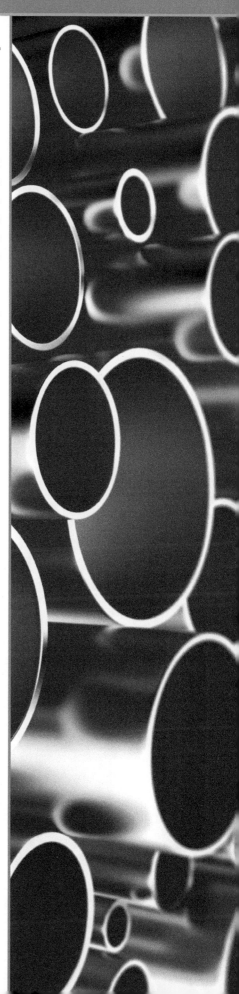

Contents (continued)

Figures and Tables

1.0.0 SERVICE-CALL SAFETY AND ETIQUETTE

Objective

Recognize and observe standards of safety and etiquette when making service calls to residential and commercial facilities.

    a. Observe safety on service calls.
    b. Observe etiquette on service calls.

Trade Terms

Etiquette: The code of polite and appropriate behavior that people observe in a public setting.

When going out on service calls, plumbers represent their company as well as their professional skills and knowledge. Customers will judge you not only by how well you do your job, but also by the impression you make. To make the best impression possible, always follow good safety practices and demonstrate proper etiquette when servicing plumbing installations in residential and commercial buildings. Etiquette means polite and appropriate behavior. This section covers the key safety practices and proper etiquette that should be observed during service calls. Before going on services calls, learn your company's policies and procedures for safety and etiquette, and follow them closely when you are in the field.

1.1.0 Observing Safety on Service Calls

Before you can begin any repair work, there are some general safety guidelines you should know and follow. It is impossible to predict what you will encounter when responding to a service call. For example, a small leak from a faucet spout into a kitchen sink, while wasteful and annoying, is not an emergency. Usually it is also not hazardous. On the other hand, a valve leaking a lot of water above a suspended ceiling is usually hazardous. In an emergency like this, you must stop the flow of water, immediately minimize the potential damage, and make the area safe before you make the repairs.

Generally, repairs to fixtures involve a stoppage in the drain fixture or a cracked or leaking fixture. If a leak in a fixture is the result of a problem with the valve or faucet, you will usually repair or replace the valve or faucet. If the problem is with the fixture itself, you will generally replace the fixture rather than repair it.

Some general safety guidelines for you to follow when responding to a service call appear below. Adapt these guidelines to the job at hand. As you progress in your career, you will add your own guidelines from your experience.

- Wear rubber-soled shoes or boots for protection from slipping and electric shock.
- Notify the owner and occupants that you will be shutting off electrical and water service within the building.
- Turn off electrical circuits.

> **WARNING!**
> Electrical shocks can happen to anyone, not just to electricians. Always protect yourself against accidental contact with energized sources when working on plumbing installations and responding to service calls.

- Shut off a valve upstream from the leak. If you think it will take you a while to locate and turn off this valve, direct the leak into a suitably sized container to minimize damage until you can turn the water off.
- Remove excess water.
- Move furniture, equipment, or any other obstacles away from the work area.
- Cover furniture, floors, and equipment to protect them from any damage that could occur as a result of your work on the valve.
- Place ladders carefully, bracing them if necessary. Wear appropriate personal protective equipment for working on ladders.
- When finished, turn the water back on to test your repair. Notify the owner and occupants that service is back on.
- Do your work neatly, and clean up when you are done.

> **WARNING!**
> Slips, trips, and falls cause most general industry accidents, according to the Occupational Safety and Health Administration (OSHA). They cause 15 percent of all accidental deaths, and they are second only to motor vehicles as a cause of fatalities. Slips, trips, and falls happen quickly, so take a few minutes to make your work area safe.

1.2.0 Observing Etiquette on Service Calls

Plumbers pride themselves on their customer-service skills as well as their technical knowledge. A professional, service-oriented plumber is prompt, prepared, thorough, and courteous on every service call. Customers appreciate plumbers who observe these guidelines. When you make an appointment, keep it. Show up for service calls on time. If you cannot make your appointment or if you are going to be late, notify the customer as soon as you can. These are courtesies you would expect from someone coming to your own home. Arrive prepared to do the job. If possible, know before you arrive what work needs to be done on site. Ensure that the proper tools and supplies are available. Customers lose confidence in plumbers who have to leave and return several times just to complete a simple task.

Construction professionals adhere to a code of professional ethics. Ethics are principles and values that guide conduct. As a plumber, you are responsible for understanding and following the code of ethics for your chosen profession. Successful plumbers exhibit all of the following qualities:

- Positive attitude
- Honesty
- Loyalty
- Willingness to learn
- Responsibility
- Cooperation
- Attentiveness to rules and regulations
- Promptness and reliability

> **WARNING!**
>
> Do not work in a crawl space or other restricted area without prior confined-space training. Toxic and flammable vapors, as well as carbon monoxide (CO) emissions, can accumulate in confined spaces. Electrical, hot-water, and steam lines also pose risks in confined spaces. Wear appropriate personal protective equipment, and wear a dust mask or respirator where needed.

1.2.1 Etiquette on Residential Service Calls

When you are on a service call to a residence, you should think of yourself as a guest in that person's home. Behave in a way that you would want other people to behave if they were working in your house. Proper self-presentation is important. Your personal habits and work ethic make powerful impressions on your customers. Once formed, impressions are hard to change—so you want to make sure that the impressions you make are good ones.

Plan to arrive at the time scheduled for the service call. If you are going to be late, call the customer as soon as possible to let them know, and tell them when they can expect you. Be polite and courteous, and have a positive attitude. When you meet the customer, introduce yourself and shake hands. Be sure to wear an identification tag or piece of clothing with your company logo to clearly identify who you are. Dress professionally. Always wear appropriate personal protective equipment when working on service calls.

Before starting to work, set up drop cloths or tarps on the floor to protect against dirt, footprints, and damage from dropped items. Cover any furniture that is too large to move. Wear boot covers to protect against scuffing or scraping floors that are not covered. To minimize trips back and forth to your vehicle, bring the tools you will need with you and set them up near the fixture or appliance you will be working on. If the customer is at home, do not play music or talk loudly while working. Never put tools on countertops, particularly those made of stone, as the tools may chip and gouge the countertops, or stain them with lubricants and oil.

If the customer is not at home, consider the neighbors before doing activities that make excessive noise. If you are working outside, be careful not to damage grass or plants. If your work requires you to go in and out of the residential building repeatedly, be sure not to track dirt and mud inside. Close doors behind you. If you are the last one in the building when you leave, be sure to close and lock the door behind you.

You should take pride in performing a job well and you should never cut corners. Never lie to your customers. Respect the property of others. Ensure

Cleaning Up is Part of the Job

Whether you are installing water supply piping or repairing a fixture, you must always clean up your work site. A clean work site is safer for you and for any trades that come in after you. Cleaning up is especially important if you are called in to repair a major leak. No one wants to pay for repairs and then have to replace furnishings or carpet because you were careless.

While working at the customer's home or workplace, keep the work area as clean as possible. Lay down protective coverings over floors, furniture, and equipment. Wear overshoes and gloves while working. These will help keep dirt and grease off the walls, floors, and furniture. When possible, remove furniture and other obstacles from the work area. Work neatly and clean up when the job is done. These simple, commonsense steps will help ensure satisfied customers and future business.

that the tools and materials belonging to your employer (or sponsor), customers, and other trades are not damaged or stolen. Treat their property with the same respect that you give to your own. Always take responsibility for your actions. Do not make excuses or blame others when you make an error.

When your work is complete, remove tools and parts and clean the work area. Remove drop cloths and tarps, and make sure the floors and walls are clean. Do not leave footprints, dust, or parts behind. When accepting payment, provide the customer with a clearly written service-call summary sheet and explain each item and the cost. Answer any questions the customer has with courtesy. If a tip is offered, decline it politely.

1.2.2 Etiquette on Commercial Service Calls

The same rules of professionalism and courtesy that you follow for residential service calls apply for service calls to commercial facilities. However, there are some important differences that you should keep in mind. Offices, factories, malls, restaurants, hospitals, and other commercial buildings are public spaces. You will be working amid people who are also trying to get their own work done. It is a professional courtesy to try to ensure that your work interferes with theirs as little as possible.

Never enter a room unless the manager or a representative of the customer accompanies you. Always give advance notice before turning off the water supply to part or all of the building. Post safety signs around the work area warning workers to stay clear until the work has been completed. Wear appropriate personal protective equipment when working, and provide appropriate personal protective equipment to customers who enter the work area.

Always follow lockout/tagout procedures to identify hazardous equipment and prevent its use until the equipment has been certified safe. Lockout/tagout devices protect workers from energy sources and from injury caused by high-pressure or high-temperature water or other liquids. In a lockout, an energy-isolating device that prevents the transmission or release of energy, such as a stop valve, a disconnect switch, or a circuit breaker, is placed in the Off position and locked. Lockout devices can be used with key or combination locks. Multiple lockout devices (*Figure 1*) are used when more than one person has access to the equipment. In some instances, as with work involving valves, a chain and lock can be used to hold the valve in place and keep it from being turned.

In a tagout, components that allow the flow of water or the flow of power to equipment and machinery, such as switches or valves, are set in a safe position, and a written warning or tagout device is attached to them (*Figure 2*). An effective lockout/tagout program should include the following:

- An inspection of equipment by a trained person who is thoroughly familiar with the equipment operation and associated hazards
- Identification and labeling of lockout devices
- The purchase of locks, tags, and blocks
- A standard written operating procedure that all employees follow

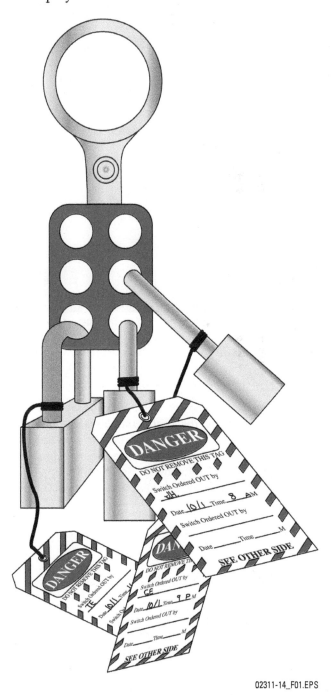

02311-14_F01.EPS

Figure 1 Multiple lockout device.

Use the PROPER Procedure for Lockout/Tagout

Using PROPER lockout/tagout procedures is the best way to prevent accidents, injury, and death.

Process (shut down process)
Recognize (identify energy sources such as electrical panels, disconnect switches, and switchboards)
Off (shut off energy sources)
Place (locks and tags in place)
Energy (release stored energy)
Return (controls are returned to neutral)

02311-14_F02.EPS

Figure 2 Placing a lockout/tagout device.

Additional Resources

Communicating at Work. 1993. Tony Alessandra and Phil Hunsaker. New York: Simon and Schuster.
Communicating in the Real World: Developing Communication Skills for Business and the Professions. 1987. Terrence G. Wiley and Heide Spruck Wrigley. Englewood Cliffs, NJ: Pearson.
Tools for Teams: Building Effective Teams in the Workplace. 2001. Craig Swenson, ed. Leigh Thompson, Eileen Aranda, and Stephen P. Robbins. Boston, MA: Pearson Custom Publishing.
Your Attitude Is Showing. 2008. Sharon Lund O'Neil and Elwood M. Chapman. Upper Saddle River, NJ: Pearson Prentice Hall.

1.0.0 Section Review

1. Before shutting off electrical and water service within a building, _____.

   a. notify the utility company
   b. notify the owner and occupants
   c. set up safety signs at the shutoff valve or switch
   d. close all valves upstream of the leak

2. When on a service call to a commercial building, enter a room only _____.

   a. if it is adjacent to the work area in order to place safety signage
   b. after the work has been completed
   c. after knocking
   d. when accompanied by the customer

2.0.0 Troubleshooting and Repairing Water Supply Systems

Objective

Explain how to troubleshoot and repair problems with water supply systems.

  a. Explain how to troubleshoot and repair leaks.
  b. Explain how to troubleshoot and repair frozen pipes.
  c. Explain how to troubleshoot and repair water pressure problems.
  d. Explain how to troubleshoot and repair water quality problems.
  e. Explain how to troubleshoot and repair water flow-rate problems.
  f. Explain how to troubleshoot and repair underground pipes.
  g. Explain how to diagnose cross-connections.

Performance Task

Troubleshoot and repair problems with water supply systems.

Trade Terms

Activated carbon filter: A water filter that uses charcoal to absorb sulfur from potable water.

Condensation: Water that collects on a cold surface that has been exposed to warmer, humid air.

Deionizer: A water filter installed on a cold-water inlet that removes metals and mineral salts.

Dynamic water pressure: The pressure water exerts on a piping system when the water is flowing (circulating water). Dynamic water pressure, measured in psi, will always be lower than static water pressure.

Electric heat tape: An insulated copper wire wrapped around a pipe to keep it from freezing.

Hard water: Potable water that contains large amounts of mineral salts such as calcium and magnesium.

Reverse-osmosis method: A method of filtering water through cellulose acetate membranes used in some water filters.

Schrader valve: A tire valve stem installed on a tank that is used to correct low pressure in the tank.

Static water pressure: The pressure water exerts on a piping system when there is no water flow (noncirculating water). Static water pressure, measured in psi, will always be higher than dynamic water pressure.

Wall hydrant: A hydrant installed in a building's exterior wall that works by draining water after the valve is closed.

Water filter: A device designed to remove small particles and organisms from the potable water supply.

Water softener: A device that chemically removes mineral salts from hard water.

Yard hydrant: A lawn spigot connected to the water main that works by having its valve located below the frost line.

Zeolite system: A water softener that filters mineral salts out of hard water by passing the water through resin saturated with sodium.

Plumbing systems require periodic maintenance. Sometimes, poor water quality can cause problems. Roots may block drainpipes. A pipe can degrade and leak. A fitting may break. Fixtures can simply wear out from prolonged or constant use. Plumbers decide whether to repair or replace worn or broken items. They need to be able to diagnose and fix problems quickly and correctly. Finding the cause of a problem is a lot like detective work.

Always take the time to look for the cause of the problem. Choose the best possible solution, not just the fastest or the least expensive. Do not put a customer's property at risk by hurrying to finish a job. These steps will reduce or eliminate callbacks and result in more satisfied customers. This means more business in the future.

Fixing plumbing problems requires knowledge, skill, and years of experience, but you have already learned many of the basic skills that you will need to perform the work according to professional standards. Beginning plumbers can develop their knowledge and experience both on and off the job. One good way to learn is to read plumbing literature, such as manuals, trade journals, and codes. Pay special attention to articles that offer practical tips. Work with experienced plumbers. Watch how they diagnose and fix problems. Do not be afraid to ask a lot of questions. The best plumbers are the ones who never stop learning.

You have learned that customers have a right to expect adequate fresh water on demand. Water supply systems meet that need. Water supply systems are made of many components such as pipe, fittings, valves, traps, and fixtures. Larger systems also use pumps and storage tanks to circulate the water efficiently. Parts can and do wear out over time. They may stop working efficiently, or they may break altogether. When either happens, the water system cannot provide fresh water on demand. In some cases, a malfunction can cause illness or even death. Plumbers are responsible for repairing or replacing broken or defective components. They are also responsible for making sure the repaired system operates as smoothly and efficiently as before.

2.1.0 Troubleshooting and Repairing Leaks

Leaks are the most common problem that plumbers encounter in water supply systems. Leaking and broken pipe can be hazardous. Pipes installed in a new building can break when the building shifts and settles. Underground pipes can break if a water-line trench is improperly backfilled. Leaks in a new system are usually caused by improper installation or damaged components. In older systems, the problem is often a worn component or a corroded supply line.

Before beginning work, find out how old the building is and determine what materials were used for the piping and fittings. This information offers clues to the nature of the problem. A leak indicates a problem with one or more components in the plumbing system. Common causes of leaks include the following:

- Physical damage to pipes
- Improper connections
- Corroded pipes
- Frozen pipes
- Excessive vibration or water hammer
- Higher-than-normal system pressure

Locating a leak is often difficult. Leaks can occur at a fitting or along the length of pipe. Water can run along a pipe or structure, showing up far from its source. Take the time to find the location of a leak by tracing the water drip or stain to its source. Fixing a leak without correcting the

problem that caused it will only result in repeated plumbing problems. Such a situation could cause the customer to lose confidence in you.

2.1.1 Repairing a Leak Along a Fitting

Use good repair practices to repair a leak near a fitting in plastic pipe. Make sure the pipe is dry before trying to repair it. Cut out the fitting using a pipe cutter or other cutting tool. A mini tube cutter is also a useful tool for cutting pipe (*Figure 3*). Test fit the new fitting. When the fit is correct, use couplings to join the new fitting to the existing pipe.

To repair a leak in a copper fitting, solder the fitting. Copper pipes can also be connected using press fittings and connections. Press systems use solderless fittings to connect pipes and fittings with water present. The system requires particular pressing tools (*Figure 4*).

2.1.2 Making a Temporary Repair of a Leak Along a Length of Pipe

Use good repair practices when making temporary repairs of leaks in a length of pipe. Make temporary repairs in plastic and copper pipe us-

02311-14_F03.EPS

Figure 3 Mini tube cutter.

Proper Flushing

After any repair to a water supply system, always purge the system when the repairs are complete. Remove all screens and filters and allow the water to run through the lines for a reasonable length of time. This step will ensure that all foreign matter is flushed out of the system. Advise customers to have water heaters and tanks flushed periodically to remove sediment and mineral deposits. Manufacturers generally recommend flushing 5 gallons of water from hot-water tanks each month for a reasonable amount of time. Make sure that the drain is fully opened. This will allow proper flushing.

02311-14_F04.EPS

Figure 4 Pressing tools and jaws for use with press fittings.

ing pipe clamps (*Figure 5*), then join them together using the skills you have learned.

2.1.3 Eliminating Pipe Condensation

Sometimes, water stains will appear on walls or ceilings where there are no leaking pipes. These stains are often caused by condensation and roof leaks. Condensation is moisture that collects on cold surfaces that are exposed to warm, moist air.

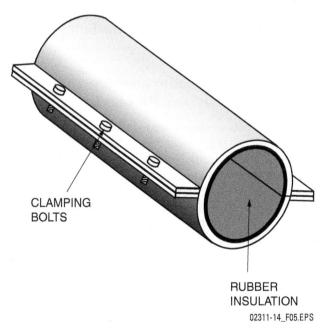

CLAMPING BOLTS

RUBBER INSULATION

02311-14_F05.EPS

Figure 5 Temporary pipe clamp.

Dripping condensation causes water stains and musty mildew odors. To eliminate condensation, add insulation to the pipe (*Figure 6*). Insulation prevents the cold pipe from coming into contact with warm air. It is the most effective way to protect against condensation. If adding insulation is impractical, increase the ventilation. Doing so will reduce the amount of moisture in the air.

2.2.0 Troubleshooting and Repairing Frozen Pipes

Water supply piping must be protected from freezing. If the piping has frozen, first determine whether the line has ruptured. If a water supply pipe has ruptured, turn off the water supply before attempting to thaw the line. Install temporary clamps to seal minor leaks. If the pipe is severely damaged, replace the damaged section immediately.

> **CAUTION**
>
> Underground supply lines can freeze if they are installed where snow is not allowed to accumulate. These areas include driveways, sidewalks, and patios. Supply lines beneath these areas are subject to freezing because there is no snow to insulate the ground. Consult the local code and building officials to identify the frost line in your region.

There are several ways to thaw frozen pipes safely. For metal pipe, an electric thawing machine (*Figure 7*) can be used. Electric thawing machines use a high-amperage direct current to warm the pipe, causing the ice that is blocking the pipe to melt. Attach the thawing machine's two clamps to the ends of the frozen section of pipe, one clamp on each end, and turn the machine on. Typically, the frozen liquid will begin to thaw within minutes.

A hot-water thawing machine uses hot water to thaw frozen pipe. Although it does not work as quickly as an electric thawing machine, it is less

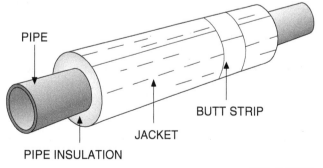

PIPE

BUTT STRIP

JACKET

PIPE INSULATION

02311-14_F06.EPS

Figure 6 Insulating a pipe.

02311-14_F07.EPS

Figure 7 Electric pipe-thawing machine.

02311-14_F08.EPS

Figure 8 Frost-free yard hydrant.

expensive. Whichever method you use, always follow the manufacturer's instructions.

When the frozen water line is exposed, an electric heat gun may be used. This is ideal when only a portion of the line is frozen. Directing hot air over the frozen line thaws the line. The running water can thaw the remainder of the line. If the frozen supply line is undamaged, a heat lamp or hair dryer can also be used to thaw it. If the pipe is indoors, heating the air can help. Also consider putting a rag over the pipe and pouring hot water over the rag. Do not use boiling water, which can cause the pipes to expand—and possibly burst—if they are excessively heated.

> **WARNING!**
>
> When thawing a pipe with an open flame, exercise caution to avoid starting a fire. Torches, lighters, and other sources of open flame are dangerous and could cause a structural fire. Using a torch on a frozen water line may damage solder joints, create hazardous steam pockets, and ignite surrounding building materials.

Repair freeze-damaged pipes with the same techniques used for fixing a leak. Note that ice usually forces the pipe to expand. As a result, repair fittings may not attach easily. Therefore, you must cut the pipe well beyond any areas that have frozen.

Where freezing is a problem, install approved frost-free yard hydrants (*Figure 8*) and wall hydrants (*Figure 9*). Consult with local experts to make sure the hydrant has the necessary approvals. A yard hydrant is a spigot connected to a pipe that runs directly to the water main. Yard hydrants work by having their valve located below the frost line. If yard hydrants continue to freeze, ensure that the water supply pipe has the proper pitch. The excess water will then be able to drain from piping when the valve is closed.

As the name suggests, a wall hydrant is installed in a building's exterior wall. Wall hydrants work by draining water after the valve is closed. The valve seat is located on the interior side of the building's wall. Wall hydrants are sized according to the thickness of the building's wall.

Wrap pipes with electric heat tape for temporary protection in case of frost. Electric heat tape is made of flexible copper wires coated in a protective layer of rubber or other nonconductive material. When wrapped around a pipe and plugged into a wall socket, heat tape provides enough heat to keep pipes from freezing. Be aware of the danger of melting plastic pipe. Install heat tape according to the manufacturer's instructions. Do not wrap heat tape over itself. Heat tape can be wrapped under insulation.

To prevent frozen pipes, consider placing the pipe in partitions. When working with outdoor faucets, install frostproof hose bibbs or install a drain or waste cock inside the building in a heated area.

2.3.0 Troubleshooting and Repairing Water Pressure Problems

Water supply systems are usually designed to operate at pressures between 35 and 50 pounds per square inch (psi). Pressure below or above the normal range can cause problems with the water supply. Too little pressure means fixtures and outlets cannot work efficiently. Too much pressure causes noise, water hammer, and excessive wear.

The allowable pressure range for noncirculating water, called static water pressure, is usually be-

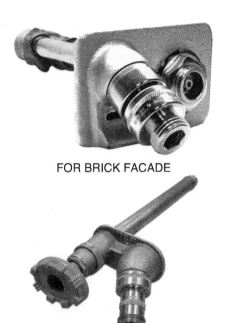

FOR BRICK FACADE

FOR NONBRICK FACADES

02311-14_F09.EPS

Figure 9 Frost-free wall hydrants.

tween 15 and 75 psi. The upper limit is the maximum setting for pressure-reducing valves (PRVs). Static pressure should be kept below 75 psi. Static pressure tends to be higher for commercial systems with flushometers, and tends to be lower for residential systems. Fifty psi is a good average pressure for homes connected to public water supply systems. Homes connected to private water supply systems operate closer to 40 psi.

The system pressure may fluctuate daily during peak and low-use periods. Fluctuations also may be seasonal. Water consumption is usually much higher in the summer than in the winter. The pressure range for circulating water, also called dynamic water pressure, is lower than that of static water pressure. Remember to distinguish between static and dynamic water pressure when designing a water supply system. Refer to your local code.

2.3.1 Low Pressure

Problems caused by low water pressure are generally harder to fix than those caused by high pressure. They also tend to be more expensive. Low pressure is caused by problems with either static pressure or dynamic pressure. Low pressure in the city main is a common static-pressure problem. Common dynamic-pressure problems include the following:

- A defective water meter
- A leak in a building main
- A defective or partly closed valve
- Improperly sized pipe
- Rust, corrosion, kinks, and other pipe blockages

Customers often mistake deposit buildups for low water pressure. Mineral buildups and corrosion in pipes restrict water flow. The result is a dynamic-pressure drop in the system. This is often a problem in older areas of a municipal water supply system. You may need to conduct a pressure test of the system to confirm the diagnosis. If the problem is extensive, a new water supply system will have to be installed.

Remember that low pressure can cause backflow if there is a cross-connection in the system. For example, opening a valve on the ground floor of a multistory building could cause back siphonage in a fixture several stories up. Low-pressure problems should be corrected quickly to prevent this. Install a pressure booster system to correct low pressure. Otherwise, the potable water supply could become contaminated.

Too many fixtures on a supply line can also cause low water pressure. In such cases, replace the supply lines to handle the additional demand. Refer to the local code before replacing any water supply pipe. Be sure to select the correct-size pipe for the anticipated demand. Remember to reconnect the water service to the house.

Tips for Preventing Frozen Pipes

The less frequently a water pipe is used, the more likely it is to freeze. Here are some easy steps to prevent frozen pipes that you can suggest to homeowners:

- Allow the water to run slowly throughout the night or during the coldest periods.
- Drain outside hose connections at the beginning of the cold-weather season.
- Insulate exposed water pipes.
- Use heat tape to protect the water line from freezing.
- Close gaps that allow cold air into spaces where water pipes are located.
- Remove all hoses from frost-free wall hydrants.

2.3.2 High Pressure

High pressure can cause vibrations and knocking in water pipes. High pressure in the water main is one of the causes of water hammer. (A quick-closing valve in the system is the other common cause of water hammer.) Simply anchoring the pipe to a wall or frame may not solve the problem. Install water-hammer arresters instead (*Figure 10*).

Another way to correct water-hammer vibrations is to install pressure-reducing valves (*Figure 11*) at the service entrance to a structure or near the source of the high-pressure problem. Install shutoff valves on both the inlet and outlet sides of PRVs (*Figure 12*). Installing shutoff valves on both sides allows water to flow through the bypass for access to the PRV for maintenance and repair.

Note that the excess pressure may be occurring in only one part of the system. For example, a system in a high-rise building could have high pressure at a branch line near the service entrance and low pressure at the lowest end. If this is the case, do not place the PRV in the main service line. Doing so will reduce the pressure at the most distant outlet. The entire piping system may have to be redesigned as a result of the excess pressure. Each specific water-hammer problem requires its own unique solution. Discuss various problems and solutions with experienced plumbers. Refer to the local code as well.

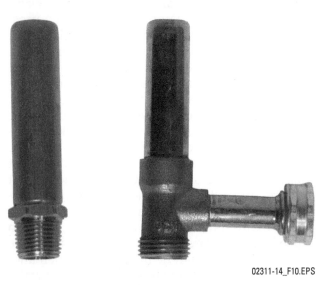

02311-14_F10.EPS

Figure 10 Water-hammer arrester.

02311-14_F11.EPS

Figure 11 Pressure-reducing valve.

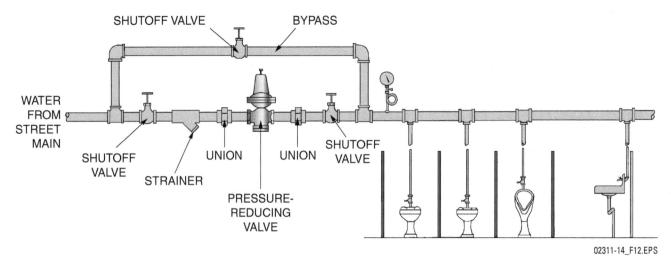

02311-14_F12.EPS

Figure 12 Installation of a pressure-reducing valve.

High water pressure is also a common cause of whistling noises in supply lines. Pipes and fixture openings that are too small can also cause whistling. Take the time to diagnose the cause of the whistling before attempting to correct the problem. Pressure-reducing devices are an effective way to eliminate whistling. Reducing the pressure in the problem area is usually easier than replacing the piping.

If the whistling is limited to a small area, check the pipe for restrictions, such as a burr that was not properly reamed. Check for defective fittings, malfunctioning valve washers, and loose valve stems. If the problem persists, install a pressure-reducing device in the branch line. If the whistling occurs throughout the entire structure, install the pressure-reducing device at the service entrance.

2.3.3 Bladder Tanks

Private water supply systems that draw on wells or springs usually store water in bladder tanks (*Figure 13*). Bladder tanks provide adequate pressure for the supply system. First, ensure that the tank's pressure switch is properly adjusted. Pressure switches can be set to different pressure ranges, so select a range that is appropriate for the building's water pressure needs. Most switches have settings for 20 to 40 pounds, 30 to 50 pounds, and 40 to 60 pounds. Consult the building's plumbing drawings to determine the pressure requirements.

A tank with low pressure cannot provide adequate water pressure throughout the system. If this is the problem, you need to pressurize the bladder tank. Begin by finding the correct pressure setting in the manufacturer's specifications. The correct pressure-switch setting is important because it determines when the pump is activated. First, empty the tank completely. Then recharge the tank to the correct air pressure, using the tank's Schrader valve (*Figure 14*). Schrader valves, which are usually located at the top of the tank, are small valve stems used to pump air into the tank. Test whether the tank can maintain a pressure range of 20 pounds. If it cannot, the chamber may have lost air pressure. Inspect the air chamber and repair any leaks that you find. If water is present at the Schrader valve, the bladder is damaged and will need to be replaced.

2.4.0 Troubleshooting and Repairing Water Quality Problems

An important part of a plumber's job is to ensure that a building's water service meets quality standards. Water quality problems can affect the health of customers. These problems can also damage the water supply system. Therefore, plumbers must be able to correct water quality problems quickly and efficiently.

When water quality may be a problem, start by testing the water. Test results will help you to determine the right purification technique. Choosing a purifier without testing may not solve the problem, and it also could force custom-

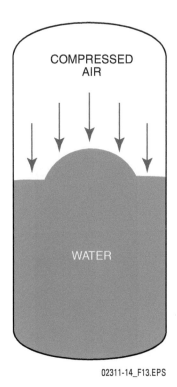

COMPRESSED AIR

WATER

02311-14_F13.EPS

Figure 13 Bladder tank for a private water supply system.

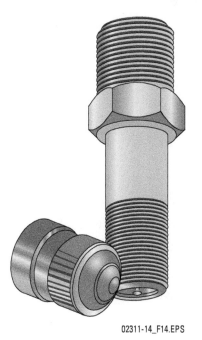

02311-14_F14.EPS

Figure 14 Schrader valve.

ers to spend much more money than necessary. Research the problem up front. It takes a lot less time and effort in the long run.

Two common causes of water quality problems are hard-water deposits and suspended or dissolved particles. Each of these is discussed in more detail in the following sections.

2.4.1 Hard-Water Deposits

Look inside an old coffeepot or teakettle. You likely will see a rough, grayish coating on the bottom. These deposits are mineral salts such as calcium and magnesium. Heat causes the mineral salts to settle out of water, called precipitating, and stick to the metal surface. Water with large amounts of minerals is called hard water. Hard-water deposits, also called scale, can collect in hot-water supply pipes and water heaters (*Figure 15*). Hard-water deposits can block pipes, reducing flow. Pipes with excessive mineral-salt buildup must be replaced. Mineral salts also collect on heating elements in water heaters. This buildup makes water heaters work less efficiently.

Plumbers usually diagnose hard-water problems using a water test. Treat hard water using a water softener (*Figure 16*). A water softener is a device attached to the water supply line that

02311-14_F15.EPS

Figure 15 Scale buildup in a water line.

chemically removes mineral salts from the water before they can enter the system.

One popular type of softener is the zeolite system. Zeolite systems consist of a resin tank, also called a mineral tank, and a brine tank. Hard water enters the mineral tank. The calcium and mag-

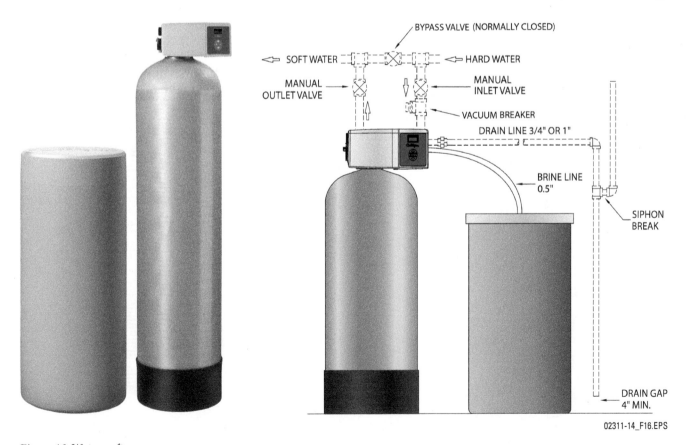

02311-14_F16.EPS

Figure 16 Water softener.

nesium salts in the water flow through resin in the tank. The resin is saturated with sodium, which reacts chemically with the salts and draws them out of the water. The minerals are then washed out through the resin tank's drain. When the cycle is complete, sodium-rich water from the brine tank recharges the resin tank. Remember that some codes require that specialists service water softeners.

Installing a water softening system is a fairly complex and costly procedure. Always discuss specific needs with the customer before installing a water softener. People may have specific requirements. For example, some people cannot drink softened water for medical reasons. In a home, the amount of water that needs to be softened may vary. For example, hose bibbs and lawn-sprinkler outlets don't need softening. If the water is slightly hard, then only the hot-water system will need to be softened. If the water is very hard, install softeners for the entire interior system.

Mineral deposits can also occur in systems where the water is not very hard. High heat can cause these mineral deposits. The water-heater settings may be too high. The solution may simply be to lower the water-heater thermostat setting, which will help keep the water below the critical temperature above which mineral deposits form. Thermostats are extremely sensitive. Therefore, adjust the thermostat in small increments. When you have made the first small change in the thermostat setting, show the customer how to make further changes. This extra step can eliminate callbacks. If the problem persists, you will have to evaluate the entire system's hot-water needs.

Common problems encountered in resin and brine systems include insufficient sodium, the presence of iron in the water, and motor problems. Sodium-level problems can be addressed by ensuring that the resin has not lost its sodium charge. If it has, follow the manufacturer's instructions for recharging the resin. If the sodium level continues to drop after recharging, check to see if the inside of the brine tank has become encrusted with sodium. This is commonly called a salt bridge. This can be caused by high humidity or the wrong type of salt. Use a broom handle or other similar tool and insert it into the tank. If a hard surface is detected well above the bottom of the tank, a salt bridge has formed. The manufacturer's instructions will specify how to break up a salt bridge. One common method is to gently poke the salt bridge with the broom handle or similar tool to break it up and allow it to dissolve. Be careful not to damage the inner lining of the tank when using this method.

Iron in the water can be treated by adjusting the setting of the water softener and cleaning the resin bed with a manufacturer-approved cleaner. If the source of the iron is bacteria or another organic source, refer to the manufacturer's instructions for cleaning, as the above methods will not solve the problem.

Electrical problems can be addressed by following the manufacturer's step-by-step testing and servicing instructions to identify the faulty component or components. Rotor seals, motor switches, and printed circuit boards are common sources of electrical malfunctions in hard-water treatment systems.

2.4.2 Small Particles and Organisms

Testing the water will indicate whether there are small particles and organisms in the water. These particles and organisms can lower water quality. Install a water filter to remove such matter from potable water. There are many different types of filters, each designed to trap certain types of materials. Be sure to consult the manufacturer's information when selecting a filter. Many filters use cellulose acetate membranes to separate out impurities. This is called the reverse-osmosis method (*Figure 17*).

To remove sulfur from water, install an activated carbon filter. This filter is filled with charcoal. The charcoal absorbs sulfur from the water as the water passes through the filter. Sediment filters trap dirt and silt. Install a deionizer on the cold-water inlet to trap metals in the water. Deionizers also filter out mineral salts.

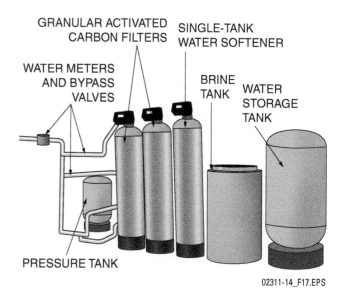

GRANULAR ACTIVATED CARBON FILTERS

SINGLE-TANK WATER SOFTENER

WATER METERS AND BYPASS VALVES

BRINE TANK

WATER STORAGE TANK

PRESSURE TANK

02311-14_F17.EPS

Figure 17 Commercial reverse-osmosis filtration systems.

2.5.0 Troubleshooting and Repairing Water Flow-Rate Problems

The flow rate in a water supply system is the volume per unit of time of circulated water in a plumbing system. It is calculated in gallons per minute (gpm). The rate of flow in a water supply system can be controlled by flow restrictors. In many cases, these devices can increase system efficiency by reducing water and energy consumption. For example, a flow restrictor installed in a typical shower can cut water consumption nearly in half without affecting the shower's performance.

Install flow restrictors at the inlet lines of individual fixtures that operate at a lower pressure than that of the system. That way, individual fixtures can have different flow rates. Follow the manufacturer's instructions when installing flow restrictors.

Many water supply systems rely on pumps to control the flow rate. These pumps may break down and need to be replaced. Systems that use pumps include the following:

- Wells
- Irrigation systems
- Hot-water recirculation systems
- Pressure booster systems
- Solar water heaters

When a pump breaks down, the entire water supply system usually has to be shut down. Plumbers must diagnose and correct pump problems quickly and efficiently.

When replacing a pump, ensure that the replacement pump is the proper size. The new pump should be able to handle the same amount of water and use the same amount of electricity as the old pump. If a pump breaks down repeatedly, review its specifications. Compare that information with the system's flow-rate calculations and plumbing drawings. Ensure that the pump is correctly matched with the system. A building engineer may need to conduct the appraisal.

Use a full port valve or gate and check valves when installing a pump (*Figure 18*). Install the pump using either a union or flange connection

so that it can be removed easily. Some water supply systems rely on more than one pump to operate (*Figure 19*). Ensure that the system can operate on one pump while the other is being repaired or replaced.

Some fixtures must be filled with water to a specific level to do their job. Water-closet tanks and washing machines are two common examples. These fixtures do not require flow restrictors. However, plumbers can adjust the water levels in these fixtures. Such adjustments allow the fixture to operate efficiently and use less water. Consult the manufacturer's installation instructions. Local codes may also provide guidance. Experience is a plumber's best guide.

2.6.0 Troubleshooting and Repairing Underground Pipes

Replace underground pipes that have been damaged beyond repair. Underground pipes that are undersized will also need to be replaced. Consult an expert plumber for guidance on how to replace pipe in your area.

In most cases, old buried supply pipe should be abandoned. Plan a new route for the replacement pipe by studying the existing site and structure. Note the location of all nearby utility lines before you start to dig. These include gas, electric, cable, and phone lines. In the structure, seal all old branches back to the old main or service pipe. Ensure that the new supply line connects with all existing branches in the structure.

2.7.0 Diagnosing Cross-Connections

Contamination caused by an improper cross-connection between plumbing systems is one of the biggest health and safety issues facing plumbers today. A cross-connection is an arrangement between two systems (for example between a potable water system and a nonpotable water system, or a medical gas system and a waste gas system) in which an accidental pressure differential causes solids, liquids, or gases from one system to enter the other system.

The Safe Drinking Water Act

Congress passed the Safe Drinking Water Act in 1974 and revised it in 1986. The act is the main federal law that ensures the quality of drinking water in the US. The act requires municipal water authorities to test and report on the quality of the water regularly. They must test the amount of minerals, organic matter, and bacteria in the water. The government can fine water authorities if the water in their area does not meet the act's standards for purity. For this reason, many water authorities have installed backflow preventers to protect the potable water supply.

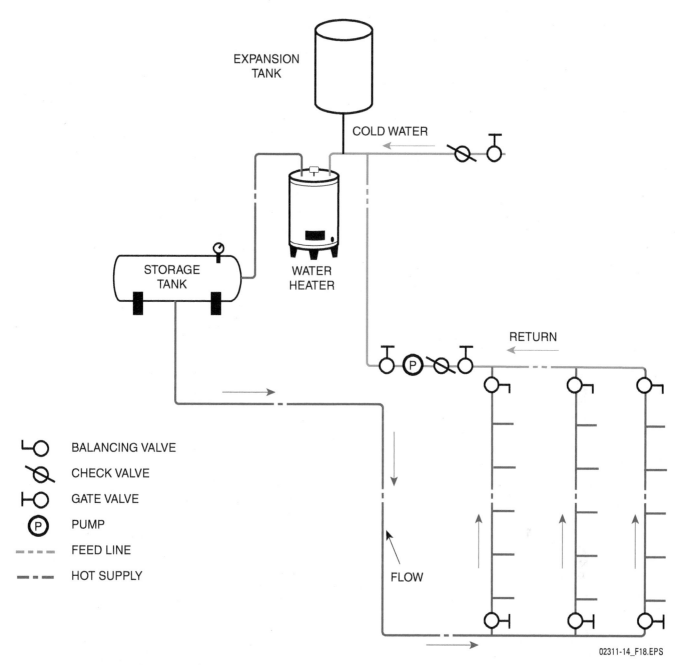

EXPANSION
TANK

COLD WATER

STORAGE
TANK

WATER
HEATER

RETURN

⊸○ BALANCING VALVE

⊘ CHECK VALVE

⊢○ GATE VALVE

Ⓟ PUMP

---- FEED LINE

-·-· HOT SUPPLY

FLOW

02311-14_F18.EPS

Figure 18 Typical pump installation in a hot-water recirculation system.

How to Recycle an Entire Building

The US Environmental Protection Agency (EPA) estimates that every year in the US, building demolition and renovation creates over 136 million tons of debris. Much of that material can be reused instead of being disposed of. Reusing and recycling building materials, fixtures, and fittings can save builders millions of dollars, can help reduce pollution from landfills and incineration, and can help save natural resources. For example, salvaging wooden beams and trusses from a demolished building can reduce the number of new trees that need to be cut down, and eliminate the pollution associated with cutting and processing those trees.

The systematic recycling and reuse of building materials that would otherwise be thrown away when buildings are renovated or torn down is called deconstruction. Deconstruction is growing in popularity in the construction industry because of the cost and environmental benefits it offers. Many companies now specialize in deconstruction, and work alongside other trades on demolition and renovation projects to identify, remove, and reuse construction materials.

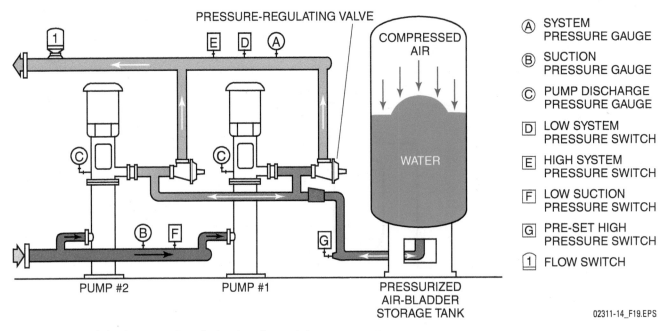

PRESSURE-REGULATING VALVE

COMPRESSED
AIR

WATER

PUMP #2   PUMP #1   PRESSURIZED
AIR-BLADDER
STORAGE TANK

Ⓐ SYSTEM
PRESSURE GAUGE

Ⓑ SUCTION
PRESSURE GAUGE

Ⓒ PUMP DISCHARGE
PRESSURE GAUGE

Ⓓ LOW SYSTEM
PRESSURE SWITCH

Ⓔ HIGH SYSTEM
PRESSURE SWITCH

Ⓕ LOW SUCTION
PRESSURE SWITCH

Ⓖ PRE-SET HIGH
PRESSURE SWITCH

☐ FLOW SWITCH

02311-14_F19.EPS

Figure 19 Typical duplex-pump installation in a domestic hot-water system.

2.7.1 Diagnosing and Servicing Cross-Connections in Water Supply Systems

Contamination of a potable water supply system through a cross-connection is called backflow. Several things can cause backflow in a potable water supply system:

- Cuts or breaks in the water main
- Failure of a pump
- Injection of air into the system
- Accidental connection to a high-pressure source

The two types of backflow are siphonage and back pressure. Siphonage occurs when negative pressure inside the DWV piping pushes the water that is normally held in a fixture trap into the DWV piping system. Back pressure occurs when a pressure that is higher pressure than the supply pressure causes a reversal of flow into the potable water piping. Siphonage and back-pressure problems are typically caused by improper venting of the DWV piping or by a blockage in the vent. To clear a vent blockage, snake the vent through the vent terminal in the roof.

Garden hoses are the most common source of cross-connections, especially in residential installations. Other common sources of cross-connection with potable water supply systems include:

- Lawn irrigation systems
- Fire suppression (sprinkler) systems
- Swimming pools, saunas, and hot tubs
- Drinking fountains
- Taps for reclaimed water
- Intakes for wells and cisterns
- Home water-treatment systems

When servicing cross-connection problems, ensure that all air gaps in the water supply system meet or exceed the minimum requirements in the local applicable code. The minimum required air gap is typically two times the size of the potable water outlet. Ensure that all faucets terminate above the flood-level rim of the sinks or basins they drain into. Ensure that vacuum breakers are installed on all hose bibbs. Install appropriate low- or high-hazard backflow prevention devices as required by the local applicable code, and follow the manufacturer's instructions when installing them.

2.7.2 Diagnosing and Servicing Cross-Connections in Medical Gas and Vacuum Systems

Cross-connections in medical gas and vacuum systems can endanger the health and safety of patients and medical professionals in health care facilities. Medical gas and vacuum systems are rigorously tested for cross-connections by qualified and certified installers and verifiers. The requirements for cross-connection testing are specified in NFPA 99, *Health Care Facilities Code*. Never attempt to service a medical gas or medical vacuum system without proper training and certification according to the local applicable code. Qualified and certified installers and verifiers must follow the steps outlined in NFPA 99 when performing cross-connection tests. You will learn more about these systems in the Plumbing Level Four module *Introduction to Medical Gas and Vacuum Systems*.

Medical Gas Systems

Any work on medical gas systems must be done by someone certified to install and service these systems. If you encounter a problem with a medical gas system, contact a certified professional.

Additional Resources

Efficient Building Design Series, Volume III, *Water and Plumbing*. 2000. Ifte Choudhury and J. Trost. Upper Saddle River, NJ: Prentice Hall.
NFPA 99, *Health Care Facilities Code*, Latest Edition. Quincy, MA: National Fire Protection Association.
Piping and Valves. 2001. Frank R. Spellman and Joanne Drinan. Lancaster, PA: Technomic.

2.0.0 Section Review

1. To repair a leak in a copper fitting without the use of press fittings and connections, _____.
   a. braze the fitting
   b. replace it with flanged fittings
   c. solder the fitting
   d. weld the fitting

2. To provide temporary protection against frost, wrap pipes with _____.
   a. fiberglass insulation
   b. duct tape
   c. heat-absorbing foam
   d. electric heat tape

3. Static pressure for noncirculating water should be kept below _____.
   a. 60 psi
   b. 75 psi
   c. 90 psi
   d. 105 psi

4. Mineral salts precipitate out of water and stick to the metal surface of pipes as a result of repeated exposure to _____.
   a. cold
   b. high pressure
   c. hard water
   d. heat

5. To allow pumps to be removed easily, install them using _____.
   a. a soldered or brazed connection
   b. a threaded connection
   c. a union or flange connection
   d. a ProPress® or SharkBite® connection

6. In most cases, old underground supply pipe that has been damaged should be _____.
   a. repaired
   b. abandoned
   c. recycled
   d. cleaned

7. Each of the following can cause backflow in a potable water supply system *except* _____.
   a. failure of a pump
   b. galvanic corrosion caused by the improper joining of dissimilar metals
   c. cuts or breaks in the water main
   d. injection of air into the system

SECTION THREE

3.0.0 TROUBLESHOOTING AND REPAIRING FIXTURES AND APPLIANCES

Objective

Explain how to troubleshoot and repair problems with fixtures and appliances.
- a. Identify common repair and maintenance requirements and procedures for fixtures, valves, and faucets.
- b. Explain how to troubleshoot and repair water heaters.
- c. Explain how to troubleshoot, repair, and replace tubs, showers, and water closets.
- d. Explain how to troubleshoot and repair other fixtures and appliances.

Performance Tasks

Troubleshoot and repair problems with fixtures, valves, and faucets using the proper tools and replacement parts.

Use manufacturer's instructions to disassemble and re-assemble a valve.

Troubleshoot and repair problems with water heaters.

Trade Terms

Corrosion: Deterioration caused by chemical reaction.

Dry well: A drain consisting of a covered hole in the ground filled with gravel.

Packing extractor: A tool used to remove packing from a valve stem.

Removable seat wrench: A tool used by plumbers to replace valve and faucet seats.

Reseating tool: A tool used by plumbers to reface a valve seat.

Screw extractor: A tool designed to remove broken screws.

Tap: A tool used for cutting internal threads.

Twist packing: A string-like material used to pack valve stems when preformed packing material is not available. It is not as durable as preformed packing.

Earlier in this curriculum you learned how to install basic plumbing fixtures, valves, faucets, and appliances. In this section, you'll learn how to service and replace the following basic plumbing fixtures:

- Fixtures, valves, and faucets
- Water heaters
- Tubs, showers, and water closets
- Dishwashers and lawn sprinklers

Many of the procedures for replacing fixtures and fittings are basically the same as for installing new ones. At this point in your training, you should know what tools and materials are required to repair and replace plumbing fixtures. Ask your instructor if in doubt.

> **CAUTION**
> A new fixture or appliance may not have the same dimensions as the one it is replacing. Always check the rough dimensions to verify measurements and placement.

Fixtures can become cracked or chipped during use. Replace damaged fixtures for both aesthetic and safety reasons. Before servicing or replacing fixtures, note the locations of the water supply lines and waste lines. Consider the ease of hookup when selecting a new unit.

3.1.0 Identifying Common Repair and Maintenance Requirements and Procedures for Fixtures, Valves, and Faucets

Once you take the necessary safety precautions, you are ready to inspect and repair the valve. While valves and faucets come in many styles, they can be divided into the following categories for repair work purposes:

- Globe valves
- Gate valves
- Flushometers
- Float-controlled valves (ball cocks)
- Tank flush valves
- Balancing valves
- Temperature and pressure (T/P) valves
- Backflow preventers
- Cartridge faucets
- Rotating-ball faucets
- Ceramic-disc faucets
- Touch faucets

A number of valves, especially those designed for commercial installations, are electronically controlled. The operation of these types of valves is controlled by a light beam.

3.1.1 Globe Valves

Internally, globe valves, angle valves (*Figure 20*), and compression faucets are all designed with the same basic parts. Therefore, the types of problems and their solutions are similar. These problems and their likely causes are presented in *Table 1*.

When repairing valves, always use the correct-size wrenches, and make sure their jaws are clean and smooth to prevent scratching finished surfaces. Do not use Channellock® pliers or pipe wrenches to repair valves.

If you have Problem 1 or 3 in *Table 1*, remove the valve bonnet, which covers and guides the stem, with a correctly sized wrench. Once you have removed the bonnet and stem assembly, inspect the seat disc (washer) and the seat. You can solve most of these cases by replacing the disc and resurfacing or replacing the seat.

A screw holds the seat disc (washer) in place. If you cannot remove the screw or if the screw breaks off, replace the stem or use a screw extractor (*Figure 21*) to remove the broken screw.

Table 1 Troubleshooting Globe and Angle Valves and Compression Faucets

Problems	Possible Causes
1. Drip or stream of water flows when valve is closed	Worn or damaged seat Worn or damaged seat disc
2. Leak around stem or from under knob	Loose packing nut Defective packing Worn stem
3. Rattle when valve is open and water is flowing	Loose seat disc Worn threads on stem
4. Difficult or impossible to turn handwheel or knob	Packing nut too tight Damaged threads on stem

> **WARNING!**
> Never attempt to extract a broken screw by attaching the extractor bit to a drill. Improper use of a screw extractor can cause injury as well as damage to the equipment. Always follow the manufacturer's instructions, and wear appropriate personal protective equipment.

Follow these steps to remove a broken screw using a screw extractor:

Step 1   Drill out a hole in the screw using a drill bit. The correct size for the drill bit is stamped on the extractor.

Step 2   Apply penetrating oil and wait for it to permeate the hole.

Step 3   With a hammer or mallet, drive the extractor firmly into the hole.

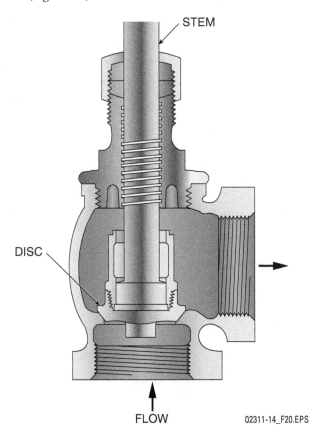

02311-14_F20.EPS

Figure 20 Basic parts of an angle valve.

02311-14_F21.EPS

Figure 21 Screw extractors with corresponding drill bits.

Step 4 Use a tee tap wrench to turn the extractor counterclockwise. Apply only the amount of torque that would normally be applied to the screw size you are trying to extract, or you may break the extractor.

> **WARNING!**
>
> Electrical shocks are not limited to electricians! Always protect yourself against accidental contact with energized sources when working on plumbing installations and service calls. Water conducts electricity.

When using a screw extractor, do not attempt to extract a broken screw by attaching the extractor bit to a drill.

Once you have removed the old screw, use a tap to recut or clean out the threads.

To replace the disc, select the correct disc and secure it to the end of the faucet stem. There are a variety of disc (faucet washer) sizes and shapes available (*Figure 22*). Plumbers generally carry an assortment of discs so that they are prepared to handle most repair jobs.

Replace the screw that holds the disc in place. Use screws with a soft metal or plastic plug near the tip. This plug locks the screw into position, eliminating the possibility that it will loosen.

To repair a damaged valve seat, fit a cutter into a reseating tool and insert it into the valve or faucet body (*Figure 23*). When you turn the handle, the cutter refaces the valve seat. After a few turns of the handle, inspect the seat. If it is smooth, stop.

If it is not smooth, give the handle another turn or two and inspect the seat again.

Some compression valves are equipped with replaceable seats. Simply replace the seat when it becomes worn or damaged. Two of the many removable seat wrenches available are shown in *Figure 24*. A few of the many replaceable seats available appear in *Figure 25*. Service and repair plumbers should make a habit of carrying an assortment of seats with them on the job.

Before you re-install the stem, check the amount of play in the threads that join the stem to the bonnet. If these threads are badly worn, the stem may rattle when the valve is opened. Replacing the stem may solve the problem. Because many different types of stems are available, you'll need to know the name of the valve manufacturer to get the right replacement. You may have to take the old stem to the supplier for a direct comparison to the replacement part. A few of the many types of stems available are shown in *Figure 26*. If the stem threads in the valve body are badly worn, you should probably replace the entire valve.

To solve Problems 2 and 4 in *Table 1.*, remove the packing nut, which holds the packing around the stem in place, and inspect the packing. Leaks around the stem generally result from wear of the packing or the stem shaft. If the stem shaft is excessively worn, replace the stem. If the problem is the packing, replacement of the packing and lubrication should solve the problem. Use a packing extractor to remove the packing (*Figure 27*). Before

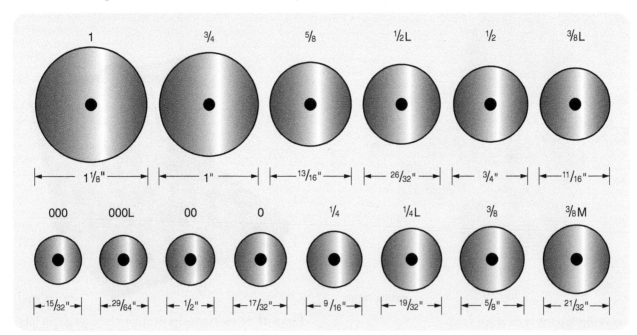

02311-14_F22.EPS

Figure 22 Seat disc sizes.

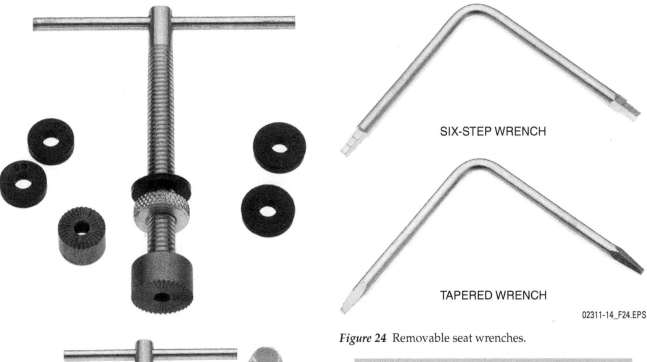

SIX-STEP WRENCH

TAPERED WRENCH

02311-14_F24.EPS

Figure 24 Removable seat wrenches.

02311-14_F25.EPS

Figure 25 Replaceable seats.

02311-14_F23.EPS

Figure 23 Reseating tools.

you select the replacement packing, examine the size and shape of the area under the packing nut. Various sizes and shapes of preformed packing are available (*Figure 28*). If the correct size and shape of preformed packing is not available, you can use self-forming graphite packing (*Figure 29*) or twist packing. Install the packing by wrapping it around the stem to fill the packing area. When

American Standard

American Standard is the world's largest producer of bathroom and kitchen fixtures and fittings. It is also a leading producer of air conditioning and heating systems.

American Standard was established when the American Radiator Company merged with the Standard Sanitary Manufacturing Company in 1929. At first, the company was called the American Radiator & Standard Sanitary Corporation. In 1967, it changed its name to American Standard.

Standard Sanitary was formed in 1899. This company pioneered improvements such as the one-piece toilet, built-in tubs, combination faucets (which mix hot and cold water), and brass fittings that don't tarnish or corrode.

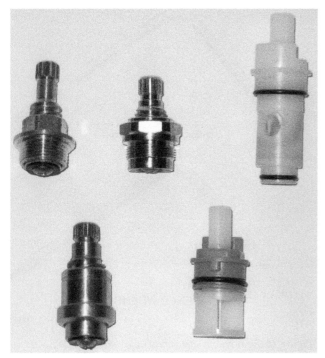

Figure 26 Stems.

you tighten the packing nut, the packing forms into the required shape. Twist packing is not as durable as preformed packing, which is made specifically to fit the valve.

If the stem is difficult to turn, the reason may be one of the following:

- *The packing nut is too tight.* Loosening the nut and lubricating the stem where it passes through the packing should solve the problem. A special lubricant that resists high tempera-

tures is available. In addition to being heat resistant, it is also waterproof.

- *The threads on the stem are damaged.* If the threads in the base of the valve are not damaged, replacing the stem should solve this problem. If these threads are damaged, replace the valve.

3.1.2 Gate Valves

Some of the problems common to globe valves are also common to gate valves (*Figure 30*). A leak around the stem is the result of either packing wear or wear on the stem. Use the same procedure described for globe valves to solve this problem.

The wedge-shaped disc in the gate valve is designed to be either fully opened or fully closed. Gate valves are not intended to throttle the volume of liquid flowing through the pipe. When gate valves are opened only partway, the gate has a tendency to vibrate or chatter, which causes the edge to erode. Repair of the valve is not likely to solve this problem. More than likely you will have to replace the valve. If a gate valve fails to stop the flow of water, one of three possible problems may be the cause:

- The gate is worn.
- The valve seat is worn.
- Some foreign material is preventing the gate from seating.

To determine which problem exists, remove the bonnet and inspect the gate and seat. You may be able to solve the problem by scraping out mineral deposits and carefully cleaning the mating surfaces. You may have to replace the gate, and if the seat area is worn, you'll have to replace the valve.

> **NOTE**
> Some companies remanufacture larger gate valves.

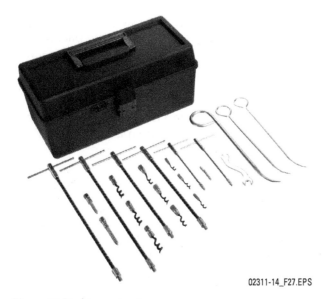

Figure 27 Packing extractors.

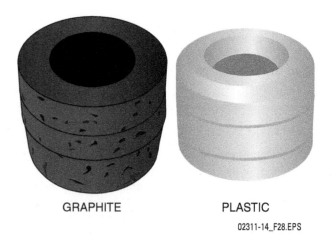

GRAPHITE   PLASTIC

Figure 28 Preformed packing.

02311-14_F29.EPS

Figure 29 Self-forming graphite valve packing.

3.1.3 Flushometers

Flushometers connect directly to the water supply and require no storage tank, which allows for repeated, rapid flushing. Manual flushometers (*Figure 31*) operate by pulling a lever. Sensor flushometers (*Figure 32*) operate automatically rather than by using a handle. An infrared sensor or electric eye automatically flushes the fixture when

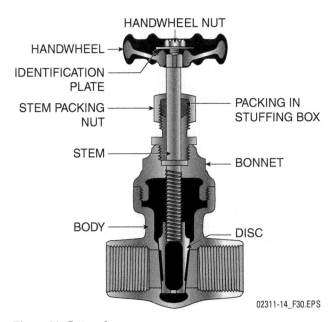

HANDWHEEL NUT
HANDWHEEL
IDENTIFICATION PLATE
STEM PACKING NUT
STEM
BODY
PACKING IN STUFFING BOX
BONNET
DISC

02311-14_F30.EPS

Figure 30 Gate valve.

it detects that a person is no longer using the fixture. Concealed-sensor flushometers (*Figure 33*) are located behind walls to protect against vandalism and damage. Servicing a concealed-sensor flushometer will require access through the wall, either through a removable panel or by cutting through the wall. The troubleshooting guide in *Table 2* shows common problems and solutions for both manual and electronic-eye flushometers. Five common problems are associated with flushometers:

- Leakage around the handle
- Failure of the vacuum breaker
- Malfunction of the control stop, which regulates the flow of water
- Failure of the flushometer to completely close
- Leakage in the diaphragm that separates the upper and lower chambers

Kits that contain the components you'll need to repair each of these defects are available. Even though all flushometers contain the same basic components, specific parts will vary from manufacturer to manufacturer. Always follow the manufacturer's specifications and instructions when installing replacement parts.

3.1.4 Waterless Urinals

Waterless urinals use a cartridge filter containing a liquid seal rather than water to remove waste and odor (*Figure 34*). The top of the cartridge is shaped like a funnel to allow wastewater to flow into it, where it passes through the liquid seal. Because the liquid is lighter than water, it floats above the wastewater entering the cartridge and thus traps odors. As the wastewater flows through the cartridge, a filter traps sediment suspended in the wastewater. The wastewater then flows through the fixture's trap and into the waste

Repair or Replace?

Often, replacing a defective valve makes more sense than trying to repair it. It can be difficult to find the right parts needed for repair. You can save yourself and your customer a lot of time and money by simply replacing defective valves. Every situation is different, so be sure to think about cost-effectiveness and customer service when deciding whether to repair or replace a valve or faucet. Determine how long service will be turned off for the repair. Notify the owner and occupants of this information, and also when the service has been both turned off and restored.

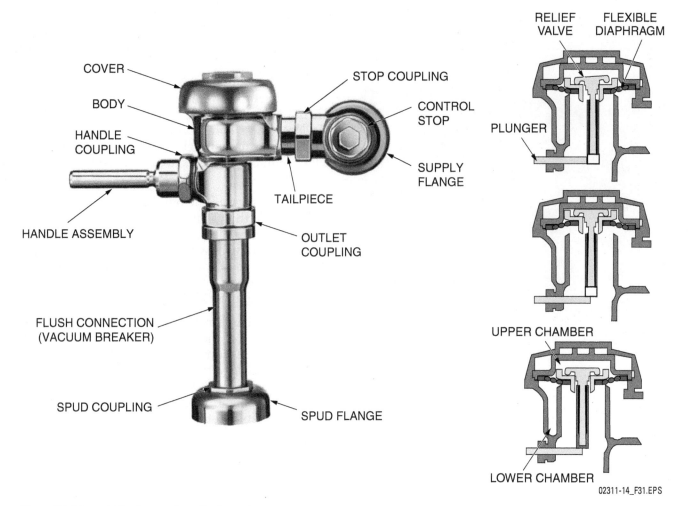

RELIEF VALVE FLEXIBLE DIAPHRAGM

PLUNGER

UPPER CHAMBER

LOWER CHAMBER

02311-14_F31.EPS

Figure 31 Manual diaphragm-type flushometer.

COVER

BODY

HANDLE COUPLING

HANDLE ASSEMBLY

FLUSH CONNECTION (VACUUM BREAKER)

SPUD COUPLING

STOP COUPLING

CONTROL STOP

SUPPLY FLANGE

TAILPIECE

OUTLET COUPLING

SPUD FLANGE

02311-14_F32.EPS

Figure 32 Sensor flushometer.

line. A sealing ring on the outside of the cartridge provides an airtight barrier between the cartridge and fixture.

The sealant liquid and cartridges must be replaced periodically according to the manufacturer's specifications. Other designs for waterless urinals rely on mechanical traps to eliminate odor. However, plumbing codes in the US prohibit the use of mechanical traps.

Most service calls for waterless urinals involve the replacement of the sealant liquid and cartridge filter. Remove the cartridge filter by inserting the key provided with the cartridge into the keyhole on the top of the cartridge. Then fill the cartridge with water. Use the key handle to pull up on the cartridge to remove it from the fixture and dispose of the used cartridge according to the manufacturer's instructions. Rinse the open trap with water to clean it, and then insert a new cartridge into the trap according to the manufacturer's instructions. Fill the new cartridge with water and pour in the required amount of sealant liquid.

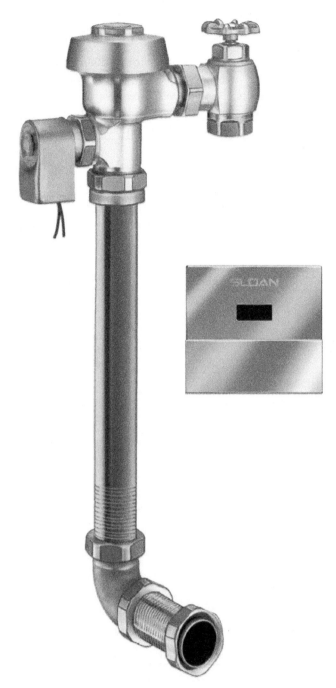

02311-14_F33.EPS

Figure 33 Concealed-sensor flushometer.

3.1.5 Float-Controlled Valves

The valve that controls the water level in a water-closet tank is a float-controlled valve, also called a ball cock (*Figure 35*). Manufacturers provide repair kits, and you must refer to their specifications and instructions when making repairs. If the stem, valve body, or float mechanism is damaged, you may have to replace the entire valve. Replacement parts are available for leaking floats or broken float rods.

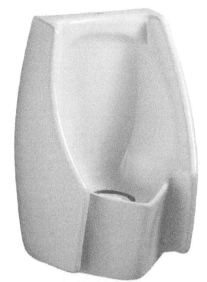

WATERLESS URINAL

CARTRIDGE INSERT, TOOL,
AND ODOR BARRIER LIQUID

02311-14_F34.EPS

Figure 34 Waterless urinal fixture.

To install a new float-controlled valve assembly, follow these steps:

Step 1 Turn off the water supply.

Step 2 Lower the threaded base through the bottom hole in the closet tank with the gasket in place against the flange.

Step 3 Place the washer and nut on the base of the assembly on the outside of the tank and tighten. Do not overtighten the nut.

> CAUTION
>
> Never overtighten the nut on the base of the assembly on the outside of the tank. Overtightening puts stress on the fixture and can cause it to crack.

Table 2 Troubleshooting Guide for Standard and Electronic-Eye Flushometers

Problem	Cause	Solution
1. Nonfunctioning valve	Control stop or main valve closed	Open control stop or main valve
2. Not enough water	Control stop not open enough Urinal valve parts installed in closet parts Inadequate volume or pressure	Adjust control stop to siphon fixture Replace with proper valve Increase pressure at supply
3. Valve closes off	Ruptured or damaged diaphragm	Replace parts immediately
4. Short flushing	Diaphragm assembly and guide not hand-tight	Tighten
5. Long flushing	Relief valve not seating	Disassemble parts and clean
6. Water splashing	Too much water is coming out of faucet	Throttle down control stop
7. Noisy flush	Control stop needs adjustment Valve may not contain quiet feature The water closet may be the problem	Adjust control stop Install parts from kit Place cardboard under toilet seat to separate bowl noise from valve noise— if noisy, replace water closet
8. Leaking at handle	Worn packing Handle gasket may be missing Dried-out seal	Replace assembly Replace Replace
9. Valve fails to close off	Damaged or dirty valve Insufficient pressure	Clean and replace, then flush lines Ensure that control stop is fully open

Electronic-Eye Flushometers

Note: These are general guidelines. It is best to follow the specific manufacturer's guidelines when troubleshooting electronic-eye flushometers.

Problem	Cause	Solution
1. Valve won't flush	Sensor not sensing presence of user; range is too short	Increase the range
2. Valve won't flush or flushes only when someone walks past	Sensor is focusing on a reflective wall or mirror; range is too long	Decrease the range
3. Unit flashes when user steps into range	Low batteries	Replace batteries
4. Valve won't shut off	Dirt or debris clogging diaphragm orifice Seal damaged or worn Overtightened solenoid valve Malfunction in electronic module	Clean Replace Loosen or replace Replace
5. Water flow to fixture too low	Control stop improperly adjusted Not enough flow or pressure	Adjust Increase flow or pressure
6. Water flow to fixture too high	Control stop improperly adjusted Improper diaphragm installed in valve	Adjust Replace with proper diaphragm

Step 4 Place the supply and coupling nut on the base of the float valve and hand-tighten.

Step 5 Align the supply with the stop valve. Remove the supply and cut to fit.

Step 6 Re-assemble the supply to the tank and the valve, and then turn on the water supply.

Step 7 Check for leaks.

Step 8 Adjust the float to achieve the water level indicated on the inside of the tank.

Step 9 Adjust the float with the adjusting screw. Do not bend the float arm.

> **CAUTION**
> Never try to adjust the water level in a water closet by bending the float arm of a float-controlled valve. If the rod works around a half-turn (180 degrees), this will cause the tank to overflow.

3.1.6 Tank Flush Valves

Tank flush valves (*Figure 36*) are available in many styles. If the valve and the lever that operates the valve are badly corroded, replace both parts. However, most problems occur with the component parts: the tank ball, the flapper tank ball, the connecting wires or chains, and the guide. In-

02311-14_F35.EPS

Figure 35 Float-controlled valve.

spect all of these parts to determine whether to repair or replace the entire assembly. If the valve seat is corroded, you can use a reseating tool to restore it.

3.1.7 Balancing Valves

A balancing valve (*Figure 37*) balances both temperature and pressure. It protects users from being scalded or getting a blast of cold water in tubs and showers. This valve allows hot and cold water to mix together so that a comfortable water temperature is delivered to the faucet or showerhead. It is designed to immediately block the flow of hot water if the cold-water supply fails. Temperature- and pressure-balancing valves are used in both residential and commercial installations.

Leaks can occur when solder or other debris gets between the washers and seat surfaces of this valve. To repair a leak, replace all washers. You should also inspect the top surface of both the hot and cold seat, and replace them if damaged.

If the hot and cold water are not properly balanced or the water temperature changes without the handle being moved, the problem is most likely debris in the pressure-balancing spindle. To correct this problem, open the valve halfway and remove the handle. Tap the spool with a plastic hammer. If this does not solve the problem, remove the spool assembly and tap the handle end against a solid object to free the piston. Soaking the assembly in vinegar will help remove small particles of lime and scale buildup.

02311-14_F36.EPS

Figure 36 Tank flush valve.

3.1.8 Temperature and Pressure (T/P) Valves

The T/P valve is designed to open and vent heated water (the temperature part of the valve) and air (the pressure part of the valve) into the atmosphere (*Figure 38*). This venting action brings the water temperature and pressure back to safe levels. Therefore, the T/P valve is the principal safety feature of water heaters.

As water is heated, its volume expands. A simple example of this principle is a teakettle filled with boiling water. The heated water rapidly expands, eventually turning into a blast of hot steam. The thermostat on a water heater maintains the water temperature to prevent a buildup of heat. However, if the thermostat fails, the water temperature will continue to increase and the rapid increase of heat inside the heater can result in an explosion. To ensure safety, codes require that T/P valves regulated for residential use be installed on water heaters.

Too much water pressure can cause the T/P valve to operate with too much force. To correct this problem, install a pressure-reducing valve on the cold-water side. Too high a temperature can cause the T/P valve to operate too frequently. To correct this problem, check the thermostat. You may have to lower the setting or replace the thermostat. Drips are typically the result of mineral buildup around the valve seat. Lift the stem lever to let the water flush out, which should also flush out debris. Rotate the lever to reseat the stem. If the valve seat is damaged, replace the valve.

Gas shutoff valves (*Figure 39*) are another way to protect against overheated water in systems that use gas water heaters. These immersion valves are equipped with a thermostat that shuts off the gas supply to the water heater's main burner and pilot if the water temperature exceeds 210°F. The

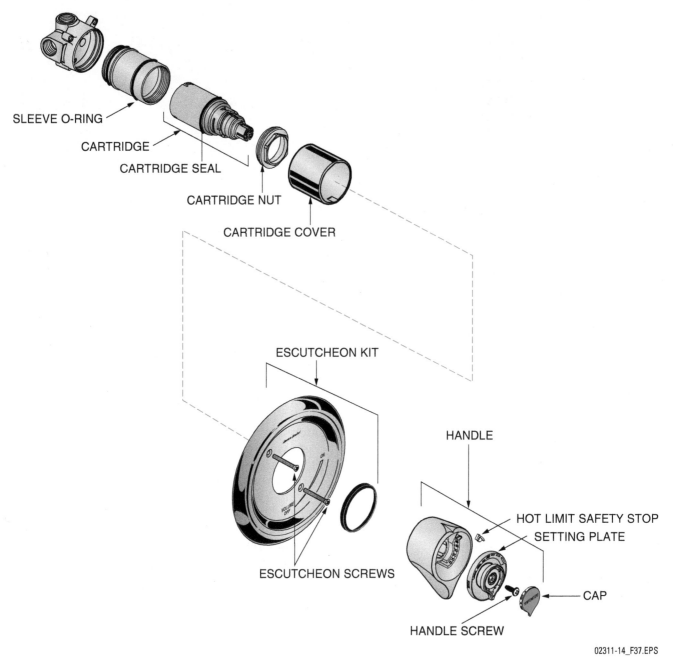

SLEEVE O-RING

CARTRIDGE

CARTRIDGE SEAL

CARTRIDGE NUT

CARTRIDGE COVER

ESCUTCHEON KIT

HANDLE

HOT LIMIT SAFETY STOP
SETTING PLATE

CAP

ESCUTCHEON SCREWS

HANDLE SCREW

02311-14_F37.EPS

Figure 37 Balancing valve for a shower.

valve is fitted with a reset button to reopen the valve after a shutoff. When servicing a gas water heater after a gas shutoff valve has tripped, be sure to shut off the main gas valve at the heater and the pilot before resetting the valve. Then re-light the pilot according to the manufacturer's instructions. Never install a gas shutoff valve on a cold-water line.

3.1.9 Cartridge Faucets

You can identify a cartridge faucet by the metal or plastic cartridge, or sealed unit, inside the valve body (*Figure 40*). The cartridge is located between the spout assembly and the handle. Many single-handle faucets and showers are cartridge designs. Replacing a cartridge is a fairly easy repair that will fix most faucet or shower drips. To replace or repair a cartridge, turn off the water supply to the faucet and open the faucet taps to let the water drain out of the fixture. Disassemble the faucet according to the manufacturer's instructions to expose the cartridge. Use a cartridge puller (*Figure 41*) for the particular model of faucet to unscrew the locking nut holding the cartridge in place, and remove the cartridge. Repair or replace the cartridge using the cartridge puller and re-assemble the faucet.

02311-14_F38.EPS

Figure 38 Temperature-and-pressure valves.

02311-14_F39.EPS

Figure 39 Immersion-type gas shutoff valve.

3.1.10 Rotating-Ball Faucets

A rotating-ball faucet has a single handle. You can identify it by the metal or plastic ball inside the faucet body (*Figure 42*). Drips usually occur as a result of wear on the valve seals. Replacement valve seals and springs are readily available. Leaks from the base of the faucet are usually caused by worn spout O-rings. You should replace these parts.

3.1.11 Ceramic-Disc Faucets

A ceramic-disc faucet is a single-handle faucet that has a wide cylinder inside the faucet body (*Figure 43*). The cylinder contains a pair of closely fitting ceramic discs that control the flow of water. The top disc slides over the lower disk. These discs rarely need replacing. Mineral deposits on the inlet ports are the main cause of drips. Cleaning the inlets and replacing the seals eliminates most drips.

3.1.12 Touch Faucets

Touch faucets (*Figure 44*) are a recent innovation. They allow users to turn the water flow on and off simply by touching or tapping the faucet's spout or handle. Touch faucets work by interrupting the faucet's capacitance, which is the ability to store an electric charge. A battery in the faucet generates a low-voltage electric charge on the faucet's

Pop-Up Stopper

A pop-up stopper consists of a stopper rod, a clevis, and a horizontal pivot rod. If the stopper gets wrapped with hair or loses its seal, water will drain from the sink even when the stopper is closed. To correct this problem, loosen the nut holding the horizontal pivot rod in place. Remove the pivot rod, mark where the clevis connects to the stopper rod, then loosen the screw holding the clevis in place. Soak the clevis and pivot rod in vinegar to clean these parts. Inspect the rubber seal on the bottom of the stopper head. If the seal is cracked or brittle, replace it. Use a small brush to scrub off stubborn debris. Rinse and dry all parts, re-assemble, and re-install.

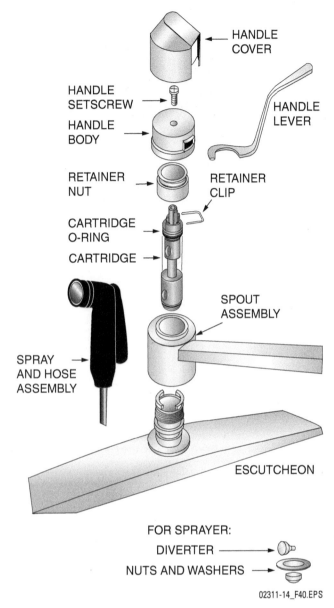

Figure 40 Cartridge faucet.

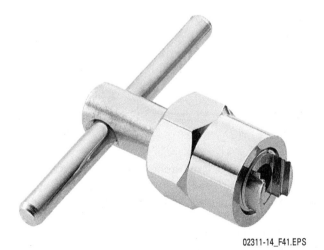

Figure 41 Cartridge puller.

spout and handle. Touching the spout or handle interrupts the flow of the current. This interruption causes a valve to open, allowing the water to flow. The electric current is far too low to cause an electric shock. This is similar to the way that touch-screen mobile devices such as smartphones and tablets work.

The user sets the water temperature and pressure using a handle similar to that used on a rotating-ball faucet. The water flow is then started and stopped by tapping the spout. The spout will not activate the water stream if the spout is held. This allows the user to rotate and position the spout without accidentally splashing water.

If the handle and the spout no longer respond to touch, or if the water flow turns on and off by itself, check the fixture's batteries to ensure they are charged, and replace them with the appropriate model of replacement battery if they are not. If the faucet is equipped with grounding clips, ensure that they are connected properly. Ensure that the spout is insulated from accidental electrical contact with the sink basin. Finally, try resetting the faucet by disconnecting the battery for 30 seconds and then reconnecting it. If the hot and cold water are cross feeding, inspect the check valves to ensure they are installed properly.

3.1.13 Faucet Water Filters

Water filters come in undercounter mountings, countertop mountings, and fixture mountings. Some faucets combine the functions of water delivery and filtration in one unit. Although styles vary, all filters work on the same basic principle. Water intended for drinking or washing passes through a cartridge, which usually contains carbon, to remove unwanted tastes, odors, and chemicals such as chlorine. Always follow the manufacturer's installation and maintenance instructions. In general, most problems result from filters not being changed regularly. In addition, customers may complain of reduced water flow. Filters do restrict the flow of water from the faucet to allow contact time with the filter. This is a trade-off for cleaner, better-tasting water.

Replacing Cartridges

Faucet cartridges come in many styles, so you may want to take the old cartridge to the supply house for comparison when you have to replace one.

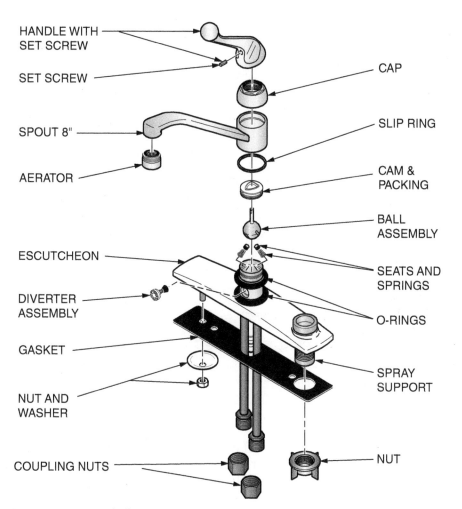

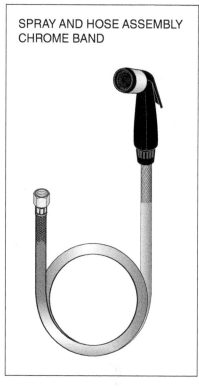

SPRAY AND HOSE ASSEMBLY
CHROME BAND

HANDLE WITH SET SCREW

SET SCREW

SPOUT 8"

AERATOR

ESCUTCHEON

DIVERTER ASSEMBLY

GASKET

NUT AND WASHER

COUPLING NUTS

CAP

SLIP RING

CAM & PACKING

BALL ASSEMBLY

SEATS AND SPRINGS

O-RINGS

SPRAY SUPPORT

NUT

02311-14_F42.EPS

Figure 42 Rotating-ball faucet.

3.1.14 *Electronic Controls*

Electronic, or hands-free, controls are available for faucets, toilets, and urinals. These controls emit a constant beam of infrared light, which operates the fixture. Water flows into sinks and lavatories when the user blocks the beam. Toilets and urinals flush when the user no longer blocks the beam. The relatively high cost of these controls generally limits their use to commercial and institutional installations, where they may be installed new or retrofitted onto existing plumbing. The controls save water and energy and are more hygienic than hand-operated controls.

Some electronic controls use battery power to operate flush valves and faucets, and most problems result when the battery power is exhausted or when a battery is improperly installed. If the problem is not with the battery, the problem may be a clogged filter or component that has been installed backwards. Clean and re-install clogged filters, and ensure that all components are installed properly.

The installation and maintenance instructions for electronic controls vary from one manufacturer to another. Always follow the manufactur-

er's installation and maintenance instructions. (For general troubleshooting guidelines for electronic flushometers, refer to *Table 2*.)

3.2.0 Troubleshooting and Repairing Water Heaters

The water heater is one of the most expensive appliances in a home. Water heaters must be se-

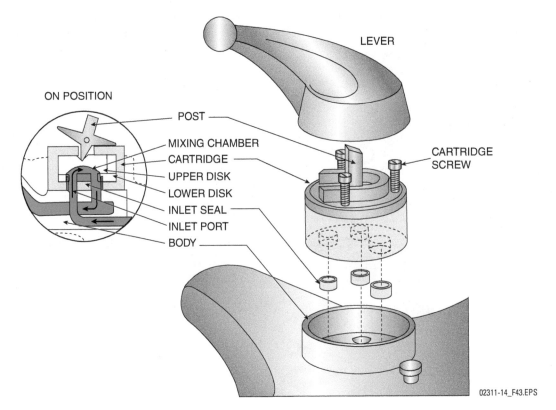

Figure 43 Ceramic-disc faucet.

ON POSITION

POST

MIXING CHAMBER
CARTRIDGE
UPPER DISK
LOWER DISK
INLET SEAL
INLET PORT
BODY

LEVER

CARTRIDGE
SCREW

02311-14_F43.EPS

lected, installed, adjusted, and maintained with care. When a water heater leaks or fails to supply enough hot water, it may need to be replaced. When there is no hot water at the tap, the water heater is the most likely cause. Common water-heater problems include:

- Failure of the pilot light or heating element
- Improper thermostat setting
- Uninsulated hot-water lines
- Overuse of hot water
- Improper size

3.2.1 Draining a Water Heater

Many water-heater problems require the water supply to the water heater to be shut off and the tank to be drained completely before servicing. Always follow the manufacturer's instructions when draining a water heater, as different models and types of water heaters will have particular requirements. To drain a typical water heater, follow these steps:

Step 1 Disconnect the water heater's electrical connection.

Step 2 Connect a drain hose to the water heater's drain valve. Run the drain hose so that it

will discharge safely. Remember, the water being drained from the water heater will be hot.

Step 3 Close the cold-water supply valve to the water heater and open a hot-water fixture to ensure that all the air has escaped the system.

Step 4 Open the water heater's drain valve and allow the tank to drain completely.

02311-14_F44.EPS

Figure 44 Touch faucet.

3.2.2 Correcting Common Water-Heater Problems

Failures of the pilot light and the heater element are the most common water-heater problems. If there is no hot water, begin by checking for an extinguished pilot light or if the electric element is burned out.

Gas water heaters (*Figure 45*) burn a mix of air and fuel gas to heat water. For a gas water heater with an extinguished pilot light, check for drafty conditions. Redirect louvers to the equipment room so that air does not blow directly on the pilot light. If that does not correct the problem, install additional shielding around the water heater. Shielding will protect the pilot light from drafts. Check the thermal couple. Check for dust and cobwebs blocking the jet. This is especially important if the pilot light has been off for a long time. If the gas jet is blocked, clean it according to the manufacturer's instructions. Inspect the vent and clear any blockages.

Gas water heaters are fitted with sacrificial anodes to protect the tank from corrosion. Anodes should typically be inspected every six months

Water-Lubricated Pumps

Many pumps are water lubricated. This means they must be immersed in water before being tested or run. Otherwise, the bearings could be ruined quickly. Pay close attention to the manufacturer's specifications when installing and testing pumps.

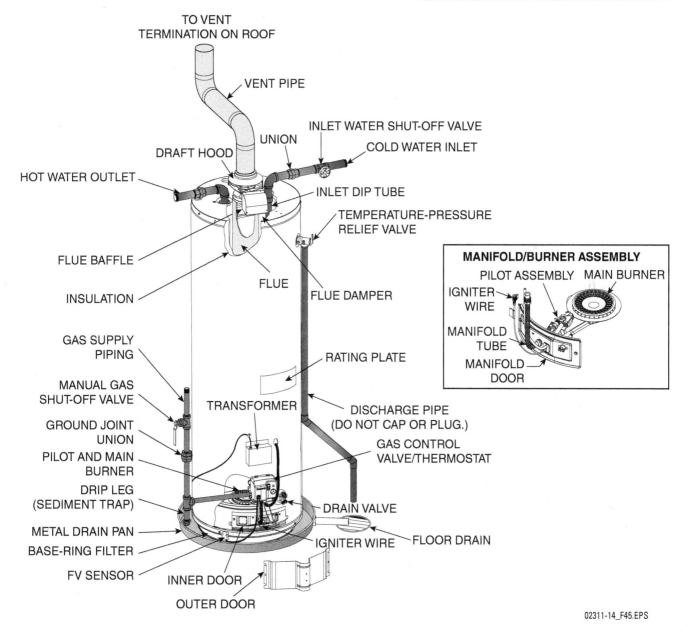

Figure 45 Typical gas water heater.

02311-14_F45.EPS

and replaced annually or when their diameter has decreased to ⅜ of an inch. To replace an anode, follow these steps:

Step 1   Close the cold-water supply line to the water heater. Turn off the burner according to the manufacturer's instructions.

Step 2   Drain and flush the water heater according to the manufacturer's instructions.

Step 3   Remove the water heater's combustion covers, access covers, and insulation.

Step 4   Use a socket wrench to remove the anode. Use a breaker bar if necessary to loosen the anode. Never use an impact wrench to loosen and remove an anode.

Step 5   Replace the anode or install a new anode according to the manufacturer's instructions. Apply pipe sealant or tape to the anode threads to ensure a seal.

Step 6   Close the drain valve and open a hot-water fixture to ensure that all the air has escaped the system.

Step 7   Open the cold-water supply to the water heater and allow the unit to fill.

Step 8   Light the water-heater burner according to the manufacturer's instructions.

Step 9   Inspect the anode for leaks and adjust the tightness of the anode accordingly.

Step 10 Replace the insulation, access covers, and combustion covers according to the manufacturer's instructions.

Electric water heaters (*Figure 46*) use an immersion element to heat water. Before replacing an electric water-heater element, disconnect the power supply, and then drain the water storage tank. When the tank has been drained, remove the element. Inspect the element for any clues as to the cause of the problem. If the element has excessive calcium buildup, install a water softener.

Replace the element with a new one when the problem has been corrected. Tighten the new element to ensure a watertight seal, and then refill the tank. You will learn how to replace an immersion element in the section *Replacing an Immersion Element in an Electric Water Heater*.

If the thermostat is out of adjustment, find the correct setting. As discussed earlier, thermostats are extremely sensitive, so adjust thermostats in small increments. Make the first adjustment and show the customer how to make further adjustments. If the hot-water pipes are losing heat, insulate the supply lines. If the thermostat does not operate, take a reading of the primary and secondary voltage leads on the thermostat's transformer

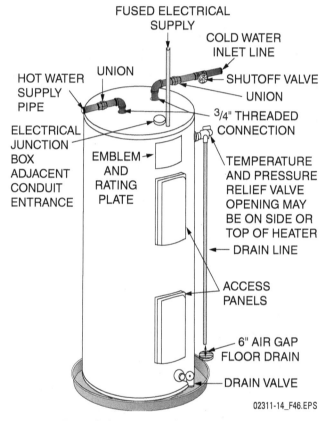

Figure 46 Typical electric water heater.

Water Heater Stands

Water heaters are often located near storage areas for flammable liquids such as gasoline and paint thinner. These liquids emit vapors that are heavier than air. These vapors thus tend to accumulate near the floor. If they are drawn into a gas-fired water heater, they can cause an explosion. Manufacturers offer water heater stands that raise the heater off the floor. The heater can then draw air from above the vapor level. Consult the local code about the use of water heater stands in your area.

to ensure there is sufficient voltage. The thermostat's sensor probe may need to be replaced if the voltage is insufficient. Otherwise, the thermostat's circuit board may need to be replaced.

Sometimes there may be not enough or too much hot water. Part of a plumber's job is to offer advice to customers on how to assess their water-heater requirements over time. For example, if the water heater is in a home, identify the family's past, current, and future usage. The family may have grown children who have recently left home. As a result, they may be using less hot water. If so, the thermostat could be lowered or the water heater could be replaced with a smaller unit. You can also suggest how the rest of the family can reduce their hot-water usage. Walk the customer through the proposed changes. Survey the family's needs and then recommend whether they should make adjustments or replace their water heater.

3.2.3 Replacing a Water Heater

If the water heater is improperly sized, replace it with one that is adequate for the system. Take the following steps when removing a water heater:

Step 1 Shut off the fuel or energy source.

Step 2 Shut off the water supply and drain the tank.

Step 3 Disconnect the heater's inlet and outlet water lines. Mark each line connected to the heater. This will help you identify the correct connection when installing the new water heater.

Step 4 Carefully disconnect the water heater's gas piping and electrical connections. If you do not have experience with handling utility connections, seek help from an experienced plumber.

Step 5 Remove the water heater according to the manufacturer's instructions. Consult with an experienced plumber on the best way to remove a water heater or check the local code for guidance. Be sure to avoid spilling or splashing any remaining water in the tank as you remove the water heater from the building.

Before disposing of an old water heater, inspect the elements (if it is an electric water heater) or the inside base of the tank (if it is gas fired). If there is excessive corrosion or calcium deposits, suggest that the water be analyzed. Also suggest to the customer that a water softener be installed because water softeners usually allow a water heater to run more efficiently and last longer.

To install a new water heater, follow these steps:

Step 1 Locate and install the new water heater according to the manufacturer's instructions. Consult the local code.

Step 2 Connect all water, gas, oil, or electric supply lines. The lines should have been marked when the old tank was removed.

Step 3 Replace the T/P relief valve with a new one before operating the new water heater.

Step 4 Fill the tank with water. The tank must always be filled before the heating elements or burners are turned on. If this is not done, the elements will burn out.

Step 5 Ensure that all air is bled through the hot-water faucets.

> **WARNING!**
>
> The T/P valve is a critical safety control feature in a water supply system. The valve must always function properly. A malfunctioning T/P valve can cause a water heater to explode, resulting in injuries or death. Install the T/P valve so that it drains into the ground as an indirect waste line. Always check to ensure that T/P valves are operating correctly and that they are clear of obstructions.

Step 6 Set the thermostat to the lowest setting that will serve the structure adequately.

Step 7 Ensure that the relief-valve drain does not create a trap. A trap will cause back pressure when steam is discharged through the relief pipe.

Step 8 Provide an air gap at the outlet end of the relief-valve pipe. The air gap should be located where the valve empties into the floor drain or other approved receptacle.

3.2.4 Replacing an Immersion Element in an Electric Water Heater

Electric water heaters use one or two immersion elements in the storage tank to heat water (*Figure 47*). Immersion elements are controlled by automatic thermostats (*Figure 48*), which sense the tank's outside surface temperature and adjust the flow of electricity to each element. In normal operation, the upper element of a typical water heater will operate first to heat the water in the upper section of the tank. When that water reaches a preset temperature, the upper element is cycled off and the lower element is cycled on to heat the remaining water (*Figure 49*).

Before replacing immersion elements, test them to ensure that they are the cause of the problem. Turn the water heater off and use a multimeter to test the resistance of the element. Set the multimeter to Ohms and connect the leads to the element as shown on the manufacturer's instructions. Usually, the black lead is connected to the Common port, while the red lead is connected to the Ohms port. Set the voltage indicator to the correct scale and read the results. If the meter goes off the scale, then it is likely that the element will need to be replaced. If the voltage is lower than the rated voltage, check to see if the element is accidentally grounded. The manufacturer's instructions will provide detailed procedures for testing an element. Always follow these procedures closely, and wear appropriate personal protective equipment when working with electrical equipment.

To replace an element in an electrical water heater, follow these steps:

Step 1 Disconnect or turn off the water heater and open the circuit breaker for the water heater.

Step 2 Use a screwdriver to remove the thermostat access panel for the element to be replaced.

Step 3 Test the resistance of both elements by setting the multimeter to read X1000 and attaching one lead to one of the screw terminals and the other to a heating-element mounting screw. If the reading is anything other than infinity, you should replace the element. Test the other element to determine if it also needs to be replaced.

Step 4 Attach a hose to the water-heater drain valve. Close the water-inlet valve on the water heater and open the drain valve to drain the tank completely. Open the temperature and pressure relief valve. Exercise caution when draining the tank, as the water is hot.

Step 5 Unscrew the element's electrical terminals. Use a breaker bar if necessary to loosen the connections. Never use an impact wrench to loosen and remove an immersion element's contacts.

Step 6 Install a replacement immersion element that is the same configuration and rating as the element being replaced. Attach and tighten the mounting bolts and screws to hold the element in place.

02311-14_F47.EPS

Figure 47 Immersion elements for electric water heaters.

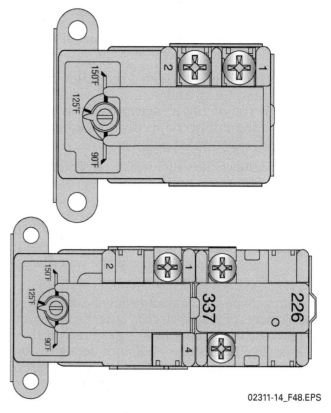

02311-14_F48.EPS

Figure 48 Automatic thermostat.

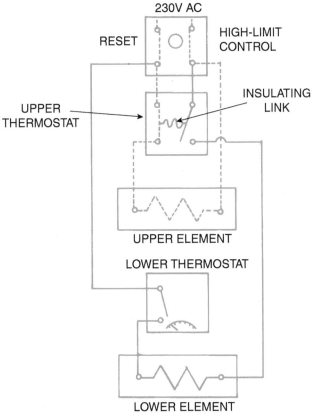

230V AC

RESET   HIGH-LIMIT
         CONTROL

UPPER
THERMOSTAT

INSULATING
LINK

UPPER ELEMENT

LOWER THERMOSTAT

LOWER ELEMENT

02311-14_F49.EPS

Figure 49 Wiring diagram of immersion elements in an electric water heater.

Step 7 Remove the drain hose from the water heater. Close the drain valve and reopen the water-inlet valve.

Step 8 Close the temperature and pressure relief valve and allow water to fill to the temperature and pressure relief valve.

Step 9 When the tank has refilled completely, connect or turn on the water heater.

Step 10 Replace the thermostat access panel and close the circuit breaker.

Step 11 Flush all faucets and fixtures.

3.3.0 Troubleshooting, Repairing, and Replacing Tubs, Showers, and Water Closets

Tubs and showers can be designed as one-piece units. Such units are generally too large to pass through doors, so they are typically installed during new construction. Lightweight fiberglass tubs and shower stalls also come in sectional units that can be installed easily. This makes it easier to replace damaged units or upgrade to fancier ones

during renovations. Plumbers repair and replace tubs and showers. Follow the manufacturer's installation directions. Take precautions to avoid damaging the units during installation.

Manufacturers often recommend laying a wet bed of mortar on the floor before installing a fiberglass tub or shower base. Set the base of the new unit into the mortar carefully and make sure the fit is correct. Let the mortar harden according to the mortar instructions. The mortar will conform to the base design as it dries. This gives maximum support to the fiberglass unit (*Figure 50*).

Shower sidewalls require a square enclosure. When installing ceramic tiles on the sidewalls, install water-resistant drywall as a backing. Remove the old wall material if necessary. In most cases, install the fiberglass unit directly against the 2 × 4 wall studs. Shim the opening to square it up. Install bracing for the tub (*Figure 51*).

The existing piping will determine which side of the shower or tub will have the faucets and drain (*Figure 52*). There are right- and left-hand units. Some tub and shower units have holes already cut for the water supply outlets. Follow the manufacturer's instructions closely when cutting into fiberglass. Wear appropriate personal protective equipment.

Showerheads in stalls and combination tub and shower units may be installed one of two ways. One option is through a hole in the fiberglass unit. The other is through the wall above the unit. Both locations are considered correct. Consult with the customer to determine the desired location.

Customers may want to replace water closets for several reasons. The bowl or tank may be cracked or damaged, or the customer might want a different style of fixture. If a bowl or tank is leaking, disassemble the unit and inspect the gaskets and seals. If they are worn, hard, or broken, replace them. One common source of leaks in water closets is the wax seal between the floor and the base of the unit. Replace a worn or damaged seal with a new one (*Figure 53*).

3.4.0 Troubleshooting and Repairing Other Fixtures and Appliances

Plumbers are often called in to work on remodeling and renovation projects. House additions need fixtures and appliances, too. In most cases, it makes sense to tie the new units into the existing water supply and DWV systems. Plumbers ensure that new lines are correctly designed and adequately sized. Plumbers are also responsible for calculating the water needs of the modified system.

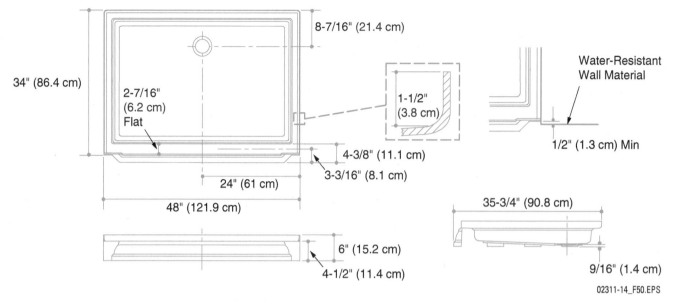

Figure 50 Rough-in for fiberglass shower and tub unit.

3.4.1 Domestic Dishwashers

When installing dishwashers (*Figure 54*), consider the following factors:

- Available space
- Access to the hot-water supply
- Location of the waste line

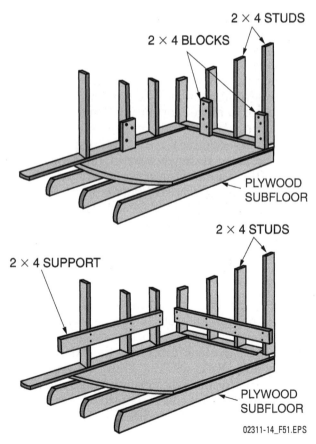

Figure 51 Bracing for the tub.

Determine the usage requirements early to help the customer select the proper-size appliance. The size of the appliance can affect the location of kitchen cabinets and countertops. The most common sizes of dishwashers are 18-inch and 24-inch units.

Dishwashers use hot water only. Usually, the dishwasher is located next to the sink. This arrangement allows easy access to the hot-water line. Install a shutoff valve on the hot-water inlet line to allow you to service the dishwasher without having to shut off the water system. Dishwashers need hotter water than other fixtures, usually between 140°F and 150°F, but most water-heater thermostats are set lower than this. To solve this problem, many high-end domestic dishwashers have their own internal reheater.

> **CAUTION**
>
> New fittings can have hidden defects. Be sure to examine new fittings for holes, cracks, or poor casting. Read and follow the manufacturer's installation instructions carefully.

Install the dishwasher's waste line so that it empties into the sink drain. If the sink has a food-waste disposer (*Figure 55*), connect the drain hose to the disposer. Check the local code before doing this. Many codes state that the drain hose should first be looped above the fixture's flood-level rim to reduce the chance of a backup into the dishwasher if the waste line becomes blocked. Some codes require a vacuum breaker or an air-gap assembly in the dishwasher discharge piping (*Figure 56*), which is another way of preventing waste from backing up into the dishwasher.

02311-14_F52.EPS

Figure 52 Tub-and-shower stall unit.

02311-14_F53.EPS

Figure 53 Location of the wax seal.

02311-14_F54.EPS

Figure 54 Domestic dishwasher.

1: SINK BAFFLE
2: VIBRATION ABSORBING MOUNT
3: VIBRATION ABSORBING TAILPIPE MOUNT
4: NOISE INSULATION
5: GRINDING RING
6: MOUNTING
7: STAINLESS STEEL GRINDING ELEMENTS
8: 3/4 HP MOTOR

02311-14_F55.EPS

Figure 55 Domestic food-waste disposer.

3.4.2 Lawn Sprinklers

Plumbers do not always install lawn sprinkler systems. However, they are responsible for hooking them up to the potable water supply. Because the potable water supply must be protected from backflow, the plumber must install an approved backflow preventer on the water supply connection. Contact the local authority having jurisdiction for which type of backflow preventer to install.

Sprinkler lines installed in colder climates will be subject to damage caused by winter freezing unless precautions are taken. Install these lines so that they are pitched (or, graded) toward the lowest point of the yard or property. This will allow the system to drain thoroughly and not retain water when cold weather begins, and will also make repairs easier if needed. Another option to protect the sprinkler line is to install an automatic drain fitting connected to an underground dry well. A dry well is simply a covered hole filled with gravel into which water can drain. Some systems may use air pressure to blow out the lines. Refer to the local code for guidance. Whatever method you choose, remember to drain the water lines before the winter season to prevent damage caused by freezing.

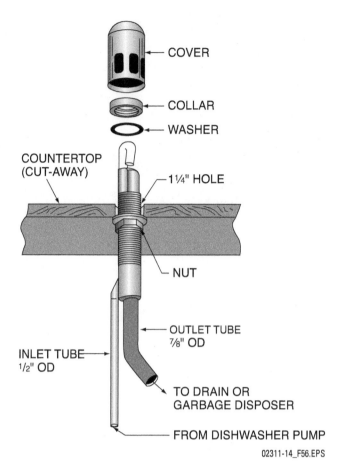

02311-14_F56.EPS

Figure 56 Air-gap assembly for a domestic dishwasher.

Additional Resources

Installing & Repairing Plumbing Fixtures. 1994. Peter Hemp. Newtown, CT: Taunton Press.
Plumbing Fixtures and Appliances. 2001. Washington, DC: International Pipe Trades Joint Training Committee.
Plumbing Fixtures and Fittings. 2003. Boca Raton, FL: Catalina Research.

3.0.0 Section Review

1. To repair a damaged valve seat, fit a cutter into a _____.

   a. reseating tool
   b. screw extractor
   c. packing extractor
   d. wrench

2. When using a multimeter to test an immersion element, it is likely that the element will need to be replaced if the meter _____.

   a. does not move
   b. reads between 1 and 3
   c. goes off the scale
   d. reads between 3 and 5

3. Before installing a fiberglass tub or shower base, manufacturers often recommend covering the floor with _____.

   a. tile
   b. a wet bed of grout
   c. a wet bed of mortar
   d. rubber sheeting

4. The water temperature required by dishwashers is usually _____.

   a. between 130°F and 140°F
   b. between 140°F and 150°F
   c. between 150°F and 160°F
   d. between 160°F and 170°F

SECTION FOUR

4.0.0 TROUBLESHOOTING AND REPAIRING DWV SYSTEMS

Objective

Explain how to troubleshoot and repair problems with DWV systems.

a. Explain how to troubleshoot and address blockages.
b. Explain how to troubleshoot and correct odor problems.

Performance Task

Troubleshoot and repair problems with DWV systems.

Trade Terms

Auger: A flexible metal wire inserted into a pipe to break up blockage. Also called a snake.

Drum machine: A large power auger designed to snake lines through a floor or other drain.

Jetter: A drain-cleaning machine that uses a stream of high-pressure water to flush away blockages while also power-washing the walls of the pipe.

Rodder: A gasoline-powered machine that uses sections of straight steel rod to clean straight runs of main lines.

DWV systems are vital for public health and sanitation. A well-designed DWV system will usually operate smoothly and efficiently. However, problems can and do occur. The most common problems are clogs, or blockages, and odors. Plumbers should address and correct drainage problems promptly. If a problem goes untreated, it could cause contamination, illness, and even death. The most common procedures for servicing DWV systems are discussed in the following sections.

4.1.0 Troubleshooting and Addressing Blockages

Blockage is the most common problem in waste drainage systems. A blockage is a pipe obstruction that interferes with the flow of waste liquids. Solid waste matter, tree roots, rust, ice, and other debris are common causes of clogs and blockage.

Blockage may occur in any part of a DWV system. Blockage in drain lines commonly happens at pipe bends, in fixture traps, and in improperly pitched sections of pipe. Vent pipes can also become obstructed. In this section, you will learn how to identify and fix different types of obstructions in DWV systems.

> **WARNING!**
> DWV lines may contain toxic and flammable vapors. The public sewer is also a biohazard. Use appropriate personal protective equipment when working with DWV systems. Wear hand and eye protection as necessary.

4.1.1 Drain and Trap Blockages

Fixture traps protect against contamination from waste products. They do this by maintaining a water seal in a U-shaped length of pipe. However, the shape of a trap makes it easy for debris to collect there. When too much debris collects in the trap, water may be slow to drain and blockages may cause water to back up. A blockage in a fixture trap can usually be reached through the fixture's drain. Obstructions in fixture traps are more of a nuisance than a serious problem.

To unblock a sink or lavatory drain, use a plunger. Note that plungers are less effective on vented pipes. This is because the pressure created by the plunger is released through the vent. The same is true for fixtures that have overflows, such as bathroom lavatories and bathtubs. To plunge a fixture with an overflow, first block the overflow with a plug, such as a wet rag. This will keep the pressure created by the plunger from escaping. When plunging does not clear a toilet blockage, use a hand or power auger (*Figure 57*). An auger, also called a snake, is a long, flexible metal wire wound on a drum. On the end of the wire is a head designed to break up a blockage. Use the following steps to auger a drain:

Step 1 Remove the drain plug from the fixture.

Step 2 Insert the end of the auger into the drain. Work the auger deeper into the drain by pushing and spinning the handle or operating the motor trigger. This will allow the auger to move through bends and fittings.

Step 3 When the auger contacts the blockage, spin the auger handle or motor until it breaks through the obstruction. Always spin the handle or motor in one direction only. Pushing and pulling the auger back and forth also helps break up the blockage.

Step 4 When it feels like the blockage has been cleared, run water through the drain. If the water flows freely, the blockage has been cleared. If the water does not flow freely, repeat Step 3.

Step 5 When the blockage has been completely cleared, remove the auger and run water through the drain to flush the debris out of the system.

Always wear appropriate personal protective equipment when using an auger. This includes eye protection and gloves that are specifically designed to protect against injury when handling spinning augers (*Figure 58*). The gloves are covered in steel studs that shield the wearer's hands. Always follow the manufacturer's safety instructions when using power augers. Never use an auger on a system for which it is not designed, or damage to the pipes and fixtures could result.

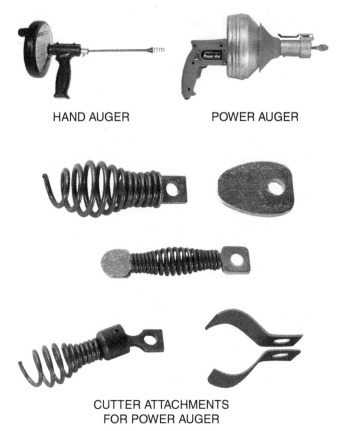

HAND AUGER   POWER AUGER

CUTTER ATTACHMENTS
FOR POWER AUGER

02311-14_F57.EPS

Figure 57 Hand and power augers.

> **WARNING!**
>
> Always wear protective eyewear and gloves designed for use with drain-cleaning equipment when using augers. The rotating metal wand of an unpowered or powered auger can cause serious injury.

If neither a plunger nor an auger removes the blockage, disassemble and thoroughly clean the trap. Remember to place a bucket or other container underneath the assembly when disassembling a trap. Couplings and washers are designed for easy removal (*Figure 59*). If the washers are stiff or damaged, replace them. Inspect the inside of the trap assembly for rough edges and projections, because these can cause a blockage. Some

traps have cleanout plugs at their base for easy removal of blockage (*Figure 60*).

Use a plunger to remove blockage in a toilet. More serious obstructions require an auger. Do not use chemical cleaners to remove toilet blockages. The water will usually dilute cleaners before they can reach the blockage. Never flush a blocked toilet until the blockage has been completely broken up and removed. Otherwise, the toilet could

The Effects of Corrosion on Metal Pipe

Exposure to air, water, and soil can cause metal water supply piping to deteriorate and leak. This is called corrosion. Corrosion happens when the metal in the pipe reacts to chemicals in the surrounding environment, causing the metal to break down. With the increasing use of polyvinyl chloride (PVC) pipe for water supplies, the problem of corrosion is less frequent today, although plumbers may see evidence of corrosion when servicing older supply pipe.

Rust, a common form of corrosion, is a powdery or flaky residue that forms on iron and steel pipes exposed to moisture and air. If steel pipe is rusting, identify any sources of moisture in contact with the pipe and take steps to correct the problem. If you find no specific water problems, replace rusted pipe with copper or plastic pipe. Use copper pipe for water supply lines and plastic for waste lines. Follow the manufacturer's specifications when installing new pipe.

Pipe buried in soil can be subject to a form of corrosion known as electrolysis. Electrolysis breaks down the chemical composition of a pipe through long-term exposure to a small electrical current. Another common form of pipe deterioration is galvanic corrosion. Galvanic corrosion can happen when two different types of metal are joined. Contact between the two metals can actually create a small electrical current. Galvanic corrosion can be prevented or stopped by using dielectric fittings or attaching a sacrificial anode to the pipe.

Figure 58 Drain-cleaning glove.

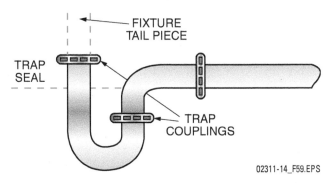

Figure 59 Location of couplings on a fixture trap.

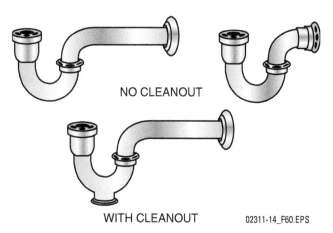

Figure 60 Examples of fixture traps with and without cleanouts.

overflow. After clearing the blockage, test the fixture by flushing several times with toilet paper. If any blockage remains in the line, the paper will catch on it and back up the flow.

For clearing difficult blockages, plumbers have the choice of several large tools that are designed specifically for DWV lines. A drum machine (*Figure 61*) is essentially a large power auger designed to snake lines through a floor or other drain. Some drum machines are equipped with a mechanism to automatically feed and retract the cable. A gas-powered rodder (*Figure 62*) is used to clean straight runs of main lines. Instead of a flexible cable, rodders use sections of straight steel rod. An auger tip is fitted to the head of the first length of rod, and additional sections of rod are added as the rod drives deeper into the line. The plumber uses a throttle to control the speed of the rodder. Plumbers can also use a jetter to clear blockages in DWV lines. Jetters use a stream of high-pressure water to flush away blockages while at the same time power-washing the walls of the pipe.

In older buildings you may find drum traps (*Figure 63*). A drum trap is a metal cylinder installed in the fixture waste line. The top of the trap sticks up through the finished floor for easy access. Drum traps are now prohibited by most plumbing codes.

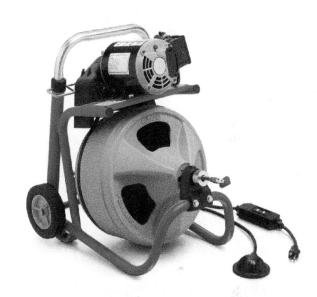

Figure 61 Drum machine.

Local plumbing authorities must approve the use of a drum trap before it can be installed. Jewelry shops are commonly permitted to use drum traps to trap stone and gem particles. Access to drum traps is usually through the basement or a ceiling access panel.

Figure 62 Rodder.

02311-14_F62.EPS

Tree roots, grease, ice, and other debris often cause blockages in the building drain. DWV systems are designed so that cleanouts are accessible to plumbers, because cleanouts allow plumbers to clean pipes that cannot be easily reached through drains and traps (*Figure 64*). Use an auger to break up blockages in the building drain. To clear tree roots, use an electric auger with a special flexible cutting blade (*Figure 65*). Tree roots will continue to cause problems as long as the trees remain and until the cracked pipe is repaired. Replace the pipe and dig a new sewer with good joints such as glued polyvinyl chloride (PVC) or cast-iron pipe. Before excavating, call utilities protection services, and follow OSHA excavation guidelines. Local horticulturalists can provide information about trees with roots that do not threaten pipes.

When the building drain has been cleaned, use a garden hose to flush the line with water. This will wash the debris out of the system. Another way to clear the line is to run several water outlets at the same time. Take precautions to avoid a cross-connection. Place a bucket below the cleanout to protect the floor from spills.

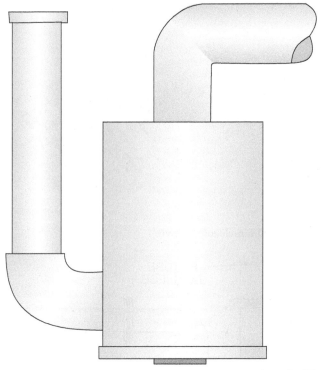

02311-14_F63.EPS

Figure 63 Drum trap.

4.1.2 Vent Blockages

Blockages can occur in vent stacks as well as in drains. Bubbling water in sink and bathtub drains indicates a blocked vent. Vent blockage can be caused by solids backing up into the vent from the drain line or by outside debris falling into the vent. In cold climates, frost and heavy snow can plug a vent opening. Plugged vents prevent air pressure in a DWV system from equalizing, which can lead to trap siphonage and the escape of sewer odors.

Sewer odors indicate dry trap seals, which can be caused by a restricted vent opening. Clean vent stacks from where they exit the roof. Insert a hand or an electric auger through the vent terminal to clear the blockage. When the blockage has been completely cleared, flush the vent using a garden hose or an inside tap. This will keep the debris from blocking the drain line at another location.

If the blockage is in a branch vent, you may have to tear out part of a wall to gain access to the vent piping. If the branch line is both a drain and a vent, use an electric auger inserted through the exposed portion of the vent. This procedure may be easier than going through a sink drain if the blockage is beyond the branch line.

During the winter, vapors from the DWV system can freeze at the vent opening when they meet cold outside air. As layers of frost continue to build up, the vent opening will become increas-

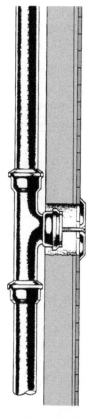

Figure 64 Building-drain cleanout installed in a wall.

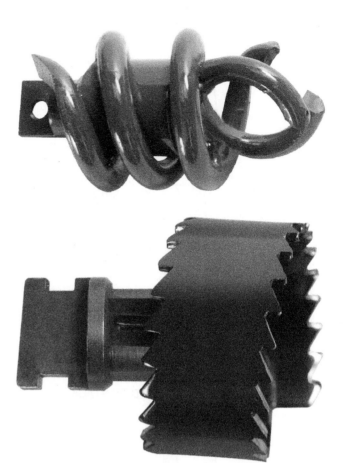

Figure 65 Root-cutting attachments for power auger.

ingly restricted. Eventually, the frost may close the vent. One way to solve this problem is to enlarge the vent opening (*Figure 66*). Note that the enlarged section of vent stack must extend at least 1 foot below the roofline to be effective.

Repeated trap siphonage means that the vent is undersized. An undersized vent prevents the air pressure in the DWV system from equalizing. When this is the problem, replace the vent.

4.1.3 Blockages in Restaurants

Many restaurants are designed with separate waste lines. Separate lines for water-closet waste, bar sinks, and kitchen waste prevent contamination of all fixtures from a backup in one. Many service calls to restaurants are the result of blockages in the grease system. When diagnosing a blockage in a restaurant, therefore, it is important to examine both the grease system and the sanitary system to identify the source of the problem.

When clearing a blockage in a restaurant waste line, make sure you are augering the correct line. You may need to consult the building's plumbing drawings. Take precautions to protect food-handling fixtures from spills and splashes.

4.2.0 Troubleshooting and Correcting Odor Problems

Sewer odors indicate dry trap seals or leaks in the waste line. Vents that exit the roof near air intakes for air conditioning or ventilating systems can also cause odors to enter a structure. In addition to siphonage, several things can cause the loss of a trap seal. If the fixture has not been used in a long time, the water in the seal may have dried up. Tight door seals or exhaust fans in public restrooms can also cause a trap to lose its seal. Install trap primers to correct dry traps (*Figure 67*).

Sewer gases can contain methane, which in large amounts can destroy copper DWV piping. To prevent the buildup of methane and the odors it causes, some state and local codes permit the installation of a house trap on the building drain. House traps on building drains can be difficult to access, however. The *International Plumbing Code*® prohibits house traps except where required by local code.

When installed, house traps must have a cleanout and a relief vent or other air intake on the inlet side of the trap. Consult the local code for other requirements related to house traps.

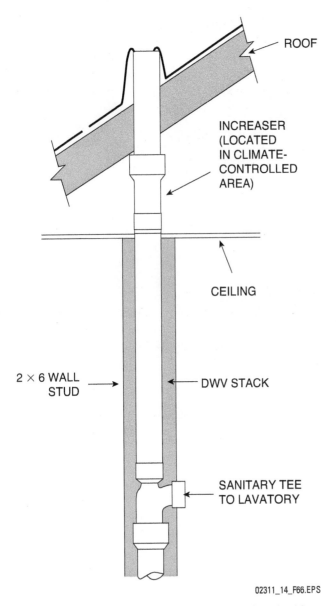

Figure 66 An enlarged vent opening prevents frost buildup.

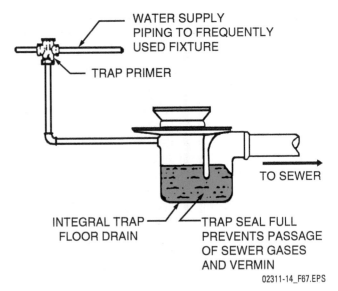

Figure 67 Trap primer installed in a DWV system.

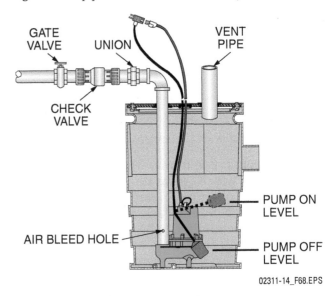

Figure 68 A typical sewage removal system.

If there is an odor in a DWV system and it has a sewage removal system, check the seating of the sump lid (*Figure 68*). Check the sump vent for blockage. Some commercial installations such as coin-operated laundries use special lint traps. Inspect the lint trap and clean it if it is the source of the odor.

Plumbers use four different kinds of tests to ensure that DWV piping is installed correctly with no leaks: the water, air, smoke, and peppermint tests. Water and air tests are typically used for installation testing. Smoke and peppermint tests are widely used to locate leaks during servicing. This section describes the latter two tests.

Always test DWV systems according to your local applicable code. Take extra precautions to ensure that all required safety measures are followed prior to and during the test.

4.2.1 The Smoke Test

In the smoke test, oily waste, tar paper, or similar materials are burned to produce a thick smoke that is blown into the DWV piping. Smoke bombs, which look like large firecrackers, can be dropped inside the stack to produce a large volume of brightly colored smoke. This procedure is much simpler than using a smoke generator. Do not create smoke by mixing chemicals such as ammonia and muriatic acid. This method produces toxic smoke and dangerous particles that are hard to see.

When the smoke begins to escape from the top of the stack, close the stack with a test plug and increase the pressure to 1 inch of water. This pressure should be maintained until the end of the test. If the pressure cannot be maintained, a leak is indicated. To find it, look for the escaping smoke. If the leak is too small for the escaping smoke to be seen, close windows and doors to retain the odor of the escaping smoke and locate the leak by scent.

4.2.2 The Peppermint Test

You can use an odor test if other tests fail to locate the leak. Use oil of peppermint to create the odor (ether is sometimes used instead). Close the outlet end of the drainage system and all vent openings except the top of one stack. Empty about 1 ounce of oil of peppermint for every 25 feet of stack (but not less than 2 ounces total) down the stack. Pour a gallon or more of warm water into the stack and immediately close the top of the stack.

Add air pressure to force the odor through the leak. Search for the leak using your sense of smell. Anyone who has recently handled peppermint should not enter the building until the search is complete. The odor test is simpler than the smoke test, but results are not always satisfactory because there may have been insufficient pressure in the pipes to force the odor through a leak. Also, it may be difficult to locate the leak after you've detected the odor.

Additional Resources

International Plumbing Code®, Latest Edition. Falls Church, VA: International Code Council.
How to Fix It Series, Vol. 1, No. 20, *Kitchen & Bathroom Plumbing*. 1998. Ron Hazelton, ed. New York: Time-Life Books.
Plumbing: A Guide to Repairs and Improvements. 1998. Jeff Beneke. New York: Sterling Publishing Company.

4.0.0 Section Review

1. Service calls to restaurants are frequently the result of blockages in _____.

   a. the grease system
   b. the vent system
   c. the DWV system
   d. the dishwasher drain line

2. When conducting a smoke test, once the stack has been closed with a test plug, the pressure should be increased to _____.

   a. 150 psi
   b. 5 inches of water
   c. 50 psi
   d. 1 inch of water

SUMMARY

Every service call presents new challenges. Plumbers are expected to do the work as professionally and efficiently as possible. To service and install plumbing fixtures, you need knowledge and experience. Service calls are an excellent chance to gain valuable experience.

Common problems with water supply systems include leaks, frozen pipes, high or low water pressure, water quality, and flow rate. Problems with water heaters may require that the unit be replaced.

Blockage in drainpipes can cause problems with DWV systems. Drain and traps can be unblocked through fixture drains or cleanouts. Odors can indicate dry traps or leaking pipe.

Forms of corrosion, such as rust, electrolysis, and galvanic corrosion, can cause pipes to deteriorate. Plumbers must know how to replace corroded pipe.

If fixtures and appliances break down, plumbers repair or replace components to fix the problem. Damaged fixtures may need to be replaced. Plumbers also install new fixtures for renovations or additions.

No matter what the service call, plumbers must always arrive promptly and work safely and efficiently. Your professional reputation depends on making sure customers receive the service they expect.

1. If a customer offers you a monetary tip when the work is completed, you should _____.

   a. accept it and thank the customer
   b. ignore it
   c. politely decline it
   d. ask the customer to donate to a charity instead

2. When a switch or valve is set to a safe position and a warning device or note is attached, the process is called a _____.

   a. tagout
   b. safety shutdown
   c. lockdown
   d. notification

3. Since it could cause the pipe to burst, do not attempt to thaw a frozen pipe using _____.

   a. a hot-water thawing machine
   b. a heat gun or hair dryer
   c. an electric thawing machine
   d. boiling water

4. The pressure exerted by circulating water is called _____.

   a. hydrological pressure
   b. dynamic pressure
   c. static pressure
   d. functional pressure

5. Water hammer is often caused by _____.

   a. too many fixtures on a branch
   b. low water pressure
   c. high water pressure
   d. scale buildup in pipes

6. Private water supply systems often store water in _____.

   a. bladder tanks
   b. sumps
   c. cisterns
   d. gravity tanks

7. Water that contains large amounts of minerals is called _____.

   a. saturated water
   b. soft water
   c. mineralized water
   d. hard water

8. A water filter containing activated carbon is designed to remove _____.

   a. rust
   b. mineral salts
   c. sulfur
   d. organic particles

9. To make sure a replacement water pump is properly matched to the system, an appraisal might have to be made by a(n) _____.

   a. inspector
   b. building engineer
   c. pump technician
   d. master plumber

10. Before digging to install a replacement water supply pipe, you must _____.

    a. determine the depth of the water table
    b. get permission from adjoining property owners
    c. hire a registered surveyor
    d. locate all nearby utility lines

11. To prevent contamination from cross-connection, hose bibbs must be equipped with a _____.

    a. relief valve
    b. low-pressure cutout
    c. vacuum breaker
    d. reduced pressure-zone principle backflow preventer

12. Some valves used in commercial installations are controlled by _____.

    a. light beams
    b. voltage drops
    c. digital pulses
    d. rheostats

13. A screw extractor should be turned using a(n) _____.

    a. locking pliers
    b. tee tap wrench
    c. electric drill
    d. pipe wrench

14. If preformed stem packing is not available for a valve, use _____.

   a. plumber's putty
   b. string soaked with creosote
   c. a wax ring
   d. twist packing

15. A flushometer that is controlled by an "electronic eye" is classified as a _____.

   a. digital flushometer
   b. manual flushometer
   c. sensor flushometer
   d. recycling flushometer

16. The float-controlled valve used in a water closet is also called a _____.

   a. pop-up
   b. ball cock
   c. butterfly
   d. bobber

17. Debris in the pressure-balancing spindle of a T/P valve may be dislodged by _____.

   a. tapping the spool with a plastic hammer
   b. removing the spindle and soaking it in denatured alcohol
   c. rapidly opening and closing the valve several times
   d. fully opening the valve and letting it run for five minutes

18. The discs in a ceramic-disc faucet _____.

   a. cannot be replaced
   b. should be replaced annually
   c. rarely need replacing
   d. can be adjusted if they leak

19. A touch faucet that is not functioning may be reset by disconnecting the battery for _____.

   a. 10 seconds
   b. 30 seconds
   c. one minute
   d. five minutes

20. Before opening a water heater's drain valve, you must _____.

   a. open all the faucets in the house
   b. be sure the pilot light is lit
   c. close the cold-water supply line
   d. place a 5-gallon bucket under the valve

21. To heat water, an electric water heater uses a(n) _____.

   a. sacrificial anode
   b. catalytic element
   c. thermocouple
   d. immersion element

22. To help a water heater operate more efficiently and last longer, suggest that the homeowner install a _____.

   a. water softener
   b. programmable thermostat
   c. water-flow monitor
   d. water-inlet filter

23. A backing of water-resistant drywall should be used when installing _____.

   a. one-piece tub/shower units
   b. ceramic tiles on tub or shower sidewalls
   c. a fiberglass shower stall
   d. a shower base

24. The most common widths for domestic dishwashers are 18 inches and _____.

   a. 20 inches
   b. 22 inches
   c. 24 inches
   d. 30 inches

25. Fixture traps protect against waste contamination by _____.

   a. active filtration
   b. maintaining a water seal
   c. creating a positive pressure zone
   d. allowing contaminants to settle out

Trade Terms Quiz

Fill in the blank with the correct term that you learned from your study of this module.

1. A(n) _____ is a tool used by plumbers to replace valve and faucet seats.

2. A flexible metal wire inserted into a pipe to break up blockage is called a(n) _____.

3. A(n) _____ is a lawn spigot connected to the water main that works by having its valve located below the frost line.

4. A tool used to remove packing from a valve stem is called a(n) _____.

5. _____ is the code of polite and appropriate behavior that people observe in a public setting.

6. The pressure water exerts on a piping system when there is no water flow is called _____.

7. A(n) _____ is a large power auger designed to snake lines through a floor or other drain.

8. The method of filtering water through cellulose acetate membranes used in some water filters is called the _____.

9. _____ is the deterioration caused by chemical reaction.

10. A tire valve stem used to correct low pressure in a tank is called a(n) _____.

11. _____ is an insulated copper wire wrapped around a pipe to keep it from freezing.

12. Potable water that contains large amounts of mineral salts such as calcium and magnesium is called _____.

13. A(n) _____ is a water filter that uses charcoal to absorb sulfur from potable water.

14. Water that collects on a cold surface that has been exposed to warmer, humid air is called _____.

15. _____ is the pressure water exerts on a piping system when the water is flowing.

16. A hydrant installed in a building's exterior wall that works by draining water after the valve is closed is called a(n) _____.

17. A(n) _____ is a drain consisting of a covered hole in the ground filled with gravel.

18. A gasoline-powered machine that uses sections of straight steel rod to clean straight runs of main lines is called a(n) _____.

19. _____ is a string-like material used to pack valve stems when preformed packing material is not available.

20. A device designed to remove small particles and organisms from the potable water supply is called a(n) _____.

21. A(n) _____ is a drain-cleaning machine that uses a stream of high-pressure water to flush away blockages while also power-washing the walls of the pipe.

22. A tool designed to remove broken screws is called a(n) _____.

23. A(n) _____ is a water softener that filters mineral salts out of hard water by passing the water through resin saturated with sodium.

24. A tool used for cutting internal threads is called a(n) _____.

25. A(n) _____ is a water filter installed on a cold-water inlet that removes metals and mineral salts.

26. A device that chemically removes mineral salts from hard water is called a(n) _____.

27. A(n) _____ is a tool used by plumbers to reface a valve seat.

Trade Terms

Activated carbon filter	Etiquette	Screw extractor
Auger	Hard water	Static water pressure
Condensation	Jetter	Tap
Corrosion	Packing extractor	Twist packing
Deionizer	Removable seat wrench	Wall hydrant
Drum machine	Reseating tool	Water filter
Dry well	Reverse-osmosis method	Water softener
Dynamic water pressure	Rodder	Yard hydrant
Electric heat tape	Schrader valve	Zeolite system

Trade Terms Introduced in This Module

Activated carbon filter: A water filter that uses charcoal to absorb sulfur from potable water.

Auger: A flexible metal wire inserted into a pipe to break up blockage. Also called a snake.

Condensation: Water that collects on a cold surface that has been exposed to warmer, humid air.

Corrosion: Deterioration caused by chemical reaction.

Deionizer: A water filter installed on a cold-water inlet that removes metals and mineral salts.

Drum machine: A large power auger designed to snake lines through a floor or other drain.

Dry well: A drain consisting of a covered hole in the ground filled with gravel.

Dynamic water pressure: The pressure water exerts on a piping system when the water is flowing (circulating water). Dynamic water pressure, measured in psi, will always be lower than static water pressure.

Electric heat tape: An insulated copper wire wrapped around a pipe to keep it from freezing.

Etiquette: The code of polite and appropriate behavior that people observe in a public setting.

Hard water: Potable water that contains large amounts of mineral salts such as calcium and magnesium.

Jetter: A drain-cleaning machine that uses a stream of high-pressure water to flush away blockages while also power-washing the walls of the pipe.

Packing extractor: A tool used to remove packing from a valve stem.

Removable seat wrench: A tool used by plumbers to replace valve and faucet seats.

Reseating tool: A tool used by plumbers to reface a valve seat.

Reverse-osmosis method: A method of filtering water through cellulose acetate membranes used in some water filters.

Rodder: A gasoline-powered machine that uses sections of straight steel rod to clean straight runs of main lines.

Schrader valve: A tire valve stem installed in a tank that is used to correct low pressure in the tank.

Screw extractor: A tool designed to remove broken screws.

Static water pressure: The pressure water exerts on a piping system when there is no water flow (noncirculating water). Static water pressure, measured in psi, will always be higher than dynamic water pressure.

Tap: A tool used for cutting internal threads.

Twist packing: A string-like material used to pack valve stems when preformed packing material is not available. It is not as durable as preformed packing.

Wall hydrant: A hydrant installed in a building's exterior wall that works by draining water after the valve is closed.

Water filter: A device designed to remove small particles and organisms from the potable water supply.

Water softener: A device that chemically removes mineral salts from hard water.

Yard hydrant: A lawn spigot connected to the water main that works by having its valve located below the frost line.

Zeolite system: A water softener that filters mineral salts out of hard water by passing the water through resin saturated with sodium.

Additional Resources

This module presents thorough resources for task training. The following resource material is suggested for further study.

Communicating at Work. 1993. Tony Alessandra and Phil Hunsaker. New York: Simon and Schuster.

Communicating in the Real World: Developing Communication Skills for Business and the Professions. 1987. Terrence G. Wiley and Heide Spruck Wrigley. Englewood Cliffs, NJ: Pearson.

Efficient Building Design Series, Volume III, *Water and Plumbing*. 2000. Ifte Choudhury and J. Trost. Upper Saddle River, NJ: Prentice Hall.

Installing & Repairing Plumbing Fixtures. 1994. Peter Hemp. Newtown, CT: Taunton Press.

International Plumbing Code®, Latest Edition. Falls Church, VA: International Code Council.

How to Fix It Series, Vol. 1, No. 20, *Kitchen & Bathroom Plumbing*. 1998. Ron Hazelton, ed. New York: Time-Life Books.

NFPA 99, *Health Care Facilities Code*, Latest Edition. Quincy, MA: National Fire Protection Association.

Piping and Valves. 2001. Frank R. Spellman and Joanne Drinan. Lancaster, PA: Technomic.

Plumbing: A Guide to Repairs and Improvements. 1998. Jeff Beneke. New York: Sterling Publishing Company.

Plumbing Fixtures and Appliances. 2001. Washington, DC: International Pipe Trades Joint Training Committee.

Plumbing Fixtures and Fittings. 2003. Boca Raton, FL: Catalina Research.

Tools for Teams: Building Effective Teams in the Workplace. 2001. Craig Swenson, ed. Leigh Thompson, Eileen Aranda, and Stephen P. Robbins. Boston, MA: Pearson Custom Publishing.

Your Attitude Is Showing. 2008. Sharon Lund O'Neil and Elwood M. Chapman. Upper Saddle River, NJ: Pearson Prentice Hall.

Figure Credits

Stacy Sabers, CO09

North Safety Products, Figure 2

Image courtesy of RIDGID®. RIDGID® is the registered trademark of the Ridge Tool Company, Figures 3–4, Figure 7, Figure 58, Figures 61–62

Woodford Manufacturing Company, Figures 8–9

Courtesy of Watts, Figure 11, Figures 38–39

Courtesy of Culligan International Company, Figure 16

Courtesy of IRWIN Tools, Figure 21

Courtesy of Superior Tool Company, Figures 23–24

Courtesy of Palmetto, Inc., Figure 27

Courtesy of Sloan Valve, Figures 31–33

American Standard, Inc., Figure 34, Figure 37

Courtesy of MultiCraft Construction LLC, Figure 36

Courtesy of Moen, Figure 41

Courtesy of **deltafaucet.com**, Figure 42

Kohler Co., Figure 44

Courtesy of A.O. Smith Corporation, Figures 45–46

Courtesy of Camco Mfg., Figure 47

Courtesy of Kohler Co., Figure 50, Figure 52

Courtesy of Whirlpool Corporation, Figure 54

InSinkErator® and Evolution Essential® are registered trademarks of Emerson Electric Co., Figure 55

General Pipe Cleaners, Figure 57 (top right and bottom), Figure 65

Little Giant Pump Co., Figure 68

Answer	Section Reference	Objective
Section One		
1. b	1.1.0	1a
2. d	1.2.2	1b
Section Two		
1. c	2.1.1	2a
2. d	2.2.0	2b
3. b	2.3.0	2c
4. d	2.4.1	2d
5. c	2.5.0	2e
6. b	2.6.0	2f
7. b	2.7.1	2g
Section Three		
1. a	3.1.1	3a
2. c	3.2.4	3b
3. c	3.3.0	3c
4. b	3.4.1	3d
Section Four		
1. a	4.1.3	4a
2. d	4.2.1	4b

NCCER CURRICULA — USER UPDATE

NCCER makes every effort to keep its textbooks up-to-date and free of technical errors. We appreciate your help in this process. If you find an error, a typographical mistake, or an inaccuracy in NCCER's curricula, please fill out this form (or a photocopy), or complete the online form at **www.nccer.org/olf**. Be sure to include the exact module ID number, page number, a detailed description, and your recommended correction. Your input will be brought to the attention of the Authoring Team. Thank you for your assistance.

Instructors – If you have an idea for improving this textbook, or have found that additional materials were necessary to teach this module effectively, please let us know so that we may present your suggestions to the Authoring Team.

NCCER Product Development and Revision

13614 Progress Blvd., Alachua, FL 32615

Email: curriculum@nccer.org
Online: www.nccer.org/olf

❏ Trainee Guide   ❏ Lesson Plans   ❏ Exam   ❏ PowerPoints   Other _____

Craft / Level: _____   Copyright Date: _____

Module ID Number / Title: _____

Section Number(s): _____

Description: _____

_____

_____

_____

Recommended Correction: _____

_____

_____

_____

Your Name: _____

Address: _____

_____

Email: _____   Phone: _____

Glossary

Activated carbon filter: A water filter that uses charcoal to absorb sulfur from potable water.

Actual cubic feet per minute: A measure of the air volume at the air-compressor inlet under actual temperature and pressure conditions. It is often abbreviated acfm.

Adhesive: A chemical used to bond thermoset pipes and fittings without fusing them together.

Adsorption: A type of absorption that happens only on the surface of an object.

Aftercooler: A device that cools air by passing it through a heat exchanger, causing moisture to condense out of the air.

Air-pressure regulator: A valve that keeps air pressure within the control range.

Air admittance vent: A valve-type vent that ventilates a stack using air inside a building. The vent opens when exposed to reduced pressure in the vent system, and it closes when the pressure is equalized.

Air dryer: A device that removes all moisture from compressed air.

Air end: A chamber at the end of the rotary screws in a rotary compressor where air is compressed.

Air lock: A pump malfunction caused by air trapped in the pump body, interrupting wastewater flow.

Air receiver: A storage tank for compressed air.

Air throttle valve: A spring-loaded valve used to control or shut off airflow in a compressed-air system.

Alum: A chemical used for coagulation. It consists of aluminum and sulfur.

Analog switch: A type of switch that is operated mechanically or electrically.

Applied mathematics: Any mathematical process used to accomplish a task.

Area: A measure of a surface, expressed in square units.

Atmospheric vacuum breaker: A backflow preventer designed to prevent back siphonage by allowing air pressure to force a silicon disc into its seat and block the flow.

Auger: A flexible metal wire inserted into a pipe to break up blockage. Also called a snake.

Austenitic stainless steel: A stainless steel alloy that contains nickel and stays strong at very low temperatures. Austenitic steel is the most popular type of steel for corrosive-waste pipes.

Backflow: An undesirable condition that results when nonpotable liquids enter the potable-water supply by reverse flow through a cross-connection.

Backflow preventer with intermediate atmospheric vent: A specialty backflow preventer used on low-flow and small supply lines to prevent back siphonage and back pressure. It is a form of double-check valve.

Back pressure: Excess air pressure in vent piping that can blow trap seals out through fixtures. Back pressure can be caused by the weight of the water column or by downdrafts through the stack vent.

Back siphonage: A form of backflow caused by subatmospheric pressure in the water system, which results in siphoning of water from downstream.

Backwashing: The cleaning process of a fully automatic water softener.

Beaded end: A raised lip around the rim of borosilicate glass pipe.

Bimetallic thermometer: A thermometer that determines temperature by using the thermal expansion of a coil of metal consisting of two thin strips of metal bonded together.

Biohazard: An organism or chemical that is dangerous to the health of living creatures.

Borosilicate glass: A pipe material made from specially reinforced glass that does not expand or contract much with changes in temperature.

Branch interval: The space between two branches connecting with a main stack. The space between branch intervals is usually story height but cannot be less than 8 feet.

Branch vent: A vent that connects fixtures to the vent stack.

Calcium hypochlorite: The solid form of chlorine. It comes in powder and tablet form.

Celsius scale: A centigrade scale used to measure temperature.

Centigrade scale: A scale divided into 100 degrees. Generally used to refer to the metric scale of temperature measure (see Celsius scale).

Centrifugal pump: A device that uses an impeller powered by an electric motor to draw wastewater out of a sump and discharge it into a drain line.

Chlorination: The use of chlorine to disinfect water.

Chlorinator: A device that distributes chlorine in a water supply system.

Circle: A surface consisting of a curve drawn all the way around a point. The curve keeps the same distance from that point.

Circuit vent: A vent that connects a battery of fixtures to the vent stack.

Coagulation: The bonding that occurs between contaminants in the water and chemicals introduced to eliminate them.

Combination waste-and-vent system: A line that serves as a vent and also carries wastewater. Sovent® systems are an example of a combination waste-and-vent system.

Common vent: A vent that is shared by the traps of two similar fixtures installed back-to-back at the same height. It is also called a unit vent.

Compressible: Able to be reduced in volume.

Condensation: Water that collects on a cold surface that has been exposed to warmer, humid air.

Conduction: The transfer of heat energy from a hot object to a cool object.

Conductor: An interior vertical spout or pipe that drains storm water into a main drain.

Constant speed control: A control that vents excess air pressure from an air compressor.

Contact time: The amount of time that a chlorine solution must remain in a water supply system in order to disinfect it, as determined by the type and amount of contamination in the system, the concentration of chlorine used, the pH (acidity or alkalinity) level of the water, and the temperature of the water.

Contamination: An impairment of potable water quality that creates a hazard to public health.

Continuous vent: A vertical continuation of a drain that ventilates a fixture.

Controlled-flow roof drainage system: A roof drainage system that retains storm water for an average of 12 hours before draining it. Controlled-flow systems help prevent overloads in the storm-sewer system during heavy storms.

Control range: The range between the highest- and lowest-allowable air pressures in a compressed-air system.

Conventional roof drainage system: A roof drainage system that drains storm water as soon as it falls.

Corrosion: Deterioration caused by chemical reaction.

Corrosive waste: Waste products that contain harsh chemicals that can damage pipe and cause illness, injury, or even death.

Coupling: A snap-fit plug connecting a tool or machine to a compressed-air system.

Coupling: Removable metal clamps with plastic seals, used to connect borosilicate glass pipe.

Cross-connection: A direct link between the potable-water supply and water of questionable quality.

Cube: A rectangular prism in which the lengths of all sides are equal. When used with another form of measurement, such as cubic meter, the term refers to a measure of volume.

Cubic foot: The basic measure of volume in the English system. There are 7.48 gallons in a cubic foot.

Cubic meter: The basic measure of volume in the metric system. There are 1,000 liters in a cubic meter.

Cylinder: A pipe- or tube-shaped space with a circular cross section.

Decimal of a foot: A decimal fraction where the denominator is either 12 or a power of 12.

Deionizer: A water filter installed on a cold-water inlet that removes metals and mineral salts.

Demand: The measure of the water requirement for the entire water supply system.

Density: The amount of a liquid, gas, or solid in a space, measured in pounds per cubic foot.

Desiccant air dryer: A type of air dryer in which chemical pellets absorb moisture from compressed air.

Developed length: The length of all piping and fittings from the water supply to a fixture.

Diatomaceous earth: Soil consisting of the microscopic skeletons of plants that are all locked together.

Digital switch: A type of switch that is operated electronically.

Disinfection: The destruction of harmful organisms in water.

Distillation: The process of removing impurities from water by boiling the water into steam.

Double-check valve assembly (DCV): A backflow preventer that prevents back pressure and back siphonage through the use of two spring-loaded check valves that seal tightly in the event of backflow. DCVs are larger than dual-check valve backflow preventers and are used for heavy-duty protection. DCVs have test cocks.

Drainage fixture unit (DFU): A measure of the waste discharge of a fixture in gallons per minute. DFUs are calculated from the fixture's rate of discharge, the duration of a single discharge, and the average time between discharges.

Drum machine: A large power auger designed to snake lines through a floor or other drain.

Dry well: A drain consisting of a covered hole in the ground filled with gravel.

Dual-check valve backflow preventer assembly (DC): A backflow preventer that uses two spring-loaded check valves to prevent back siphonage and back pressure. DCs are smaller than double-check valve assemblies and are used in residential installations. DCs do not have test cocks.

Duplex pump: Two centrifugal pumps and their related equipment installed in parallel in a sump.

Duplex stainless steel: A stainless steel alloy mix of austenitic and ferritic stainless steels that offers very good corrosion resistance.

Dynamic water pressure: The pressure water exerts on a piping system when the water is flowing (circulating water). Dynamic water pressure, measured in psi, will always be lower than static water pressure.

Ejector: A common alternative name for a sewage pump.

Electric heat tape: An insulated copper wire wrapped around a pipe to keep it from freezing.

English system: One of the standard systems of weights and measures. The other system is the metric system.

Equilibrium: A condition in which all objects in a space have an equal temperature.

Equivalent length: The length of pipe required to create the same amount of friction as a given fitting.

Etiquette: The code of polite and appropriate behavior that people observe in a public setting.

Fahrenheit scale: The scale of temperature measurement in the English system.

Ferritic stainless steel: A stainless steel alloy that is very resistant to rust and has a very low carbon content.

Field effect: A change in an electrical state that occurs when a conductive material such as water passes through an electric field.

Float switch: A container filled with a gas or liquid that measures wastewater height in a sump and activates the pump when the float reaches a specified height.

Floc: Another term for precipitate.

Flow rate: The rate of water flow in gallons per minute that a fixture uses when operating.

Friction: The resistance that results from objects rubbing against one another.

Friction loss: The partial loss of system pressure due to friction. Friction loss is also called pressure loss.

Fulcrum: In a lever, the pivot or hinge on which a bar, a pole, or other flat surface is free to move.

Furan: Any resin that is highly resistant to solvents. Furans are used to make thermoset pipe.

Fusion coil: A device that uses heat to join thermoplastic pipe.

Gallon: In the English system, the basic measure of liquid volume. There are 7.48 gallons in a cubic foot.

Grains per gallon: A measure of the hardness of water. One grain equals 17.1 ppm of hardness.

Hard water: Potable water that contains large amounts of mineral salts such as calcium and magnesium.

Hazard communication: A workplace safety program that informs and educates employees about the risks from chemicals.

Head: The height of a water column, measured in feet. One foot of head is equal to 0.433 pounds per square inch gauge.

High hazard: A classification denoting the potential for an impairment of the quality of the potable water that creates an actual hazard to the public health through poisoning or through the spread of disease by sewage, industrial fluids, or waste. This type of water-quality impairment is called contamination.

Horizontal vent: A vent that runs at an angle no greater than 44 degrees to the horizontal.

Hose-connection vacuum breaker: A backflow preventer designed to be used outdoors on hose bibbs to prevent back siphonage.

Hydraulic gradient: The maximum degree of allowable fall between a trap weir and the opening of a vent pipe. The total fall should not exceed one pipe diameter from weir to opening.

Hydraulics: The set of physical laws governing the motion of water and other liquids.

Impeller: A set of vanes or blades that draws wastewater from a sump and ejects it into the discharge line.

In-line vacuum breaker: A specialty backflow preventer that uses a disc to block flow when subjected to back siphonage. It works the same as an atmospheric vacuum breaker.

Inclined plane: A simple machine that has at least one flat surface inclined to the horizontal.

Indirect or momentum siphonage: The drawing of a water seal out of the trap and into the waste pipe by lower-than-normal pressure. It is the result of discharge into the drain by another fixture in the system.

Individual vent: A vent that connects a single fixture directly to the vent stack.

Infrared thermometer: A thermometer that measures temperature by detecting the thermal radiation emitted by an object.

Intercooler: A heat exchanger that removes heat from air as it moves through a multistage air-compressor system.

Ion: An atom with a positive or negative electrical charge.

Ion exchange: A technique used in water softeners to remove hardness by neutralizing the electrical charge of the mineral atoms.

Isosceles triangle: A triangle in which two of the sides are of equal length.

Jetter: A drain-cleaning machine that uses a stream of high-pressure water to flush away blockages while also power-washing the walls of the pipe.

Kelvin scale: The scale of temperature measurement in the metric system.

K value: A mathematical variable used to determine contact time based on the highest pH level and lowest temperature of the water in a water supply system.

Laminar flow: The parallel flow pattern of a liquid that is flowing slowly. Also called streamline flow or viscous flow.

Leader: An exterior vertical spout or pipe that drains storm water into a main drain.

Lever: A simple machine consisting of a bar, a pole, or other flat surface that is free to pivot around a fulcrum.

Lift station: A popular name for a sewage removal system. The term is also used to refer to a pumping station on a main sewer line.

Liquid thermometer: A thermometer that measures temperature through the expansion of a fluid, such as mercury or alcohol.

Liter: In the metric system, the basic measure of liquid volume. There are 1,000 liters in a cubic meter.

Loop vent: A vent that connects a battery of fixtures with the stack vent at a point above the waste stack.

Low hazard: A classification denoting the potential for an impairment in the quality of the potable water to a degree that does not create a hazard to the public health but does adversely and unreasonably affect the aesthetic qualities of such potable water for domestic use. This type of water-quality impairment is called pollution.

Lubricator: A device that injects a fine oil mist into the air supply line, lubricating tools and machines.

Manufactured air gap: An air gap that can be installed on a reduced-pressure zone principle backflow preventer to prevent back pressure and back siphonage.

Martensitic stainless steel: A stainless steel alloy that is very strong but is mostly used to handle mild corrosive wastes.

Mercury float switch: A float switch filled with mercury. It activates the pump when it reaches a specified height in a basin.

Metric system: A system of measurement in which multiples and fractions of the basic units of measure are expressed as powers of 10.

Network control: A device that controls multiple compressors by linking them together into a single network controlled by a computer.

Osmosis: The unequal back-and-forth flow of two different solutions as they seek equilibrium.

Oxidizing agent: A chemical that is used for oxidation and is made up largely of oxygen.

Packing extractor: A tool used to remove packing from a valve stem.

Part-load control: A device that allows an air compressor to operate at a given percentage of its normal output.

Pasteurization: Heating water to kill harmful organisms in it.

Percolate: To seep into the soil and become absorbed by it.

Permeability: A measure of the soil's ability to absorb water that percolates into it.

Pneumatic ejector: A pump that uses compressed air to force wastewater from a sump into the drain line.

Pneumatics: The set of physical laws governing the motion of air and other gases.

Point-of-entry (POE) unit: A water conditioner that is located near the entry point of a water supply.

Point-of-use (POU) unit: A water conditioner that is located at an individual fixture.

Pollution: An impairment of the potable water quality that does not create a hazard to public health but affects the water's taste and appearance.

Ponding: Collecting of storm water on the ground.

Pounds per square inch: In the English system, the basic measure of pressure. Pounds per square inch (psi) is measured in pounds per square inch absolute (psia) and pounds per square inch gauge (psig).

Precipitate: A particle created by coagulation that settles out of the water.

Precipitation: The process of removing contaminants from water by coagulation.

Pressure-type vacuum breaker: A backflow preventer installed in a water line that uses a spring-loaded check valve and a spring-loaded air inlet valve to prevent back siphonage.

Pressure: The force applied to the walls of a container by the liquid or gas inside.

Pressure drop: In a water supply system, the difference between the pressure at the inlet and the pressure at the farthest outlet.

Pressure float switch: An adjustable float switch that can either be allowed to rise with wastewater or be clipped to the sump wall.

Prism: A volume in which two parallel rectangles, squares, or right triangles are connected by rectangles.

Probe switch: An electrical switch used in pneumatic ejectors that closes on contact with wastewater.

Pulley: A simple machine consisting of a rope wrapped around a wheel.

Radar: A system that determines distance by bouncing radio waves off an object and measuring the time it takes for the waves to return. Radar is an acronym of "radio detection and ranging."

Re-vent: To install an individual vent in a fixture group.

Reciprocating compressor: A type of air compressor in which air is compressed by a piston.

Rectangle: A four-sided surface in which all corners are right angles.

Reduced-pressure zone principle backflow preventer assembly (RPZ): A backflow preventer that uses two spring-loaded check valves plus a hydraulic, spring-loaded pressure-differential relief valve to prevent back pressure and back siphonage.

Refrigeration air dryer: A type of air dryer in which air is cooled as it passes through a heat exchanger cooled by a liquid refrigerant.

Relief vent: A vent that increases circulation between the drain and vent stacks, thereby preventing back pressure by allowing excess air pressure to escape.

Removable seat wrench: A tool used by plumbers to replace valve and faucet seats.

Reseating tool: A tool used by plumbers to reface a valve seat.

Retention pond: A large man-made basin into which stormwater runoff drains until sediment and contaminants have settled out.

Retention tank: An enclosed container that stores stormwater runoff until sediment and contaminants have settled out.

Reverse-flow pump: A duplex pump designed to discharge solid waste without allowing it to come into contact with the pump's impellers.

Reverse-osmosis method: A method of filtering water through cellulose acetate membranes used in some water filters.

Right triangle: A three-sided surface with one angle that equals 90 degrees.

Rodder: A gasoline-powered machine that uses sections of straight steel rod to clean straight runs of main lines.

Rotary compressor: A type of air compressor in which rotary screws compress the air.

Rotary screw: Helical devices in a rotary compressor that compress air by passing it down the screw path.

Run dry: To operate a pump without water.

Schrader valve: A tire valve stem installed in a tank that is used to correct low pressure in the tank.

Screw: A simple machine consisting of an inclined plane wrapped around a cylinder.

Screw extractor: A tool designed to remove broken screws.

Scupper: An opening on a roof or parapet that allows excess water to empty into the surface drain system.

Secondary drain: A drain system that acts as a backup to the primary drain system. It is also called an emergency drain.

Self-siphonage: A condition whereby lower-than-normal pressure in a drainpipe draws the water seal out of a fixture trap and into the drain.

Sequencing control: A device that adjusts air pressure in a compressed-air system by activating and deactivating individual compressors.

Sewage pump: A device that draws wastewater from a sump and pumps it to the sewer line. It may be either a centrifugal pump or a pneumatic ejector.

Sewage removal system: An installation consisting of a sump, pump, and related controls that stores sewage wastewater from subdrains and lifts it into the main drain line.

Simple machine: A device that performs work in a single action. There are six types of simple machines: the inclined plane, the lever, the pulley, the wedge, the screw, and the wheel and axle.

Siphonic roof drain system: An engineered drainage system that blocks air from entering the drain system during the drainage process, resulting in greater flow volume at full bore than can be achieved by conventional drainage systems.

Sodium hypochlorite: The liquid form of chlorine, commonly found in laundry bleach.

Solenoid valve: An electronically operated plunger used to divert the flow of water in a water conditioner.

Sovent® system: A combination waste-and-vent system used in high-rise buildings that eliminates the need for a separate vent stack. The system uses aerators on each floor to mix waste with air in the stack and a de-aerator at the base of the stack to separate the mixture.

Square: A rectangle in which all four sides are equal lengths. When used with another form of measurement, such as square meter, the term refers to a measure of area.

Square foot: In the English system, the basic measure of area. There are 144 square inches in 1 square foot.

Square meter: In the metric system, the basic measure of area. There are 10,000 square cen-timeters in a square meter.

Stack vent: An extension of the main soil-and-waste stack above the highest horizontal drain. It allows air to enter and exit the plumbing system through the building roof and is a termi-nal for other vent pipes.

Standard conditions: An air pressure of 14.7 psia, temperature of 68°F, and barometric pres-sure of 29.92 inches of mercury.

Standard cubic feet per minute: A measure of the air volume at the air-compressor inlet un-der standard conditions. Standard cubic feet per minute is often abbreviated scfm.

Static water pressure: The pressure water exerts on a piping system when there is no water flow (noncirculating water). Static water pressure, measured in psi, will always be higher than dy-namic water pressure.

Stop/start control: A device that turns an air compressor off when the control range is exceeded.

Stormwater removal system: An installation consisting of a sump, pump, and related con-trols that stores stormwater runoff from a subdrain and lifts it into the main drain line.

Subdrain: A building drain located below the sewer line.

Sump: A container that collects and stores wastewater from a subdrain until it can be pumped into the main drain line.

Sump pump: A common name for a pump used in a stormwater removal system.

Tap: A tool used for cutting internal threads.

Temperature: A measure of relative heat as measured by a scale.

Test cock: A valve in a backflow preventer that permits the testing of individual pressure zones.

Thermal expansion: The expansion of materials in all three dimensions when heated.

Thermometer: A tool used to measure temperature.

Thermoplastic: A type of plastic pipe that can be softened and reshaped when it is exposed to heat.

Thermoset: A type of plastic pipe that is hardened into shape when manufactured.

Throttle: A device that controls airflow into an air compressor by adjusting the compressor's intake valve.

Transient flow: The erratic flow pattern that occurs when a liquid's flow changes from a lam-inar flow to a turbulent flow pattern.

Turbidity unit: A measurement of the percentage of light that can pass through a sample of water.

Turbulent flow: The random flow pattern of fast-moving water or water moving along a rough surface.

Twist packing: A string-like material used to pack valve stems when preformed packing material is not available. It is not as durable as preformed packing.

Ultraviolet light: Type of disinfection in which a special lamp heats water as it flows through a chamber.

Uniform Laboratory Hazard Signage: Signs placed on laboratory walls that use pictographs to indicate hazards.

Unloading: The process of relieving excess air pressure from an air compressor.

Use factor: Expressed as a percentage of an average hour of use, the amount of time that a tool or machine will be operated.

Vent: A pipe in a DWV system that allows air to circulate in the drainage system, thereby maintaining equalized pressure throughout.

Vent stack: A stack that serves as a building's central vent, to which other vents may be connected. The vent stack may connect to the stack vent or have its own roof opening. It is also called the main vent.

Vent terminal: The point at which a vent terminates. For a fixture vent, it is the point where it connects to the vent stack or stack vent. For a stack vent, it is the point where the vent pipe ends above the roof.

Viscosity: The measure of a liquid's resistance to flow.

Volume: A measure of a total amount of space, measured in cubic units.

Wall hydrant: A hydrant installed in a building's exterior wall that works by draining water after the valve is closed.

Water conditioner: Device used to remove harmful organisms and materials from water.

Water drip leg: A vertical pipe on an air supply line that acts as a drain for condensed water in the line.

Water filter: A device designed to remove small particles and organisms from the potable water supply.

Water softener: A device that chemically removes mineral salts from hard water.

Water supply fixture unit (WSFU): The measure of a fixture's load, varying according to the amount of water, the water temperature, and the fixture type.

Wedge: A simple machine consisting of two back-to-back inclined planes.

Wet vent: A vent pipe that also carries liquid waste from fixtures with low flow rates. It is most frequently used in residential bathrooms.

Wheel and axle: A simple machine consisting of a circular wheel attached to a central shaft that spins with the wheel.

Whip check: A cable attached to the ends of an air supply hose and a tool, or to the ends of two hoses that are connected to each other, to prevent the hose from whipping uncontrollably when charged with compressed air.

Work: A measure of the force required to move an object a specified distance.

Yard hydrant: A lawn spigot connected to the water main that works by having its valve located below the frost line.

Yoke vent: A relief vent that connects the soil stack with the vent stack.

Zeolite system: A water softener that filters mineral salts out of hard water by passing the water through resin saturated with sodium.

Index

A

ABS. *See* acrylonitrile butadiene styrene
absorption, (02303):11
AC. *See* carbon, activated
accidents, (02308):12, (02309):6, (02311):1, 42. *See also* explosion; fire
acids and acidity. *See also specific acids*
  lime to counteract, (02303):8
  and pipe, (02303):17, (02308):12, 15
  testing for, (02303):17
  waste, (02308):17
  water, (02303):7, 8, 12, 17
acrylonitrile butadiene styrene (ABS), (02306):17
actual cubic feet per minute (acfm), (02309):1, 2, 20
adapters, (02308):14
adhesives, (02308):7, 12, 14, 15, 24
adsorption, (02303):7, 11, 25
aerators, (02303):8, (02305):13, (02311):31
aftercooler, (02309):4, 13, 20
agriculture, (02306):9, (02308):2, (02312):23, 24
air
  compressed (*See* compressed air)
  in the DWV system, (02305):1–2, 11, 12
  pressure (*See* pressure, atmospheric)
  properties, (02309):1
  in the siphonic roof drainage system, (02306):16
  test, for glass pipe, (02308):10
  volume, (02309):2
  in the water supply system, (02311):16
air conditioners, (02312):25
air end, (02309):8, 9, 20
air gap
  dishwasher system, (02311):38, 40
  installation, (02312):18–19
  manufactured, (02312):15, 24, 25, 36
  pressure relief valve, (02311):35
  to prevent backflow, (02311):16, (02312):16
  reverse osmosis system, (02303):13
  water softener, (02303):9, 10
air lock, (02307):2, 13, 14, 21
alarms, (02307):3, 9, 10, 11
alkalies and alkalinity, (02303):3, 4, (02308):12
alternator, electrical pump, (02307):3, 4
alum, (02303):7, 12, 18, 25
aluminum, (02301):20, (02303):12, (02307):6, (02308):12
Amazon.com, (02305):7
American National Standards Institute (ANSI), (02309):12
American Society for Testing and Materials International (ASTM), (02312):12
American Society of Mechanical Engineers (ASME), (02309):10
American Society of Plumbing Engineers (ASPE), (02306):16
American Society of Sanitary Engineers (ASSE), (02305):7, (02312):17
American Standard, (02311):21
American Water Works Association (AWWA), (02311):31, (02312):12, 17, 27
ammonia, (02303):8, (02308):12, (02311):46

anodes, sacrificial, (02311):33–34, 42
ANSI. *See* American National Standards Institute
anthracite, granulated, (02303):11
apartments or condominiums, (02305):10, (02306):5
appliances, (02303):7, 9, (02311):37–38
apps, mobile, (02305):7
area
  calculations in plumbing, (02301):13
  definition, (02301):5, 37
  measuring, (02301):5–9, 13, 41
  and rainfall weight calculations, (02306):12
  units, (02301):1, 39–40
arrester, water hammer, (02301):23, (02309):1, (02311):10
asbestos, (02308):11
ASME. *See* American Society of Mechanical Engineers
ASPE. *See* American Society of Plumbing Engineers
ASSE. *See* American Society of Sanitary Engineers
ASTM. *See* American Society for Testing and Materials International
atmospheric vacuum breaker, (02312):15, 19–20, 36
atomic structure, (02303):9
auger, (02308):16, (02311):41–42, 43, 44, 45, 52
auto servicing facilities, (02308):2, (02312):16, 23
AWWA. *See* American Water Works Association
axle, (02301):26, 29, 38

B

backfill, (02308):11, 12, (02311):6
backflow
  definition, (02311):9, 16, (02312):5, 15, 16, 36
  due to bypass of backflow preventer, (02312):24
  in pneumatic ejector, (02307):5
  in stormwater pump, (02307):6
backflow preventers
  air gap (*See* air gap)
  atmospheric vacuum breaker, (02312):15, 19–20, 36
  bypass is prohibited, (02312):24
  classification, (02312):18
  containment, (02312):20
  double-check valve assembly, (02312):15, 21–23, 36
  dual-check assembly, (02312):15, 21–22, 36
  intermediate-atmospheric-vent, (02312):15, 26, 36
  isolation, (02312):20
  for lawn sprinklers, (02311):40
  overview, (02311):14, 16, (02312):16
  pressure drop due to, (02312):10
  pressure-type vacuum breaker, (02311):44, (02312):15, 20–21, 22, 36
  reduced-pressure zone principle assembly, (02312):15, 23–25, 36
  selection, (02312):26
  specialty, (02312):26, 27
  standards, (02312):17
backhoe, (02301):27
backwashing, (02303):7, 9, 25
bacteria and viruses
  diseases from, (02303):1, 3, 5, (02307):5, (02312):24, 25
  Escherichia coliform, (02312):24, 25

Plumbing & Mechanical Magazine, (02305):7
PlumbingNet, (02305):7
PlumbingWeb, (02305):7
plunger, (02311):41, 42
pneumatic ejector. *See* ejector, pneumatic
pneumatics, (02309):1, 2, 20
pneumatic tools, (02305):14, (02306):7–8, (02309):6, 11–12
POE. *See* point-of-entry unit
point-of-entry unit (POE), (02303):1, 2, 25
point-of-use unit (POU), (02303):1, 2, 7, 25
pollution
    building materials in landfill/incineration, (02311):15
    Clean Water Act, (02306):16
    from combined sewer overflow, (02306):16
    definition, (02303):1, 2, 25
    nonpoint-source (*See* runoff)
    NPDES permit to control, (02306):9
    point-source, (02306):9
polyesters, (02308):12
polyolefin, (02306):17
polyphosphate, (02303):17
polypropylene (PP), (02308):12, 13
polystyrene, (02308):11
polytetrafluoroethylene (PTFE), (02309):12
polyvinyl chloride (PVC)
    pipe, (02301):21, (02306):3, 17, (02308):12, (02311):42, 44
    thermal expansion, (02301):21
polyvinyl film, (02308):11
polyvinylidene fluoride (PVDF), (02308):12, 13
ponding, (02306):11, 17, 23
ponds, retention, (02307):1, 2, 9, 21
potable water. *See* water supply, potable
POU. *See* point-of-use unit
pounds per square inch absolute (psia), (02301):22, (02309):2
pounds per square inch gauge (psig), (02301):22, (02306):17, (02309):1–2, 9
pounds per square inch (psi), (02301):17, 37, (02311):8–9
power failure, (02307):2, 6, 9, 10, 13
power supply
    backup, (02307):6, 9, 10
    shutoff, (02311):1
    solar, (02311):14
PP. *See* polypropylene
precipitate, (02303):7, 12, 25. *See also* scale
precipitation (chemical process), (02303):7, 9, 12–13, 25
precipitation (weather). *See* rainfall; snow
pressing tool, (02311):7
pressure
    air compressor, (02309):9
    atmospheric (air), (02301):22, 23, (02305):1–2, 5, (02307):11, (02309):1–2
    back (*See also* backflow preventers; siphonage, back)
        contamination caused by, (02311):16, (02312):16, 18, 22
        creation in vent system, (02305):2, 3, 11
        definition, (02305):1, 19, (02312):16
        in pumps, (02307):3
    bladder tank, (02311):11, 16
    calculations, (02301):22–25
    compressed-air system, (02309):12–13, 14
    definition, (02301):17, 37
    within the DWV, equalization, (02305):1–2, 3
    dynamic water, (02311):5, 9, 52
    effect on boiling point, (02301):25
    flow, for fixtures, (02312):5
    and force, (02301):22–25
    negative, (02305):2, 3
    relationship with height of water column, (02301):23–24
    relationship with temperature, (02301):22, (02309):1

relationship with volume, (02309):2
    standard conditions, (02309):1, 2
    static water, (02311):5, 8–9, 52
    units, (02301):1, 2, 39–40
    water, (02301):22–25, (02311):8–11, (02312):6
    in water heaters, (02301):25
pressure drop, (02312):1, 3, 10, 13, 18, 36
pressure gauge, (02309):13, (02311):16
pressure loss. *See* friction loss
prism, (02301):5, 9–10, 37, 41
professional considerations
    certification (*See* personnel, certified or licensed)
    cleanliness, (02311):2, 3
    ethics, (02311):2–3
    never stop learning, (02311):5
    payment and tips, (02311):3
    promptness, (02311):1, 2
    reputation, (02303):8
    responsibility for public safety, (02303):2, (02307):9
    safety during service calls, (02311):1
    service-call etiquette, (02311):1–3
propane, (02312):18
PROPER mnemonic, (02311):4
PRV. *See* valves, pressure-reducing
psi. *See* pounds per square inch
psia. *See* pounds per square inch absolute
psig. *See* pounds per square inch gauge
PTFE. *See* polytetrafluoroethylene
public health issues
    acidic wastewater, (02308):18
    backflow and backflow preventers, (02312):15, 16–17, 18, 21, 22, 23–24
    proper venting, (02305):1
    restaurants, (02312):21
    sewer leaks, (02306):4
    vaccinations, (02303):5
    water contamination, (02303):1, 2, (02312):6
puller, cartridge, (02311):28, 30
pulley, (02301):26, 27–28, 37
pulp and paper mills, (02308):2
pumps
    backup, (02307):3, 6, 9, 10–11, 13
    bilge (*See* pumps, sump)
    capacity, (02307):6, 14
    centrifugal, (02307):1, 2–3, 6, 21
    cycle rate, (02307):6, 8
    discharge capacity, (02306):9
    duplex, (02307):1, 3, 4, 21, (02311):14, 16
    failure, (02311):16, (02312):16
    history, (02307):4, 11
    pasteurization system, (02303):4
    priming, (02307):2, 14, 15
    rating, (02307):14
    replacement, (02307):15, (02311):14
    reverse-flow, (02307):1, 3–4, 21
    self-priming, (02307):15
    sewage (ejectors), (02307):1, 2–5, 6, 9–11, 14, 15, 21
    steam, (02307):11
    stormwater (*See* pumps, sump)
    sump, (02307):1, 2, 6, 8, 9–11, 15, 21
    troubleshooting and repair, (02307):13–14, (02311):14
    water-lubricated, (02307):2, 15, (02311):33
    well, (02303):12, (02311):14
PVB. *See* vacuum breaker, pressure-type
PVC. *See* polyvinyl chloride
PVDF. *See* polyvinylidene fluoride
Pythagorean theorem, (02301):8